The Theory of Superconductivity in the

High-T_c Cuprates

PRINCETON SERIES IN PHYSICS

Edited by Sam B. Treiman

11/97

The Theory of Superconductivity in the High-T_c Cuprates

P. W. ANDERSON

Princeton Series in Physics

PRINCETON UNIVERSITY PRESS • PRINCETON, NEW JERSEY

Library of Congress Cataloging-in-Publication Data

Anderson, P. W. (Phillip W.), 1923–
The theory of superconductivity in the high-T_c cuprates
/ P. W. Anderson.
p. cm. — (Physical series in physics)
Includes bibliographical references and index.
ISBN 0-691-04365-5 (alk. paper)
1. Copper oxide superconductors—Congresses.   2. High temperature
superconductors—Congresses.   I. Title.   II. Series
QC611.98.C64A54   1997
537.6′23—dc20   96-43338

This book has been composed in Times Roman with Helvetica

Princeton University Press books are printed on acid-free paper, and meet the
guidelines for permanence and durability of the Committee on Production
Guidelines for Book Longevity of the Council on Library Resources

Printed in the United States of America

10   9   8   7   6   5   4   3   2   1

Contents

PART III. A Critical Postscript

Introduction

GENESIS OF THE BOOK

This book has been "in preparation" for over six years as I write this intro-duction. It started out as a joint venture of the entire Princeton group, based on a course of lectures in the spring term of 1988 which was contributed to by many other people besides myself. (I have listed all of my collaborators elsewhere.) Chapter 1 is not far from the corresponding part of that original venture. However, the glow of optimism of early 1987, when we seemed to see an almost immediate conclusion to the problem of high T_c in the "RVB"—resonating valence bond—idea was rapidly fading and much of what we wrote and said then was false or misleading.

Again, in 1989 I organized a summer school in Cargése for NATO, and again I promised that it would result in a book, which I saw as resembling the same Princeton University Press book which I had already contracted for. But I found myself, again, convinced that the result would be a travesty, in that the theory had clearly not reached final form, and the outcome, if I simply collected the sets of lectures, would resemble any of innumerable meeting proceedings volumes then being published, with perhaps a more selective assortment of opinions represented, but no uniform point of view. From that time the central chapter 2—the "central dogma"—dates. It was badly needed to organize our diffuse thinking.

I have waited to write the book until I felt that the theory is in sufficiently final form that major revisions will not be necessary. This does not mean that the subject is finished: there are many exciting directions for further development, both in the elaboration and precision of the theory of high T_c, and in modifica-tions appropriate to other systems, but no change in the central structure can be anticipated: there are no outstanding inexplicable anomalies.

I must apologize to my originally projected coauthors,* who expended con-siderable labor on this project. Several of their contributions appear as parts of the book. But I found in the end that it was essential to take full personal responsibility for the basic material and for the point of view expressed, since

---

*G. Baskaran, Z. Zou, B. Doucot, S. Shastry, S.-D. Liang, T. Hsu, J. Wheatley, and later also X.-G. Wen, T. M. Rice, P. Stamp, F.D.M. Haldane, Y. Ren, E. Siggia, S. John, and P. Wiegmann. Baskaran and Doucot, particularly, were very helpful in note-taking and organizing.

there are so many rather subtle points of judgment which must be decided, occasionally on intuition or taste (which is really a synonym for experience and for complex but logical deduction) rather than rigorous argument, and I cannot ask others to commit themselves to my personal convictions on such matters, no matter how deep these may be.

HISTORICAL INTRODUCTION

During Christmas vacation, 1986, the world of many-body theory first became aware of $(La - Ba)_2 Cu O_4$, a superconductor with the now modest T_c of $33°K$, but still far above previous values. In the same few months in which this T_c was pushed to $45°$, $52°$, and finally higher than $90°K$ by the discoveries of $(La - Sr)_2 Cu O_4$ ("2 1 4") and $Y Ba_2 Cu_3 O_x$ ("YBCO"), I proposed (correctly) that the appropriate model was a doped Mott insulator based on a single band of hybridized $Cu\ d$ orbitals; and groped toward a connection with the "resonating valence bond" liquid of singlet pairs of which Bethe's linear spin chain was the only physical exemplar, and which also had a close relationship to the Gutzwiller projected Fermi sea. The first insight was fully correct, the second gave us the vital concept of the "spinon," and in fact does describe, in the sense of a "fixed point," the magnetic state of the normal metal. The spinon remains the central notion of the theory: the basic magnetic excitation in the Luttinger liquid (which is our name for the anomalous normal state) is a "semion" or Majorana Fermion, with spin $1/2$, no charge, a Fermi surface, and with a restricted Hilbert space.

From this true insight several false trails led, and I followed all or most of them before finding the right one. One false direction was indicated by the "flux phase" of the 2d Heisenberg model, which is unfortunately a slightly better variational guess at the correct magnetic state (which Chakravarty et al. conclusively demonstrated to be always Néel-like). In the Mott insulator, one may Gutzwiller project whatever half-filled electronic state one likes, including one with a magnetic field of half a flux quantum per plaquet. The corresponding states for doped Hubbard models are disastrously poor states for the kinetic energy, and do not occur, as is clear from an argument of P. Lee, but leave traces in the tendency to chirality for very low doping.

From this early, hopeful insight developed a massive literature whose distance from physical reality became increasingly evident. It was unfortunate that the apparent seductiveness of the mathematics blinded so many able physicists to this fact. Perhaps the oddest result was the claim of "anyon superconductivity" which is an incorrect description of a nonexistent physical state.

A second false direction was also partly my doing; Zou, Baskaran, and I picked up Kivelson, Rokshar, and Sethna's apparently self-evident invention of the "holon"—at first, a boson to add to the spinon to make an electron. This correctly led us in the direction of gauge fields to enforce the constraint which follows from expressing one particle in terms of two others. But we disastrously oversimplified in several ways: our boson was quasi-free, our gauge field Abelian and treated in mean field theory, etc. Unlike the "anyon" theories which resulted from flux phases, there seemed to be much real physics here, especially when we assumed our holons to have very flat dispersion relations. It was, for instance, in this form that we first expressed the "interlayer" superconductivity mechanism, and Lee-Nagaosa as well as Wiegmann found workable high T theories of resistivity. And the gauge principle explains the basic fact of "confinement," i.e., dimensional reduction, which is the most important experimental observation about high-T_c superconductors. But it remained difficult to deal in detail with the very many suggestive anomalies which could be seen in experiments.

In the year 1989 genuine progress beyond the initial insight gradually began to appear. In the course of a series of gatherings of many of the important actors— the Cargése School at which I first realized the explicit reason for the breakdown of Fermi liquid theory; the IBM meeting at Mt. Fuji where the photoemission data were first discussed, and the Raman spectrum clarified; and the Los Alamos Symposium where we first proposed the "Tomographic Luttinger Liquid"—the central outstanding problem was solved. The main insight was that the soluble 1-dimensional Hubbard model must be taken very literally as the template for the structure of a Luttinger liquid, because in the limit of low energies the Fermi surface curvature can be neglected, and scattering becomes "non-diffractive"— purely forward—and every point on the Fermi surface supports a bosonized 1d model. The holons, as in 1D, have a Fermi surface equivalent at every $2k_F$, each of the spanning vectors of the original Fermi surface. This is the only way in which the spinon's Majorana symmetry can be fleshed out to give the full structure of the electron.

Converting this primitive insight into a full theory of the phenomenon of high-T_c superconductivity has taken the intervening four years. The theory which was constructed, or at least as much as exists, is given in chapters 6 and 7 of this book. A major milestone in this final labor has been the development of a transport theory which could account for the strikingly anomalous temperature-dependent Hall effect, an observation the importance of which has been emphasized to me by my colleague Nai-Phuan Ong since very early in the story. The explanation of this data was the first essentially new prediction of the theory, and caused me to announce, in Kanazawa in 1991, that it was effectively complete. A second milestone was the development of a serviceable gap equation, which took place only in summer 1992 and allowed detailed calculations of gap anisotropy and isotope effect to be added to the preexisting theory of the variations of T_c with chemical structure. This period also contained the explicit proofs of the non-Fermi liquid character in 2D and of the confinement process, accompanied by a theory of c-axis resistivity and the Hall effect.

Almost all of these theories follow a new calculational procedure which is unfamiliar to many and is probably the source of more misunderstanding than any other aspect. Since conventional perturbation theory fails, it becomes essential to express the response we need entirely in terms of the response functions of the *interacting* system, preferably *without* vertex corrections. If (as for the pair susceptibility, or the conductivity in the dirty limit) this can be done, and the result expressed in terms of interacting Green's functions, we can have considerable confidence in the result. Another milestone, actually the first in time, was the wonderfully clearly resolved photoemission data of Arko, Olson, and their collaborators, which cried out for an interpretation in terms of a Luttinger liquid spectrum: this was not quite a prediction, since it is input to the theory, but is a unique and binding constraint, if one only interprets the data without prejudice, as we have seen very clearly only recently in the explanation

*a   of "ghost" Fermi surfaces.

There are still many problems with the theory. What is known theoretically does not allow direct calculations of all processes. We know, approximately, only the Green's function and certain other response functions for a Luttinger liquid; and all processes must be reduced to simple diagrams involving these entities by ingenuity and physical arguments. We do not know in detail how to insert an energy gap into the bosonized spectrum, only that its pair response functions can be rather sharply bounded and that the gap removes charge-spin separation. And most of all, we have not yet a complete formal theory which relates the electron to the elementary bosons explicitly, though its outlines are

*b   reasonably clear.

Finally, we have only vague ideas about how to deal with impurity scattering

*c   as with many other responses.

The most exciting feature of the "Luttinger liquid" phenomenon is that it cannot be expected to be confined to high-T_c cuprates alone. It has its genesis in the theory of one-dimensional metals which applies to a large class of organic and other 1D or quasi 1D materials such as polyacetylene, Bechgaard salts, etc. The phase transitions in the superconducting systems, and their anomalous anisotropic resistivity, both suggest a dimensional reduction as the basic phenomenon as in the cuprates. In spite of the sophistication of theory for the 1D model, little serious effort has gone into applying it to these substances, possibly partly because the "confinement" phenomenon is a new idea. There are also many puzzling transport measurements.

There are also many 3D materials which remain essentially mysterious. The occurrence of non-Fermi liquid states in 3D materials such as $(BaK)BiO_3$, fullerides, chevrels, heavy electron materials, etc. is suggested by experimental anomalies, but there is as yet no theory. This book will not attempt to go on to these problems, except to include a reprint of my remarks at the '91 Goteborg Nobel symposium.

ACKNOWLEDGMENTS

I think it entirely appropriate first to acknowledge my debt to the community of experimental physicists, since they have been, in general, extraordinarily generous in sharing their data with me, and much more open-minded than the theorists in listening to new ideas. Above all, I wish to thank N.-P. Ong, without whose constant consultation this book could not have been written. It was Ong's conviction that there was a generic behavior pattern in the transport properties of the cuprates when they are "optimal" in some sense which sustained me through the years. His and his students' many skillful measurements were invaluable. A small selection of the experimentalists whom I can mention are: Laura Greene and John Rowell, who first noticed the strange resistivity and called it to my attention. Bertram Batlogg has been a constant consultant on materials questions, and when there was doubt he (in addition to Ong) tended to be most nearly right. Zack Schlesinger and Ruben Collins were very helpful; their infrared data and judgment as to the meaning and quality of others' work were invaluable. In photoemission, I have benefited from many discussions with Cliff Olson and I am most impressed by his and his group's open-mindedness and absolute honesty. Jim Allen has also been very helpful; and later on also Z.-X. Shen, Bill Spicer, and Juan Campuzano. Of groups with which I have not been personally close, Iye's and Uchida's groups in Japan have produced much outstanding data, and, of course, the experimentalists, Don Ginsberg, Myron Solomon, and Miles Klein at Illinois. In NMR, discussions with Walstedt were useful. Neutron scattering data has never reached the level of precision that I felt could be useful, specifically never allowing the necessary focus on detailed Fermi surface excitations. Given these limitations, I am indebted for early discussions with Birgeneau and later ones with Keimer. *d

I must emphasize that literally hundreds of experimentalists have helpfully shared their data, and any selection is intrinsically unfair. A more complete list follows: Dynes, Kapitulnik, Geballe, Takigawa, Arko, Malozemoff, Chaudhuri, Penney, Mehran, Bishop, Junod, Phillips, Thomas, Tanner, Ott, Fisk, Smith, Chaikin, Z. Z. Wang, Mitzi, Petroff, Simon, Mihaly, Kawai, Cooper, Fleury, Crabtree, Heeger, Cieplak, Takagi, Sawatsky, List, Aeppli, Hardy, Bozovic, van der Marel, Timusk, and Bontemps.

I have listed my putative theoretical coauthors but it would be unfair not to mention some of those who have invested the most time, thought, and energy in my ideas. These include, above all, G. Baskaran, and along with him my original group of students, Hsu, Wheatley, Zou, and Doucot. Other Princeton students directly involved were Ren, Stafford, Hristopolos, Clarke, and Strong; and, peripherally, Wen and Marston. Others among the Princeton group who contributed substantively were Liang, Shastry, John, Wiegmann, Tsvelik, Sarker, Haldane, Georges, and Khveshchenko. Also, for special mention as a collaborator, Sudip Chakravarty and his associate Asle Sudbo.

I have benefited greatly from discussions with a number of theorists, often (as with some of the above list) in substantive but constructive disagreement: above all E. Abrahams, but also D. H. Lee, Dan Rokshar, S. Kivelson, S. Girvin, R. Laughlin, T. M. Rice, P. A. Lee, A. Millis, M.P.A. Fisher, R. Shankar, E. Fradkin, S. Mele, A. Larkin, Y. Yakovenko, Yu Lu, O. K. Andersen, W. Metzner, M. Norman, M. Randeria, and C. M. Varma.

I must thank a number of institutions for their hospitality during this long slog. The work was started during a Fairchild Professorship at Cal Tech. Also central were the IBM Watson Labs and the IBM corporation, who hosted my constant visits to their laboratory and my visit to Japan. Parts of lectures were written at the Italian Physical Society's Varenna site, at the NATO School at Cargése, at the ICTP's Nepal school as well as at the ICTP itself, at Bell Labs, and at Los Alamos National Labs. I enjoyed the hospitality of the Aspen center for Physics for a very fruitful month in 1988 and of the George Eastman Professorship and of David Sherrington and the Dept. of Theoretical Physics at Oxford, where much of the final version took shape. Much of the work was financed by NSF Grant # DMR 8518163 and # DMR 9104873 at Princeton.

Finally, I very gratefully acknowledge the enormous effort put in by my secretary, Eva Zeisky, in typing and organizing chapters, circulating copies of these, and even preparing many of the figures.
Postscript added in proof (Dec. 1995).

The theory is a living, evolving entity, and new problems are being solved and old misconceptions being cleared up regularly. I have added a lettered asterisk in the margins to indicate statements which are to one degree or another out of date. Usually there is a recent reprint giving further information in this area. A notes section at the end of each chapter explains each of these asterisks.

I also added an epilogue to chapter 2 because I felt that nowhere else in the book was there a synthesis of all the main results.

NOTES

*a "Ghost" Fermi surfaces have turned out to be mainly experimental artifacts primarily cleared up by the efforts of J-C Campuzano et al. The theoretical work referred to here is still valid qualitatively but does not have as much of a burden to carry.

*b See reprint of PWA and Khveshchenko for a more complete story.

*c A recent preprint begins the work on this, though it is quite heuristic and qualitative in nature.

*d Aeppli and, particularly, Keimer (thanks to samples from Aksay) are beginning to produce quantitatively useful results.

PART I
The Cuprate Superconductors:
Basic Theory

1

High-T_c Superconductors as a Magnetic Mott Phenomenon

INTRODUCTION TO THE HUBBARD MODEL

Even a cursory glance at the structure of the two best-studied high-T_c oxide superconductors focuses the attention on the striking Cu-oxide layers. Many experimental investigations, for instance NMR or neutron scattering, also suggest the existence of a precise relationship of complementarity between superconductivity and antiferromagnetic ordering in both $(La.\binom{Sr}{Ba})_2CuO_4$ and $YBa_2Cu_3O_x$. Superconductivity is exhibited by the same structures, and literally the same electrons, which show simple magnetic ordering, on what must quite clearly be seen as an "either-or" basis: either magnetic and insulating, or superconducting and metallic, with probably a frozen disordered magnetic phase intervening over a small intermediate range of electron concentration.

It requires extraordinary perversity, of which many physicists are capable, not to search for the physics of high-T_c superconductivity primarily in the only slightly better understood physics of metal oxide insulators.

T. M. Rice[1] has pointed out the wide variety of electronic states available to transition metal oxides, from simple metals through CDWs, SDWs, ferroelectrics, ferri, ferro, and antiferromagnets, etc. Fortunately, the high-T_c superconductors seem to be based on what is fundamentally one of the simplest cases, the nondegenerate, approximately magnetically isotropic, single orbital d-ligand lattice. This case was rather exhaustively discussed in the 1959 superexchange paper,[2] and we will here repeat and flesh out the arguments of that paper, which have been recently clarified and experimentally supplemented by the impressive work of Sawatsky and Allen[3] in particular in the matter of understanding the microscopic electronic energy levels, and of T. M. Rice and collaborators in generalizing and formalizing the theory. This work has been confirmed by explicit cluster calculations by Schluter, Hybertsen, and Stechel.[4]

Before doing so, let us start with a brief background discussion of Wannier functions and of "ultralocalized" Wannier functions or "chemical pseudopotentials."[5] Some important theorems due to Kohn, Blount, and others, the ideas of Heine, and the methods pioneered by Bullett and Anderson, are important in underpinning the local point of view. There is a tendency nowadays among solid-state physicists to believe that only band (preferably density-functional) calculations bear any respectability, ignoring the fact that the apparently primitive L.C.A.O. methods favored by organic quantum chemists have a much longer and more successful history, and give a much clearer account of chemical bonding. Chemical bonding is often much better described in terms of atomic orbitals and atomic states: for instance, in ionic crystals it is much more intuitive to think in terms of filled states on the individual ions than in terms of filled bands. But by use of the Wannier transform one may describe any band in terms of local orbitals.

All of these methods referred to above are based on the idea that one can successfully describe a set of isolated bands by discussing them in terms of the appropriate set of symmetry-related local functions. More precisely, given a set of Bloch wave functions

$$\phi_k(r) = e^{-ikr} u_k(r) \ , \qquad (I-1)$$

the corresponding Wannier functions are obtained from

$$W(r - R_i) = \sum_k e^{ikR_i} u_k(r) e^{-ikr} e^{i\phi(k)} \qquad (I-2)$$

where u_κ has the periodicity of the crystal lattice and $\phi(k)$ is an arbitrary real function of k. This freedom in the choice of ϕ has a very important consequence. As shown by Kohn and Blount, if the set of Bloch functions describes a well-isolated band in the spectrum, it is possible to choose ϕ in such a way that $W(r - R_i) \sim e^{-\kappa|r-R_i|}$ at large distances. The decay rate κ depends on the widths of the gaps which separate the band from the remainder of the spectrum. When the gaps vanish, the decay of the Wannier functions is only polynomial $W(r - R_i) \sim \frac{1}{|r-R_i|^\alpha}$. In such a case, it is better to add up more bands until we come to a well-defined gap, and we get as many Wannier functions as there are bands. This set of orthogonal localized functions define the relevant "symmetry orbitals" in the quantum chemist's terminology, indicating that they will exhibit the point group symmetries of the lattice.

For instance, the bonding structure of diamond is quite well described, starting from a set of four lattice functions, with a tetrahedral symmetry around each carbon site. In the case of covalent silicon, Bullett has shown the necessity of taking into account both valence and conduction bands, by an L.C.A.O. based on sp^3 hybridized orbitals.

For the high-T_c oxides, the oxygen $2p$ and copper $3d$ orbitals are well separated from the oxygen $2s$ below and copper $4s$ above. These O_{2p} and Cu_{3d} levels provide then a good starting point for a microscopic description of these compounds.

Once we get the relevant set of symmetry orbitals, and if we can estimate a reasonably good one-electron potential, say using local density functional theory, or the more local and straightforward "Wigner trick" of a correlation hole removing the electron's charge from that atom on which it finds itself, one may write down an effective local Schrödinger equation which in turn leads directly to an optimally simple L.C.A.O. secular equation. From the exponential decay of Wannier functions, the matrix elements $(\phi_i^\alpha|H_{eff}|\phi_j^\alpha)$ of the effective Hamiltonian between two local symmetry orbitals fall off also exponentially with the distance between sites i and j. However, because of the orthogonality of Wannier functions, the wave functions exhibit oscillations at long distance which might induce long-range correlations in the off-diagonal part of the effective one-particle Hamiltonian. Eventually, it can be shown that many cancellations occur in the secular equation so that only nearest neighbor off-diagonal matrix elements are relevant. Such a cancellation actually indicates that it may be better to relax the orthogonality constraint in order to get more localized orbitals, with very short-range off-diagonal matrix elements. This is the main motivation for the method of localized pseudopotentials which we briefly discuss in Appendix I.[4] Using them, local symmetry orbitals are no longer orthogonal, but the method provides a very simple and local secular equation. This is the physics behind the "downfolding" method used so successfully by O. K. Andersen.[6]

All of this has been just to show that an atomic L.C.A.O. representation is intellectually viable and likely to be very accurate. We have shown that it is meaningful to consider the bands to be made up as L.C.A.O.s, and that the atomic orbitals satisfy a local equation of simple, locally symmetric form. These methods take care of such problems as the appropriate definition of O^{--} wave functions which, atomically, are unbound. They allow us to use the natural atomic orbital description of transition metal-ligand complexes.

ONE BAND MODEL FOR THE CUO_2 LAYERS

In the case of the Cu-0 structures (planes and chains) in the canonical high-T_c materials, one may consider a schematic energy level diagram, in which, to zeroth order, we neglect kinetic energy within the bands and ask how hybridization affects the band center energies which start out as the various "E_0's" of the different orbitals.

The unperturbed energy of the oxygen p orbitals lies slightly below that of the d's—perhaps no more than one volt. The hybridization matrix elements V_{dp} are of the order of a couple of volts, and split $d - p$ hybrids into rather widely separated pairs. In an octahedral cluster, the d_{xy}, d_{yz}, and d_{zx} orbitals hybridize less strongly by a factor of 2 or so, so the splitting between bonding and antibonding orbitals is reduced correspondingly.

The energy difference between d_σ and d_π orbitals is of the order of $10Dq$— typically 1–2 eV. In the case of Cu^{++}, the energies may be as shown (see figure 1.1).

It is important to show that not many of the oxygen p orbitals are left over as "non-bonding." Leaving out the Z-direction oxygens, we have two oxygens per Cu, having 6 p orbitals and 5 d symmetry orbital linear combinations, so we have only one nonbonding p, which is the symmetric linear combination of σ orbitals which hybridizes with $(Cu)_{4s}$, driving that state up and the p states down. The c-axis oxygens may provide another 3 nonbonding linear combinations, basically uninteresting ones of p or f-like symmetry about Cu.

In the absence of doping, Cu^{++} is in a $(3d)^9$ state, and the physics is dominated by the $Cu_{(3d)}x^2 - y^2$ or $z^2 - O_{2p}$ antibonding orbitals. All the nonbonding d and p orbitals seem to be dead because they are doubly occupied.

Furthermore, for Cu^{++}, in cuprates, a striking feature is the tendency to a Jahn-Teller distortion, bonding strongly to the $d_{x^2-y^2}$ orbitals in a square planar distortion of the local octahedron. It is noteworthy how stable the square-planar structure seems to be; the existence of this stability alone is absolutely clear evidence that the $d_{x^2-y^2}$ is the only relevant d orbital in the planes, if not in the chains.

The above picture makes clear the important but "slaved" role of the O_{2p} orbitals: these hybridize strongly with Cu 3d's but it is the antibonding linear combinations of the two which are relevant.

The Jahn-Teller distortion seems, according to band theory estimates, to split the d-spectrum by another 1–2 eV. This is probably an important reason for the favorable values of t and J in the Cu planes.

It is important to realize that doping on the planar Cu's with, say Ni or Zn, lowers T_c severely and decreases the planar distortion, while other metallic doping, if anything, stiffens the planes and pulls in the $Cu - O$ bond. This means that, in the latter case, we empty a second antibonding $d_{x^2-y^2}$ orbital, further strengthening the $Cu - 0$ bonding and stiffening the planes. Undoubtedly this will turn out to have a large effect on the phonon spectrum when that is analyzed.

Here is one of the few instances where one is not unequivocally driven to a single possible theory in all cases. There do exist many less covalent compounds—e.g., K_2CuF_4 and many nickel analogues—where the second orbital emptied is d_{z^2} rather than $d_{x^2-y^2}$ and the d^8 ion is octahedral with $S = 1$, or where the degeneracy is not lifted and the d^9 system is metallic. It is precisely because of the strong covalent bonding that "Cu^{+++}" in oxides prefers to stay planar and to empty a second antibonding $d_{x^2-y^2}$.

Almost the only viable alternative theory, then, might involve some kind of use of the nearby d_{z^2} band, which will indeed couple very strongly to phonons because it wants to be octahedral. In fact, however, we find that Cu^+ carriers (electrons) self-trap much more easily and do not lead to metallic conduction, and one suspects that $S = 1$ nickel-like holes might well do the same. Also,

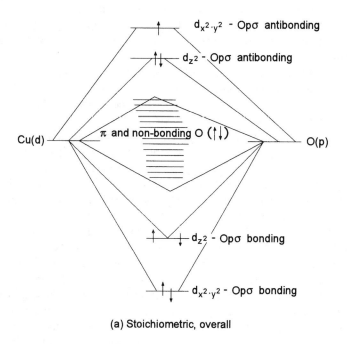

(a) Stoichiometric, overall

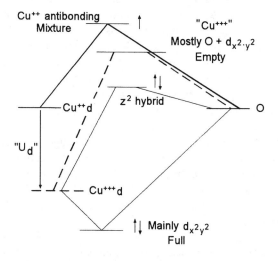

(b) Detail of top of diagram

Figure 1.1. For Cuprates (*a*) Stoichiometric, overall (*b*) Detail of top of diagram

K_2CuF_4 is a ferromagnetic metal, exactly in the same way as Ni metal for instance. Here, the exchange integral between $d_{x^2-y^2}$ and d_{z^2} orbitals is ferromagnetic, since the degeneracy is not lifted in the absence of Jahn-Teller distortion.

Throughout this chapter—and in fact most of the book—I will not discuss very thoroughly the "electron-doped" materials like $(Nd-Ce)Cu_2O_4$. They are anomalous in many ways. They conform to the above remarks in that very heavy doping is required to metallize them; isolated carriers do self-trap. They are also suspiciously Fermi-liquid-like. But their photoemission spectrum, and the highish T_c, suggest that they are members, if not very typical ones, of the cuprate class. Few reliable single crystal transport measurements exist.

Let us now mention the work of Lee and Read and of Gros et al. in showing that a careful treatment of the L.C.A.O. problem leads, in the end, simply to an effective Hubbard model. Actually, these considerations resemble an old line of reasoning by Haldane and Anderson, which was very completely checked out by detailed calculations and spectroscopy by George Watkins.

The problem we were interested in was the peculiar behavior of d transition metal impurities dissolved in Ge and Si. Watkins, years before, had made an excellent EPR and NMR study of the magnetic states of these impurities, from Mn to Cu and even Ag and Au, and found that they exhibited a whole sequence of strange valences, which were very dependent on the doping of the semiconductor, i.e., the Fermi level. There were as many as four "valences" for Mn for instance. The model we studied contained the key ingredients of the Lee-Read-Gros picture: a large U on the d-level atom, and a strong hybridization with a valence band with a high density of states (the nonbonding and bonding p-d hybrids in the high-T_c case). It is found that, rather than dipping into the Ge or Si valence bands, the d-impurity levels are strongly repelled and simply increase their level of hybridization, while going onto various magnetic states corresponding to Mn, Mn^+, Mn^{++}, and Mn^{+++}. In each of these states, the actual d level occupancy varies hardly at all, remaining at the extremely stable level of 5. All that happens is that the degree of hybridization increases. Watkins demonstrated this phenomenon by measurements and detailed calculations for a number of cases.

In that case, the hybridization became so extreme, and the corresponding d-symmetry wave functions so extended, that the effective atomic d-level properties such as hyperfine coupling, U, spin-orbit splitting, etc. became much smaller as one pushed the level toward the valence band. This is not really going to happen for Cu^{+++}, but may well be the reason why Cu^{+++} is reasonably mobile and does not self-trap.

This strong effect of admixture has been studied in the beautiful work of Allen and Sawatsky, for NiO and CuO, using photoemission and BIS spectroscopy. The canonical Mott insulator, which inspired the work of deBoer and Verwey, which in turn caused Peierls to invent the Mott insulators, was NiO. In local density functional theory, NiO is a metal, whereas one observes experimentally a gap of the order of several eV. If one leaves admixture out of account, U is of course 9 to 10 eV. By combining XPS and BIS spectra (x-ray photoemission and Bremsstrahlung-isochromat-spectroscopy), Sawatsky and Allen showed that the gap is 4.3 eV and is between an occupied state of character primarily ligand-p with a significant admixture of d, to an empty state which is all d. The mean occupancy is d^8. To remove an electron, one takes it from the ligands, which costs covalency energy, and to add one, one puts it onto d^9, so the gap is a mixture of covalency and coulomb energy. What the XPS and BIS reach is a weighted ImG for holes and electrons, respectively. From that information, it is then possible to pick out the d or ligand pieces and determine the amount of admixture. The corresponding curves for Cu in 123 have not been published yet, but according to J. Allen (Aspen, Jan. 1988), the weights for Cu^{+++} in CuO were 60% of O_{2p} hole, 30% of d hole and 10% of d^9 and two O_{2p} holes, while Cu^{++} was mostly d. The average occupancy does not change much from Cu^{++} to Cu^{+++} d^9–$d^{8.8}$ roughly). He would estimate that the Hubbard gap, if there was one, was at least 2.5 eV. The insulating state is called a "Charge Transfer" rather than a "Mott" insulator, for some reason. I believe this is an almost meaningless distinction, since all gradations between the two cases occur, and there is no sharp, rigorous test for which a given substance is.

EFFECTIVE HAMILTONIAN NEAR THE MOTT TRANSITION

The purpose of all these considerations is to produce an effective Hamiltonian to describe correctly the relatively low-energy excitations. Let us start from the Cu^{++} compound La_2CuO_4 or $YBa_2Cu_2O_{6.5}$.* Here, we have a Fermi level somewhere in the band gap ΔE which separates the effective energy $E_{d^9 \to d^8 L}$ for removal of a $d_{x^2-y^2}$ symmetry orbital from the energy for adding one further d electron $E_{z^2 d^9} \to d^{10} + d^8$. At this occupancy, there is a cusp in the curve of energy vs. particle number, representing a "commensurability gap" for charge, as discussed in the Varenna notes. It does not matter physically whether the value of this gap is controlled by the $d^8 + d^{10} - 2d^9$ U value or the hybridization energy V_{dp}, or even if a higher band such as Cu $4s$ intervenes; it only matters that it exists and what its value is; and that the symmetry orbital which represents the hole state is of $d_{x^2-y^2}$ shape, in order to get mobile holes (by contrast, in $NiO(d_{x^2-y^2})(d_{z^2})$, removing an electron destroys the octahedral symmetry, and one gets strong self-trapping polaronic effects which account for the low mobility in doped NiO).

---

*Nominally; actual insulating behavior ceases at $\sim O_{6.4}$ because of chain Cu^+ occupancy.

We can then approximately represent the system by an effective Hubbard Hamiltonian:

$$H = t \sum_{\substack{nearest \\ neighbors}} c_{i\sigma}^+ c_{j\sigma} + U \sum n_{i\sigma} n_{i-\sigma}$$

$$+ t' \sum_{\substack{next\ nearest \\ neighbors}} c_{i\sigma}^+ c_{j\sigma} \qquad (I-3)$$

with U a rather complex effective repulsive energy. In Anderson's 1959 paper was pointed out a relationship which is still true—that in a rough way, the matrix element for hopping, t, is proportional to the amount of semicovalent splitting, which determines the so-called "crystal field splitting" between d_{xy} and $d_{x^2-y^2}$ orbitals. This is because both are proportional to the degree of $d-p$ hybridization times the $d-p$ matrix element V_{dp}. The very strong splitting in the Jahn-Teller distorted Cu square planar configuration gives us a rather large value for the total d bandwidth $8t$ of order 2–3 eV. It seems then likely that the $Cu - O$ planes are in an intermediate coupling regime with $U \sim 8t$.

The exchange interaction must not be taken too literally from the 1959 paper, which was aimed at systems with considerably less covalency. In the first place, there is always a reduction of order <50% due to direct Coulomb exchange which is always ferromagnetic. The basic energy is, of course, of order $J = \frac{2t^2}{U}$ and comes from the lowering of kinetic energy which is obtained by forming a singlet pair of electrons in two neighboring orbitals. Nonetheless, this expression leaves us with a number of order $J \sim 0.1eV$ or $1000^0 K$ which (according to P. Fleury's Raman spectra, to detailed neutron data on La_2CuO_4, and to the observed spin susceptibility) is not too far from the truth. Furthermore, it is important to realize that the entire qualitative and in fact semiquantitative empirics of exchange in the oxides and other ionic d-band compounds is explained straightforwardly in the 1959 paper. It is particularly clear that J_{eff} of CuO is exceptionally large because of the Jahn-Teller distortion, that $Fe^{++} - Fe^{+++}$ shows a factor of 2 in favor of Fe^{+++} because of the stronger covalency; that the series KXF_3 and K_2XF_4 all fit very well, that ferromagnetism is predicted for certain special compounds and observed, etc. Unlike the d band metals, magnetism in these compounds is completely explicable—the consequence of the existence of a basic smallish dimensionless parameter $\frac{t}{U}$.

The many papers which deal with the so-called "2-band model"—which is not our 2-band possibility mentioned above, namely, the $d_{z^2} - d_{x^2-y^2}$ problem—seem to us to miss the point. The big, effective energy in the problem is the d-p hybridization which causes the structure as well as the big energy "U." U can be thought of as an effective Brueckner pseudopotential, the consequence of summing ladder diagrams involving d-d scattering and d-p hopping, as first discussed for magnetism problems by Kanamori. But there is simply no possibility that the bonding hybrid of $d_{x^2-y^2} + 2p_\sigma$ will play any dynamic role at energies less than several volts. U is of order t_{d-p}, t of order $t^2_{d-p/Ud}$ and hence somewhat smaller.

Again, the motivation is to find an effective Hamiltonian for low-energy phenomena. So long as $T \ll U$, we assume that no processes costing an energy U exist. This may be formalized, as Rice did long ago, and as also in the '59 paper in a special case, by a canonical transformation formally eliminating matrix elements into the high-energy subspace. If we have a subspace (1) with a projection operator P_1 for which the Hamiltonian may be schematized as:

$$H = (1-P_1)H_0(1-P_1)+P_1 U P_1+P_1 V_{12}(1-P_1)+(1-P_1)V_{21}P_1$$
$$U \gg H_0, V \qquad (I-4)$$

it is trivial to see that the correct Hamiltonian in the subspace (2) is:

$$H_{eff} = H_0 - V_{21}\frac{1}{U}V_{12}. \qquad (I-5)$$

This may be used as the start of a systematic perturbative elimination of the subspace (1) if U is not infinitely large. It is this systematic elimination which is repeatedly used in this problem. It is tantamount to the assumption that once a true Hubbard gap exists, the renormalization flows toward infinite U, so that at sufficiently small T, we must project on the subspace of no double occupancy. First we eliminate the true atomic d.d U, obtaining an effective—if smaller—covalency U; then eliminate all double occupancy obtaining an effective Rice Hamiltonian.

$$H_{eff} = H_{Rice} = P \sum t c^+_{i\sigma} c_{j\sigma} P + J \sum S_i \cdot S_j \qquad (I-6)$$

where P projects on zero or single occupancy. One can repeat the same manipulation using slave boson or slave fermion representations, but the physical idea is the same. (6) is called the "$t - J$" model and, it must be emphasized, is simply physically equivalent to the Hubbard Hamiltonian (3) except for detailed, basically irrelevant corrections. It is unfortunate that often (6) is used without taking into account the projection "P" in the first term, or by treating the second term as of primary importance and P by mean-field theory. This is fundamentally incorrect: "P" has much more drastic effects than "J" and cannot be neglected or averaged. This is the source of many incorrect papers, some of them by me or collaborators.

One may well ask, since it will turn out that this projection process has enormous physical consequences: "When does the baby go into the bath water?"— i.e., is there a sharp symmetry-breaking phase transition between Mott-Hubbard and conventional states? The answer is the same as for the metal-insulator, that there is no sharp distinction except at absolute zero, in ground states. At any finite temperature, there are a few real (as opposed to virtual) double occupancies, just as any insulator has a finite conductivity at any finite temperature. The possibility of a first-order transition is real, so also is a crossover, and undoubtedly, both happen. This subject is fairly seriously approached in the Varenna notes: it has just as much or as little importance for the fractional quantum hall effect, where the fractionally charged solitons depend on the existence of energy gaps and energy cusps which are not really sharp at finite temperatures, yet the fractional states persist to a crossover at quite high temperatures. We will find much the same physical behavior here. Especially strange is the fact that, in almost complete generality, the gap forms already for small U, and we would contend that already the renormalization flows toward $U \to \infty$. This is to be contrasted with the Anderson Model where intrinsically U flows to zero in the low temperature limit: the system renormalizes to a Fermi sea and U_{eff} decreases. The key to the theory (variously called "RVB" or Luttinger Liquid) is this renormalization to a non Fermi liquid fixed point $U \to \infty$. It is worthwhile to note that RVB and high-T_c phenomena occur in a single isolated band, whereas mixed valence is the case where a free electron band crosses the Hubbard band. So, the main requirements are:

— one isolated Mott-Hubbard band, insulating at half-filling
— spin $\frac{1}{2}$ (otherwise, we would get the usual magnetic phenomena, for instance magnetic polarons)
— Isotropic spin interaction (Anisotropy does not favor quantum fluctuations).

This leaves only a few cases: Cu, Bi, Ag, and maybe Ti or Nb; also favorable are the organics BEDT, TTF, etc.

The effective Hamiltonian just discussed describes the physics in the planes and in the chains at low energy. One would have to include in a full quantitative treatment:

— long range Coulomb repulsion
— phonons
— and, especially, interlayer tunneling.

In the case of phonons, the local BCS-type electron-phonon interaction mechanism does not work unaided, because the interactions here are fundamentally repulsive and put very strong constraints on the local charge fluctuations. In a metal, because of the extended nature of the wave functions, there is always an appreciable amount of ionicity at any site, which is responsible for the local nature of the interaction, and for the attraction mechanism. No such thing can happen once U has been renormalized to infinity. We are left then with the simplest $S = \frac{1}{2}$ Hubbard model insulator, which strikingly enough appears to be the last well-posed problem to be understood in many-body physics! That the $t - J$ Hamiltonian is an excellent approximation has been confirmed by detailed calculation on clusters by Schluter, Hybertsen, and Stechel. The lowest energy levels were fitted to a one-band Hubbard Hamiltonian and to a $t - J$ Hamiltonian; both gave excellent fits, the Hubbard slightly better. After this calculation it simply seems nonsensical to continue using 2- or 3-band models, or other constructed, artificial models.

O. K. Andersen et al. have fitted LDA bands to a tight-binding model containing t, t', and $t_\perp$, with a great deal of care and sophistication also; their numerical results will be valuable later on. *a

Appendix I

Chemical Pseudopotentials

The simplest form for an effective one-particle Hamiltonian is the sum of a set of atomic potentials and the kinetic energy

$$H = T + \sum_i V_i \qquad\qquad (i: \text{ atoms})$$

$$T = -\frac{h^2}{2m} \nabla^2.$$

More generally, we can write:

$$H \text{ (for atomic wave function i)} = T + V_i + (\mathcal{H}_{out})_i.$$

The idea is to introduce a set of local orbitals ϕ_i which are eigenfunctions of associated Schrödinger equations in which we remove as much as possible the component of H_{out} which lives in the subspace generated by the ϕ_j's for $j \neq i$. The ϕ_j's are then defined from the pseudo potential equation:

$$(T + V_i)\phi_i + H_i^{out}\phi_i - \sum_{m \neq i} \left\{ \int \phi_m(r') H_i^{out}(r')\phi_i(r')\, dr' \right\} \phi_m = E_i \phi_i.$$

These equations are to be solved self-consistently for the E_n's and ϕ_n's. Note that if we take $H^{out} = H$, this is simply an identity for a generalization of Wannier functions:

$$(H - E_0) W_0(r) = \sum_{n \neq 0} (n|H|0) W_n(r)$$

which again can be thought of as a pseudopotential self-consistent equation for W_0. By contrast, if we set $H^{out} = 0$, $V_{tot} = V$, this is an equation for the eigenfunctions. But it is most useful when we choose an atomic local potential as V_n, so approximately:

$$(T + V_n)\phi_n(r) = E_n^0 \phi_n(r).$$

Then ϕ_n is very nearly a solution, in most cases, of the pseudopotential equation, since the pseudopotential term:

$$V_{ps} = V_n^{out} \left[\delta(r - r') - \sum_{m \neq n} \phi_m(r')\phi_m(r) \right]$$

tends to cancel insofar as $\{\phi_m\}$ is approximately complete in the region of the potential V_n^{out}. In any case, the wave function can be improved perturbatively.

It is important to stress that the solution of the pseudopotential equation is unique, which is not the case for Wannier functions, due to the extra freedom in the choice of the phase $e^{i\phi(k)}$. For a reasonable choice of H^{out}, it can be shown that the pseudopotential functions are more localized than the corresponding Wannier functions:

$$\int r^2 \phi_i^2(r - R_i)\,dr = \langle R^2 \rangle_{\phi_i} \leq \langle R^2 \rangle_{Wannier}.$$

However, these local wave functions are by no means the final solution. They are now to be used to generate, with great simplicity, the matrix elements which hybridize them to make up the secular equation. We have:

$$H\phi_n = E_n\phi_n + \sum_{m \neq n} \left\{ \int \phi_n(r') H_n^{out}(r')\phi_m(r')\,dr' \right\} \phi_m(r).$$

The secular equation takes the usual form:

$$|(E - E_n)\delta_{nm} + V_{nm}| = 0$$

where the V_{nm}'s are very short ranged. Indeed, it is only the overlap between ϕ_n and both the potential and the wave function at m which gives any contribution to the matrix element. Core potentials, three center integrals, and all the fussiness of usual L.C.A.O. are absent in this formalism.

Here, the electron interaction has been only taken into account on the average. However, it may be shown that since the secular equation is the same as that for Wannier functions, the Coulomb interactions appropriate to the Wannier functions may be added as though the ultralocalized functions were Wannier functions.

Appendix II

Some Basic Facts about Normal Fermi Liquids: Conventional View

I give here some basic facts about conventional Fermi liquid theory in order to contrast with the behavior we will find in the CuO_2 planes and with our approach (which is equally valid for Fermi or Luttinger liquids).

It may be that the "normal" Fermi liquid is a figment of the imagination, in a sense, being unstable to the formation of one ordered state or another at low enough temperature. Kohn and Luttinger have claimed to show that the coulomb repulsion alone gives rise to an instability toward anisotropic superconductivity (with an extremely low T_c!). Nonetheless, we can easily imagine the state which continues adiabatically from the weakly interacting Fermi gas and which has long since been nicely characterized by Landau himself, Migdal and the Russian heroes, as well as Luttinger and Nozieres.

The key ideas of the normal Fermi liquid which we need here are those of the Green's function and of the quasiparticle. So long as quasiparticles can be defined, the exclusion principle guarantees that their decay rate is proportional to T^2 or ϵ^2, whichever is bigger, so that, focusing on a given energy ϵ, the interaction may be turned on at a rate which is slow compared to h/ϵ but fast compared to the decay rate τ, so that quasiparticles retain the identity they had in the noninteracting gas.

The Green's function—say for $T = 0$, ground state type—may be defined as:

$$G_{k\sigma}(t) = \langle 0|T c_{k\sigma}(0) c_{k\sigma}^+(t)|0\rangle.$$

T is defined in such a way that

$$i\frac{\partial}{\partial t} G_{k\sigma}(t) = \delta(t) + \langle 0|T c_{k\sigma}(0)[H, c_{k\sigma}^+(t)]|0\rangle$$

so that for a linear, one-electron Hamiltonian:

$$H = \sum_{k\sigma}(\epsilon_k - \mu)n_{k\sigma}$$

$$G_k(\omega) = \int_{\infty}^{\infty} e^{i\omega t} G_k(t)\, dt = \left(\frac{1}{\omega - \mathcal{H}}\right)_{k,k} = \frac{1}{\omega - (e_k - \mu)}.$$

Redefining $\epsilon_k \to \epsilon_k - \mu$, note that $Im G_k = i\pi\delta(\omega - \epsilon_k)$ and that $Tr Im G_k$ is proportional to the density of states $\rho(\omega)$. This latter identity is very general.

The apparent simplicity masks a vital role of the chemical potential μ in dividing positive from negative frequencies. When t is positive, the definition of G is such that we first empty the state k and later we measure the amplitude with which we can refill it. Thus we have created a hole, and any incoherent scattering or propagation away from the locality will cause gradual decay of the hole amplitude, so as $t \to +\infty$, G decays as $e^{-\eta t}$. Equally, for negative t, we create an electron, which propagates away and is incoherently scattered, so we get $e^{-\eta|t|}$. This, in terms of the inverse Fourier transform of $G(\omega)$, translates into a requirement giving small imaginary parts to the frequencies of the poles of $G(\omega)$, such that when t is positive, we pick up the hole poles, and the electron poles when t is negative.

When there are interactions, in a normal metal, we have the phenomenon of quasiparticles: the actual poles of G are a quasicontinuum representing the true many-body eigenstates which, for the interacting case, are everywhere dense and very slowly decaying; but these are meaningless for one interested in the coherent part of the states. As we approach μ, in general the free particle state comes more and more to resemble an eigenstate because of the exclusion principle prohibition on real final states—for electron-electron scattering, $\hbar/\tau$ goes as ω^2, and for electron-phonon, $\hbar/\tau$ goes as ω^3 because of the phonon density of states.

This means, as Landau pointed out, that there is an intermediate adiabatic turn-on time which gives us a labelling of states from those of the free system, so in the limit we can define quasiparticle states. These states can be described by representing the exact states approximately by an array of poles further from the real axis:

$$G_k = \frac{1}{\omega - \epsilon_k - \Sigma(k, \omega)}$$

where $Im \Sigma(k, \omega) \to 0$ as $\omega \to 0$. For our purposes we write

$$G_k \sim \frac{1}{\omega - \epsilon_k - \Sigma_R - i\Sigma_{im}} \simeq \frac{Z}{\omega - E_k - i\Gamma}$$

where $E_k = \epsilon_k + \Sigma_R(E_k)$ and $Z = [1 - \frac{\partial\Sigma}{\partial\omega}]^{-1}$. $\Sigma_{im}(\omega) \to 0$ as $\omega^2 + |\omega|^3$. This leads, after Kramers-Krönig transformation, to Σ_R like $\omega^3 \ln\omega$ which is visible in many metals and ^3He, in e.g., the specific heat. In ^3He, an important contribution of this kind is coming from long wavelength spin fluctuations. Re Σ is continuous and differentiable at least once, establishing the existence of the Fermi level, an absolutely basic requirement of Fermi liquid theory, which means that hole and electron quasiparticles merge smoothly into each other through the Fermi surface. Of course, the quasiparticle part is not the whole Green's function, but in fact, the quasiparticle representation, as Landau has shown, carries the whole physics, in that the quasiparticles are all of the low-energy excitations of the system.

A lot of high-energy physics of course is missed—e.g., with high-energy probes, one sees the momentum distribution—but that is not relevant to metallic properties because the quantum numbers e and p are not renormalized.

Kohn recently made a very important point about the question of the renormalization of e. In the early days of Fermi Liquid Theory, certain physicists were ridiculed because they seemed determined to introduce a renormalized charge e^* into transport and Fermi Liquid Theory. Nonetheless, it is easy to see that any electron scattering process conserves total charge so that the quasiparticle has true charge e (e is not renormalized, as a consequence of gauge invariance). Kohn's argument involves the key notion of Fermi-Liquid coefficients. In the Fermi Liquid theory, the presence of a quasiparticle modifies the energy of all the others in such a way as to induce an appropriate polarization or backflow. For instance, F_0^a is the exchange constant which reduces the magnetic susceptibility per spin, effectively inducing a cloud of reversed polarization. F_0^s plays the same role for particle number or charge. Kohn and his student pointed out that a wave packet of electrons in the quasiparticle state k will be automatically accompanied by a charge cloud $e(-\frac{F_0^s}{1+F_0^s})$, so that it is reasonable to suppose that there really is a renormalized charge $e^* = \frac{e}{1+F_0^s}$. F_0^s determines the compressibility of the Fermi gas, i.e., $\kappa = \frac{(free)}{1+F_0^s}$. Nonetheless, conductivity is not renormalized by this factor because of Luttinger-Ward identities which relate the self-energies and vertex corrections.

An incompressible Fermi liquid would be described by taking the limit $F_0^s \to \infty$, with the effective mass (controlled by F_1^s) remaining finite. Actually, in liquid ^3He, F_0^s is fairly large ($F_0^s \sim 10$), and this is for instance manifested by the fact that zero sound and first sound have almost the same dispersion relation. It seems then quite natural to consider this strongly renormalized Fermi liquid as an almost solid. In a true solid, the existence of an underlying lattice leads to the opening of an energy gap in the quasiparticle spectrum, and a cusp in the curve of total energy versus particle number. In the case of ^3He, there is no underlying lattice, so there is no gap and F_0^s remains finite. However, our old friend, the Mott Hubbard insulator, may be approached by letting F_0^s go to infinity. In this limit, we would expect Z to go to zero, so there are no longer true quasiparticle poles in the Green's function. However, we can imagine that even though the poles disappear, the Fermi liquid bookkeeping device remains valid to describe the quantum numbers of the low-lying excitations. In such an incompressible Fermi liquid, exchange interactions still give the possibility for the particles to move, and it is clear that we have to keep a finite (possibly strongly renormalized) effective mass m^*.

m^* finite and F_0^s going to infinity seems to be the reasonable limit to consider, in a first attempt at describing a novel type of fixed point in the large U limit.

There is still a kind of Fermi surface, which separates particle and hole states. In this limit, we expect the effective mass m^* to scale like $1/J$ rather than $1/t$, and the quasiparticles to become heavier.

A very important theorem about the Fermi Surface is Luttinger's Theorem, that its volume is equal to that of the non-interacting system. A perturbative proof is given in AGD, but it holds in a wider domain than perturbation theory. It is closely related to Friedel's theorem.

One of the most interesting theoretical descriptions of the metal-insulator transition in the half-filled Hubbard model has been provided by Brinkman and Rice[7] using Gutzwiller's variational approach. In this theory, F_0^s diverges at the insulating transition, and m^* diverges as well. This last feature occurs because the exchange interaction is not taken into account there, and nothing prevents the effective mass from being renormalized to infinity instead of $1/J$. One of the motivations for developing the original RVB was the need for an improved Brinkman-Rice theory, where the quasiparticles would keep a finite mass, in spite of the absence of charge excitations.

It is worthwhile to stress that Fermi Liquid Theory is very tricky in dirty systems, for instance, where localization effects give rise to dramatic transport anomalies. $Im\Sigma$ may become then very large at low temperature, proportional to $T^{3/2}$ compared to the T^2 of a standard Fermi liquid. In the presence of electron-electron interactions, things may become even worse, as, for instance, $Si P$ which behaves in an extremely anomalous way. We might say that this was the first non-Fermi liquid to be observed. These situations show clearly the need to go beyond Fermi Liquid Theory.

As we will see, $Z \equiv 0$ and Z_{small} are very different systems, even though, as we have emphasized above, it is sometimes useful to think of the spinons as the $Z \to 0$ limit of a Fermi liquid. For instance, the conductivity of a Fermi liquid is unrenormalized by interactions and contains no factors of Z (by the argument that e is unrenormalized, or the argument using $\int v_\perp \, dk_\perp = \int d\omega$ which was used by Schrieffer in explaining tunneling conductivity—these are equivalent). But when $Z \to 0$ there can be major interaction effects, as we shall see. When $Z \equiv 0$ the Fermi liquid literally has no compressibility per se, and charge and current are carried *only* by collective excitations. There are *no* true quasiparticles.

We shall see in chapters 3 and 6 how different experimentally the Luttinger liquid of the cuprate layers is from a Fermi liquid.

NOTES

*a I realize that there is now quite good data on at least two Fermi surfaces and, to a lesser extent, band structures, *NdCCO* and *BISCO* (2-layer), while *YBCO*, in the planes, seems to closely resemble the latter. These are compatible with (2), which has its maximum at "X" (π, π) because of the sign reversal of t in the O 2p state. Both are large hole surfaces around this point.

REFERENCES

1. T. M. Rice, unpublished; P. C. Canfield, J. D. Thompson, and G. Gruner, Phys. Rev. B **41**, 4850 (1990).
2. P. W. Anderson, Phys. Rev. **115** 2 (1959).
3. G. A. Sawatsky and J. W. Allen, Phys. Rev. Lett., **53**, 2339 (1984).
4. M. Schluter and M. S. Hybertsen, Physica C **162–164**, 583 (1989).
5. P. W. Anderson, Phil. Trans. Royal Soc., London **A334**, 473 (1989).
6. O. K. Andersen, O. Jepsen, A. I. Liechtenstein, (1991), and I. I. Mazim, Phys. Rev. B**49**, 4145 (1994); also M2S-HTSC IV Poster IR-PS 466.
7. W. F. Brinkman and T. M. Rice, Phys. Rev. **B2**, 4302 (1970).

2

The "Central Dogmas"

At a certain point in the process of unravelling the "Secret of Life"—for which read the mechanisms of reproduction and transcription of biological information—F. C. Crick propounded what he called "The Central Dogma" which constrained the overall structure of any description of the actual mechanism. The Central Dogma was determined by logical deduction from the overall experimental facts of biology. The very important conceptual function which was played by the "Central Dogma" was to limit serious discussion of mechanisms and theories to those which were consistent with logic and with the overall burden of experimental fact, while allowing a great deal of freedom in working out specific mechanisms, and leaving the overall structure of the theory immune to changes in specific processes. For instance, it was "dogma" that a genetic code existed, but the theory was independent of details of that code.

The main function of this chapter is to convince the reader that such a system of "dogmas" is useful for the field of high-T_c superconductivity, a field which has the same kind of complexity and confusion as microbiology had at that time. As in molecular biology, there is enough irrelevant complexity that an unwitting theorist may never reach the neighborhood of the actual problem, even though he is working along a line which is widely represented in the literature. Understanding high T_c involves not one, but a multiplicity of steps, and it is vital to provide a map through the maze of alternative paths, almost all of which can be eliminated by simple logic using basic and well-founded experimental or theoretical reasoning, of a sort which should be immediately persuasive.

The word "dogma" does not imply that these are principles laid down from on high; they are mostly empirical generalizations so direct and pervasive that it would seem perverse or frivolous to ignore them. "Dogma" is used entirely in Crick's sense. A second reason for propounding these dogmas is to correct the general misapprehension that there is no viable theory of high-T_c superconductivity. There is in fact an evident path indicated through all the complications. The problem is not the lack of a theory but its complexity and the fact that it consists of a number of steps of widely different completeness, involving different types of arguments, published in a bewildering variety of places and versions, and embedded in a literature of great complexity which is beset with controversy. The student may not be familiar with the process of rigorous deduction from theoretical concepts combined with a broad range of experimental facts, the process which is the primary source of the picture we now have. (As it was the primary source of the Fermi liquid theory of real metals which it replaces.) Relatively few theorists have been through the process of actually solving a puzzle like superconductivity or the Kondo effect.

The problem is not primarily finding a theory, but clearing away the underbrush of too many theories, some of which contain germs of truth because of tying in to a few valid experimental facts or theoretical concepts, but which do not take into account the key requirement of overall consistency with the complete picture. This is the key requirement because of the mature state of condensed-matter physics, which leaves theoretical problems extremely *over*determined: finding a theory is a redundantly posed problem. Any correct theory must be consistent with anomalous behavior of a bewildering variety of experimental probes, in addition to the very basic requirement of being internally consistent and embodying the entities which are really there: the known crystal structure, the outermost electronic bands and their interactions, the lattice vibrations, and *nothing else*. The theory must touch its base in the quantum theory of these entities. There remains very little flexibility, and one would feel there was no hope of finding such a theory except for one's knowledge that physics actually *does* work and there has to be one—and only one—solution. This then, is the final methodological clue—that one must retain one's faith that the solution exists. Thus when one has found *a* way through the maze of conflicting requirements, that is certain to be *the* way, no matter how many deep-seated prejudices it may violate and no matter how unlikely it may seem to those trained in the conventional wisdom.

Aside from theoretical papers which have some germs of truth in them, there is a larger group which is completely inconsistent with the basic realities of the subject. Many of these belong sociologically in the mode of particle theory, where there is no a priori foundation and speculations about the underlying physical model are acceptable, but many others come from naive or careless thinking in which extraneous independent entities like "anyons" or "spin fluctuations" or "bipolarons" are introduced but not tied down to the actual physical model and/or experimental observations. Many of the papers written in this field are not just easily falsified, in the Popperian sense, but actually falsified by preexisting evidence. I think the problem is one of a new kind of scientific sociology: physics has become so specialized and fragmented that an attempt at overall consistency with the observed facts and fundamental restrictions of theory is not seen as a necessary precondition for publication. Pure speculation, or at the other extreme, simple-minded applications of standard formulas, with no discussion of validity, seem to be seen as acceptable.

We feel the best presentation of the complete picture is to give the overall view, step by step, postponing as far as possible the details, such as precise justifications, alternative techniques, and detailed critiques. We give the resulting steps in the reasoning in the form of "dogmas," which take us, step by step, through the process of solving a difficult problem in quantum condensed matter physics. Each dogma is a stage at which the observed facts limit our choices to one possibility. Typically, one has to go through several stages of "renormalization," which is a fancy word for abstracting the relevant parts of the problem and eliminating the irrelevant high-energy degrees of freedom. This is why it is a canard that such a problem can be solved by enough computing power: it is hopeless until one has gone through this process. The most important terms by far are those which open up gaps in the spectrum, because states above any such gap can always be eliminated exactly and replaced by effective interactions among the remaining low-energy, low-frequency degrees of freedom: a process discovered by Van Vleck many decades ago. The whole problem of high T_c is a lesson—almost a poem—in restricting Hilbert space.

To summarize what we shall do, the first three of the six dogmas have to do with such eliminations. The first two restrict our attention to a single band of the one-electron spectrum, the antibonding, "$d_{x^2-y^2}$" symmetry band on the CuO_2 planes. Other bands are separated by large energy gaps from the relevant degrees of freedom near the Fermi level, and can be eliminated: the only relevant Hilbert space for electrons is this band. The third dogma tells us that only one of the interactions of the electrons in this band is so large as to open up yet another gap and further restrict Hilbert space. The next two dogmas are descriptions of the state of the "normal" nonsuperconducting metal which we encounter when we enforce these restrictions, both of which are almost equally supported by experimental and theoretical arguments. The first states that the resulting metal is in an unconventional state which we call a "Luttinger liquid," possessing a Fermi surface but no conventional electron quasiparticles; and the second that this state is strictly two-dimensional in this case. Only the last dogma, then, has to do with the superconducting state: it tells us which of the residual interactions that are left can be strong enough to give us the unconventional high transition temperatures which are observed. Again much of the argument is from experiment.

Let us then set out this list of "dogmas" with some discussion of the basis behind each and of alternatives which have been proposed.

Dogma I: All the relevant carriers of *both* spin and electricity reside in the CuO_2 planes and derive from the hybridized $O_{2p} - d_{x^2-y^2}$ orbital which dominates the binding in these compounds.   *a

The main alternative was the "chain" school, but now the one compound, YBCO, which has chain coppers, is in a tiny minority among some dozen compounds, none of the rest of which have chains and all of which behave with remarkable similarity. The infrared data on single untwinned crystals of Schlesinger and Collins show that chain conductivity is qualitatively different and relatively little affected by superconductivity. A more persistent and subtle fallacy, which this dogma excludes, is the literal acceptance of band theory results which often give bits of Fermi surface attached to other parts of the structure: the notorious "bismuth pockets" in BISCO, for instance, predicted by Freeman et al. and discussed in ARPES papers. The c-axis resistivity in BISCO, 10^5 times that in the planes, means unequivocally that no essentially 3-dimensional pockets of carriers can exist in that case. The band calculations, based as they are on an idealized, stoichiometric structure, are also incompatible with the graphite-like cleavage between the Bi layers, which shows that no bands near the Fermi surface are occupied at Bi. No band calculations are to be trusted at this kind of level of accuracy, and the real band structure must be deduced from experiment. Whether there are ever pockets of other carriers, or chain carriers, is not important but may help explain some results.

*b

The most commonsensical approach to Dogma I is to recognize the many anomalous properties of the cuprate layer materials, which are unique to those materials; and the rather unique chemistry of the cuprate layers, involving an unusual valence of Cu and very strong Jahn-Teller distortion and semicovalent bonds. The "Anderson mystery story principle," which is the original source of much of the dogma, then operates: this principle is that we must associate all truly unusual events with each other and with the basic problem.

Excitations outside of the planes, and probably even nonbonding or bonding bands in the planes, play very little role and can be ignored except for minor renormalizations. There is a remarkable similarity in behavior of materials with widely different chemistry outside the planes: for instance, the ab plane normal state resistivity per plane in optimally doped materials doesn't vary by more than a factor 2 among 5 or 6 compounds, as Batlogg observes. Normal state infrared and tunneling data are also very similar. Only T_c varies widely, a fact which we will discuss later.

Dogma I in summary: look at the planes only (a great and welcome simplification.)

Dogma II: Magnetism and high-T_c superconductivity are closely related, in a very specific sense: i.e., the electrons which exhibit magnetism are the same as the charge carriers.

The initial source of this was the "generalized phase diagrams" of state vs. doping, which can be traced out in several compounds. $\delta = 0$ (pure Cu^{++} oxidation state) is an antiferromagnetic insulator, with relatively high T_{Neel} if not frustrated: it is a straightforward Mott-Hubbard insulator with at least a 2 volt charge-transfer gap. The present view is that a relatively sharp transition (which is sometimes first order) occurs at $\delta \sim .1$ to a metallic state which almost always is superconducting, with usually a T_c which is initially finite. T_c seemed originally to rise continuously with δ from the insulator but this has not been demonstrated clearly in most cases. The metallic state is always peculiar as we shall later discuss; when it turns into a more normal metal, with excessive doping, T_c goes *down*.

Since in (almost) all other substances low carrier number, metal-insulator transitions, and antiferromagnetism decrease T_c, the mystery story principle requires the association of the magnetism with the phenomenon of high T_c. A theoretical point: the effect of doping on Mott insulators has been a controversial and unsolved problem for decades; again the overwhelming temptation is to associate difficulties.

More straightforward, and equally logically compelling, are optical, photo-electron spectroscopy, and NMR data. From optics and PES it is clear that the carriers appear in the Mott-Hubbard gap in proportion to the doping. NMR data show that the hyperfine couplings of the metallic carriers are identical with those of the spins responsible for magnetism.

Theory—so long as optics, PES, and other probes confirm the presence of the new carriers in the same orbitals as the magnetism—is equally compelling. The strength of the semicovalent bond due to $O_{p\sigma} - d_{x^2-y^2}(Cu)$ hybridization is responsible for the great integrity of the square planar configuration of CuO_2 planes. This suggests that, next to the Cu^{++} "U" repulsion, the second-largest parameter is the $2p\sigma - d\sigma$ hybridization t_{dp} which must be of order $2 - 3\,ev$, leading to the observed $\sim 6ev$ splitting of bonding and antibonding bands. Between these two levels are a spaghetti of weakly bonding $Cu - O$ hybrids which form a very large hump $\gtrsim 1-2ev$ below the Fermi level. The antibonding band has only one state per Cu ion, and therefore the appropriate Hilbert space for leaving the magnetic state intact and introducing a new set of carriers does not exist.

A considerable body of reliable electronic structure calculations by Schluter and others confirms this picture and gives us reasonably reliable values for the Hubbard "U" and "t" parameters.

The most persistent fallacy evading Dogma II is the "extended Hubbard Model" and variants thereof, which are at least formally correct in that they can be reduced to the right model unless they have the wrong parameter values, but are an unnecessary detour of no physical value. More naive theories simply don't question where, electronically, the magnetism comes from, and use coupled magnetism-carrier physics. Various probes show that the magnetic and charge form-factors of the carriers differ somewhat: they are $\gtrsim 60\%\,d$, at least, magnetically, and $\gtrsim 60\%\,p$ electrically. This can be understood as different polarization of the background bands by exchange and Coulomb interactions, by a single band of renormalized carriers. Thus the conclusion is:

II: We must solve the old problem of doping a single Mott-Hubbard band before we can *begin* the problem of high T_c. After renormalizing away high-energy excitations, the physical particles live in a single band. The problem is the very old problem of reconciling their magnetic structure and their charge transport.

Dogma III: The dominant interactions are repulsive and their energy scales are all large. Clearly the existence of a Mott half-filled band insulator implies large repulsive interactions whose scale may be bounded below by the Mott-Hubbard gap for charge transfer. A second scale is set by the exchange parameter which, by various accurate experimental measurements, especially spin wave velocities and Raman spectra, is at least $1200°$K, leading to spin wave bands .2–.3 ev wide and a spin wave velocity comparable ($\sim 1/4$ at least) to Fermi velocities. In Hubbard model terms, we are in the case of large but not infinite U.

Many clear indications place the intrinsic electron-phonon coupling at normal to moderately strong. Most striking are the shifts of optical phonons associated with the gap, shown in Fano resonances with the anomalous electronic background in the Raman effect. There is no reason to doubt various direct electronic calculations of these couplings. What is striking is that they have so little effect. We believe the resistivity which would have been caused by phonons and by static lattice distortions and defects in the (nonstoichiometric) normal metal, is *larger*, at least at low T, than the observed ρ_{ab}, which,

*c again, does not vary from substance to substance as much as a phonon resistivity would. Equally, "phonon bumps" do not show up strongly in tunneling, infrared, or ARPES spectra. This situation reveals one of the crucial anomalies of the high-T_c materials. We can and will describe the physics by saying that the strong repulsions dominate and restrict the response of the charge and spin density fluctuations to phonon and static potentials.

A little more may be said. We are accustomed to the fact that collective, bose-like modes (phonons, plasmons, spin waves) are much less easily scattered at low frequencies and long wavelengths than particle wave functions. In the "Luttinger liquid" type of theory, for charge transport the particle modes have been replaced by the collective motions, by the simple construction dating back to Tomonaga: these experimental facts support this kind of electronic theory. Thus we have, to the lowest order,

III: Restrict your attention to a *single-band, repulsive* (not too big) U Hubbard Model.

Dogma IV: The "normal" metal well above T_c is the solution of the planar one-band problem resulting from Dogma III, and is not a Fermi liquid, in the sense that $Z = 0$. (Z being the quasiparticle wave-function renormalization constant.) But it retains a Fermi surface satisfying Luttinger's theorem at least in the highest T_c materials. We call this a Luttinger Liquid.

This has several sound experimental and theoretical bases. The most vital experimental evidence lies in the giant anisotropy of resistivity ρ_c/ρ_{ab}. The resistivity perpendicular to the planes extrapolates to ∞ at T=0 and is in all cases well above the Mott limit $\frac{h}{e^2 k_F}$. This means that there is no coherent electronic transport in the c-direction: all motions are inelastic. Fermi liquids cannot localize in one direction and extend in a second, because simple localization is a question of coherence: are the electrons coherent in extended or in localized states? Thus a fortiori the normal state is a two-dimensional metal with only inelastic processes connecting the layers.

In these experimental considerations and also in later ones a theorem due to Schrieffer (as far as I know) is very important: single-particle tunneling, and transport are not renormalized by the wave-function renormalization constant "Z," in any conventional—or even unconventional, as in the case of weak localization—Fermi liquid theory. This was discovered in relation to superconducting tunneling, and in the verification of phonon interactions in BCS theory it played a vital role, but it was equally important in the hands of Mott, Thouless, and other early workers in localization theory, who recognized that the conductance e^2/h is a universal boundary between metallic and insulating states, independently of dynamic effects, because conductivity is not renormalized. The modern ideas on conductivity using the S-matrix, pioneered by Landauer, give us considerable understanding of the universality of this theorem. The essential idea is that conductivity contains no *dynamical* corrections: it may be written entirely in geometrical terms as

$$\sigma_a = \frac{e^2}{h} \, (c_D) \, A_F^a \cdot \ell_a \tag{1}$$

where c_D is a dimensionless, D-dependent constant of order unity, A_F is the Fermi surface area perpendicular to the direction a, and ℓ_a is the mean free path in direction a. (1) is proved by essentially the same technique as Schrieffer's proof for tunneling conductivity, which gives it as an integral over the entire quasiparticle spectrum which reduces to a contour integral around the pole with no "Z" correction.

Several authenticated cases of "insulating" transverse conductivity exist in the literature in other systems (e.g., TaS_2) and we would suggest that these be reexamined, since this behavior will not normally occur in a Fermi liquid. If the matrix elements are genuinely tiny, it is possible that inelastically assisted conductivity could dominate, but in all high-T_c materials but BISCO the observed $3d$ superconductivity rules this option out (see later).

A second very strong argument is the small value of ρ_{ab} and the relative sharpness of the features in the ARPES spectrum. Reasonable estimates of impurity scattering by the large nonstoichiometry in (2 1 4) or BISCO leads to a mean free path $l \sim$ a few $a_0 \sim 20$Å while ρ_{ab} corresponds to roughly $l = 50$–100Å at T=100°K. Any reasonable estimate of phonon scattering also lead to a bigger ρ_{ab} than observed. The ARPES peak *widths* are in good agreement with ρ_{ab} while the feature sharpness is even smaller. Hence, giant concentrations of charged impurities have no effect on ρ_{ab}; tiny percentages of substitutional uncharged impurities in the planes, on the other hand, which carry free bound spins according to several measurements, lead to reasonable residual resistances and to T_c lowering. This point of view is carefully worked out in my Science article.

We can only conclude that current is carried by some collective or soliton excitation whose motion is controlled by uncharged entities, i.e., $Z = 0$, $F_{OS} = \infty$, Fermions as we shall shortly discuss. A good model for such a state and such excitations is given by appropriately reinterpreting the exact Lieb-Wu solutions of the one-dimensional Hubbard model.

$Z = 0$ is confirmed by the several measurements, all of which agree, of inelastic scattering τ's: infrared, Raman background, and ARPES. All tell us that $\frac{1}{\tau} \propto \omega$, and if we treat the carriers as quasiparticles with a self-energy $\sum$, $\sum_{im} \propto \omega$ and $\sum_{re} \propto \omega \ln \omega$ by Kramers-Kronig transform, hence $\frac{\partial \Sigma}{\partial \omega}$ diverges.

In fact, it is likely that the true ω-dependence is not $\omega \ln \omega$ but $\omega^{1-\alpha}$ and both real and imaginary parts are power laws. In either case $\frac{\partial \Sigma}{\partial \omega}\big|_{\omega=0} \to \infty$ and $Z = (1 - \frac{\partial \Sigma}{\partial \omega})^{-1} \to 0$. One's first response to $Z = 0$ is to abandon entirely the Fermi liquid quasiparticle theory on which this derivation is based, as well as the Fermi surface. However, we shall see that both the one-dimensional model and the experimental facts show that the Fermi surface and the Fermion-like excitations may remain even when electron-like quasiparticles are absent.

There is a strong theoretical argument and motivation for this peculiar non-Fermi liquid normal state. The one-band Hubbard model has the property of having the "upper Hubbard band," a separated band of states thought of as comprising the motions of electrons on doubly occupied sites. This can be given a precise meaning in at least two ways: as a band of separated "anti-bound" particle-particle scattering states, or by the Rice-Kohn-Anderson canonical transformation procedure to the "t-J" model in which the kinetic energy term is *exactly* projected onto an equivalent singly occupied subspace. The key operative word is "projective": the Hilbert space of the new low-energy problem is smaller than that of the corresponding Fermi liquid states, because projection operators have zero eigenvalues. This change of Hilbert space means that the states of the $N + 1$ body problem live in a new Hilbert space and are necessarily orthogonal to those of the N body problem, hence Z—which is a ground state to ground state overlap integral—is zero.

Intrigued by this problem of doping of the Mott-Hubbard insulator, theorists have rather ingeniously found at least three viable alternative approaches, and we take it as an optional premise that one should follow only one of these—but as dogma what the final result must be like.

Two reasonable but somewhat indirect approaches are twisted antiferromagnetic order parameters and "flux" or "anyon" phases. A few carriers doped into the antiferromagnetic insulator can be shown rigorously to generate a co-moving distortion ("twist") of the antiferromagnetic order parameter, and the resulting soliton may be a model for our $Z = 0$ object—it is an excellent one in 1-dimension. A second concept is to model the Mott insulator with an

"RVB" liquid of short-range singlet pairs, and to study solitons in this; again, a limiting process starting this way is a way of thinking about 1d. The short-range RVB may be based on a "flux phase" and give solitons which are anyons. The resulting particles carry gauge fields in order to implement the projective transformations. But experiment seems firmly to reject the short-range picture. Yet another method is to follow the original suggestion of Anderson and Zou modeled on the short-range RVB ideas and to implement the projection with a "slave boson" (or "slave Fermion") theory with spinons, holons, and gauge fields, calculating directly using the full gauge theory. Lee and Nagaosa, and Yoffe and Wiegmann, calculate in this way without assuming short-range RVB, and in particular Lee assumes a Fermi surface for his spinons. Wilczek and Zou carry this system still further. Many properties can be approximately calculated in this way, but as yet it is not accurate, as far as one can tell. A fourth concept, closest to the original picture of Anderson and BZA, is a $Z = 0$ liquid of spin 1/2 Fermi-like particles, and charged S = 0 excitations or holons. Perhaps the most straightforward way to describe such a fluid is as a limiting case of the Fermi liquid, such a system as one might find if—as in 3 dimensions—there is a U_{crit} dividing Fermi liquid from Mott insulating states. As one approaches U_{crit} $Z \to 0$ implying $\frac{\partial \Sigma}{\partial \omega} \to \infty$. However, we recognize that there remains a finite spin velocity in this limit, hence the Fermion mass does not go to zero. This is only possible if $\frac{\partial \Sigma}{\partial \omega} / \frac{\partial \Sigma}{\partial k} = \frac{1}{v}$ remains finite, hence the compressibility which is proportional to $(\frac{\partial \Sigma}{\partial k})^{-1} \to 0$ and the Landau parameter $F_0^S \to \infty$. Thus charge cannot be carried by the Fermions, any Fermion being perfectly surrounded by compensating charge in the medium. It must reside in collective sliding motions of the Fermion liquid, or bose-like charged excitations which we think of as a second, charged soliton or"holon." All of this picture is precisely modeled on the one-dimensional case which is exactly soluble.

Several experimental facts drive us to this type of theory. The Pauli-like spin susceptibility (except in the anomalous case of $Y(Ba)_2Cu_3O_{6.65}$ and other spin-gap materials which have many unique properties) is that of an equivalent spinon liquid. Most strikingly, the existence of a Fermi surface in the ARPES measurements on the normal state can only be understood this way (the interpretation of ARPES energy distribution data certainly does not contradict, and possibly strongly supports, the "Luttinger liquid" picture).

One of the methodological strengths of condensed matter physics is the overdetermination by data. It is often the strongest evidence for a theory that none of the wide variety of possible probes contradicts it; let us assure you this is the case here.

The main alternative theory, which is excluded by the evidence for Dogma IV, is any of the many versions of conventional or unconventional Fermi liquid theory. The data require that the electron excitation be composite, not elementary, and, in fact, that it decay at a rate given by the available phase space. This type of behavior of ρ_c is seen elsewhere *only* in conjunction with some kind of "unusual" condensed state such as SDWs or CDWs in dichalcogenides or organics, where the charge carriers are coupled to an order parameter. And the existence of the upper Hubbard band, which drives the $Z = 0$ process, is clear in optical data.

The "marginal Fermi liquid" theory is an alternative path which uses much of the above experimental argument but attempts to distance itself from the Hubbard model. It is true that attractive models can also have varieties of non-Fermi liquid behavior in 1 and 2d systems. The primary distinction is the rejection of the concept of spinons and of the spinon Fermi surface without any consistent alternative being proposed. The theory as it is normally given rejects our argument that v_F and m^* remain finite, so has no second excitation for charge. The evidence which is relevant is that for Dogma II, and for the upper Hubbard band. Our frank assessment is that the most valid results of MFLT are those which are equally well explained by the main line of reasoning. There is one interesting question: assuming, as we must, that $Z = 0$ implies charge-spin separation and the spinon Fermi surface, is it still necessarily true that there is a true holon excitation, or could the charge be carried by a collective resonance near the $2k_F$ edge of the spinon pair spectrum? Calculations aimed at proving the existence of abound $2k_F$ charged collective mode of this sort have not succeeded. We tentatively reject such an alternative, but this is a serious question for study.

We summarize Dogma IV by the statement: the normal metal is a two-dimensional Luttinger liquid: i.e., Fermion-like spin excitations—spinons—establish a Luttinger Theorem Fermi surface. Charge is carried by an alternative excitation which necessarily itself exhibits a Fermi surface. Charged excitations cannot be thought of as having fixed statistics, and none of the three—electron, spinon, holon—is a simple bound state of the other two (in contradiction to short-range, KRS RVB and to PWA's Varenna notes!).

Dogma V: The above state is strictly two-dimensional and coherent transport in the third dimension is blocked. We feel it is also, but less, evident that this two-dimensional state is not superconducting and has no major interactions tending to make it so, at least near the usual T_c. For the dogma, the often cited c-axis resistivity would be argument enough (the most common data being those in the infrared; see Uchida et al. and Timusk et al.), but there is a second important experimental one. The ARPES data on BISCO 2212 resolve a very sharp cusp feature near the Fermi surface, and an even sharper quasiparticle peak below T_c in a variety of directions presumably moving through the Fermi surface. Because of the pair of close CuO_2 planes in BISCO 2212, the Fermi surface in the planes should split in two by an amount equal to the effective interplane hopping integral $t_\perp$, and this effect is indeed to be seen in the calculated band structures. The relevant splitting is a couple of tenths of an ev. At some symmetry points, the splitting is small because the hopping effectively takes place via the Sr^{++} ions between the CuO_2 planes, and hence the effective matrix element can be frustrated. Nonetheless, enough directions have been probed to indicate strongly that this odd-even splitting of the CuO_2 planar states doesn't exist.   *d

Theoretically this is true—here we have only the one-dimensional model to use as an analogy. Several workers in the heyday of one-dimensional physics pointed out that interchain hopping, if weak, renormalizes to irrelevant at $\omega, k \to O$. (More recently, D.-H. Lee and our group have come to the same conclusion.) Thus there exists a critical value of the hopping, $t_\perp$, between CuO_2 layers, below which the physics of the Luttinger liquid can remain strictly lower-dimensional. [More recently, this seems to have become controversial, but as far as I can see this is purely a semantic misunderstanding (see JETP Letters).]   *e

This is, however, strictly a one-dimensional result depending on the small exponent α in the Green's function, while we feel that the physical effect is larger and more obvious. If the Luttinger liquid dogma is correct, the electron is not a stable excitation in the plane any more than a quark is in free space: its charge and spin move off immediately at different velocities. Thus it has no $\omega = 0$ amplitude: it cannot find a stable eigenstate with the same quantum numbers of charge, spin, and momentum, which is the prerequisite for coherent motion. Thus theoretically Dogma IV seems to lead automatically to Dogma V.

The impact of Dogma V, then, is that the two-dimensional state has separation of charge and spin into excitations which are meaningful only within their two-dimensional substrate; to hop coherently as an electron to another plane is not possible, since the electron is a composite object, not an elementary excitation.

The assumption that the two-dimensional system is not superconducting is based on the scale and the variability of T_c. T_c varies from 0 to 6 to 130° depending on the overall 3-dimensional structure: except for ρ_c, the only highly variable experimental parameter. This is a scale which is much smaller than the scale at which the Luttinger liquid properties set in, which is not less than $\sim 1000°$K. The properties of the planes have become reasonably uniform, and dominated by strong interaction effects, from this temperature downward. The question is, what could possibly make T_c be of an entirely different, and widely varying, scale? It seems almost required that T_c itself depends radically either on interactions between planes or with the substrate, and is *not* a purely 2-dimensional effect. When superconductivity in fact ensues, the properties seem indeed to become *more* isotropic, and the penetration depth λ_c^2 which is basically a measure of the c-axis plasma frequency changes to a value which is much too low vis-à-vis the c-axis resistivity.

There are simply no indications of a unique, purely two-dimensional type of superconductivity. We go into this point more fully, since it *is* true that 2d fluctuation effects of a fairly conventional type exist above T_c in YBCO and in fact in all the "multilayer" systems such as Bi and Tl 2212. We propose that *pairs* or *triplets* of layers *do* become superconducting by dint of their interlayer coupling and behave like a single conventional superconducting layer. This is quite different physically from "anyon superconductivity" within a single layer. The Kosterlitz-Thouless behavior often seen is consistent with this idea.

It is possible that the various 2-dimensional superconducting states proposed by others are in fact nothing but the $Z = 0$ Luttinger Liquids of Dogma IV in a new guise. No attempts at a fluctuation or thermodynamic theory of "anyon superconductivity," for instance, have been made, only a ground state is discussed. Thus a real possibility is that T_c is zero or the state is sensitive to impurity scattering of one type or another. The anomalous ρ_{ab} conductivity seen in experiments extrapolates to zero and hence, in some sense, the normal state really is "superconducting"; with $T_c = 0$; but some other T_c mechanism intervenes, probably that of Dogma VI.

***f*

Dogma VI: Interlayer hopping together with the "confinement" of Dogma V is either the mechanism of or at least a major contributor to superconducting condensation energy.

There is better than a rough identity between the conductivity per electron in a given direction and the total kinetic energy: each are proportional to a velocity-velocity correlation function. Kohn has shown that in a single tight-binding band the frequency integral of the conductivity is proportional to the mean kinetic energy per electron

$$\int \sigma(\omega)\, d\omega \propto \langle t_{ij} c_i^+ c_j \rangle.$$

Along the c-axis there is a great defect in conductivity: there is no coherent motion of electrons in the c-direction. This means that there is, in the normal state, a missing energy of order $t_\perp^2/t_{\parallel}$, *which is regained in the superconducting state* (since, experimentally, the superconducting response function in this direction appears to be consistent with band structure, and this measures the restoration of the low-frequency part of $\int \sigma(\omega)\, d\omega$).

***g*

There is, therefore, a contribution to the condensation energy which is not a theoretical but an experimental fact and which comes from the interlayer tunneling energy. It is compelling to identify this as the source of the condensation energy, but, of course, pairing has other consequences and may well gain energy from phonon and other modes as well.

The heuristics of transition temperatures lend strong support to this view. We have published a study of relative T_c's of *Bi* and *Tl* 2- and 3-layer materials on this basis. Pressure coefficients are very favorable: these are large of the right sign in all 1-layer materials, but for two or more strongly coupled layers one cannot be sure what pressure will do. The one mystery is *Tl* one-layer, which has a very mysterious chemistry, fluctuating from zero to 60° with physical state. We suspect that Tl orbitals are close to the Fermi level. The fact that this T_c is so sensitive to interlayer chemistry is in itself mysterious if T_c is intralayer. We also point out (stimulated by a remark of S. Trugman) that the fact that the top layer of an YBCO crystal seems not to be superconducting is very strange on any intralayer theory. We suspect that YBCO surface problems are caused by the lack of a satisfactory cleavage in this crystal separating the pairs of layers at the chain lattice planes.

The mathematics of "confinement"—the blockage of interlayer coherent motion—is not complete but is becoming so.

On mechanism, there are no plausible competing theories. It is hardly necessary to detail this. It may be necessary to point out that the "BCS mechanism" has quite a subtle heuristics which was well understood by Morel and Anderson following Bogobliubov within three years of BCS. No other "theory" of high T_c is beyond the level of the Cooper model calculation, using vaguely described "attractive forces" of some hypothecated kind with no chemical or structural heuristic whatever. What is more, no hint of these mechanisms appears in well-designed experiments.

The mystery story principle requires that a unique consequence—high T_c— be associated with the only other truly unique properties: $Z = 0$ Fermi liquids, and exact two-dimensionality. In summary, then, Dogma VI tells us that whatever the normal state physics, we can count on the mechanism of interlayer tunneling deconfinement to account for T_c.

A parenthetical remark: the anomalous case of $YBa_2Cu_3O_{6.65}$ (the "60° superconductor") has attracted a lot of attention. It may well be that this spin liquid is a real example of one of the other alternative states, with spirals or fluxes. However, the tendency to a *reduction* of all spin fluctuations with temperature suggests that we are actually getting a singlet-paired CDW forming, not an SDW. All such liquids have the $Z = 0$ syndrome and charge-spin separation, so all are equally subject to the same interlayer mechanism.   *h

These six dogmas lead us through a rather intricate maze of alternatives to a consistent, coherent view of the whole fascinating phenomenon of the superconducting cuprates. The data-overdetermination of condensed matter physics tells us, actually, that if a crucial countervailing experiment existed we would surely not be able to sustain such a heavy structure.

There is, however, a set of experiments which can confirm it beyond the shadow of a doubt: and, in preliminary fashion, partially they have done so. These are of three kinds.

(1) Careful infrared and superconducting studies of the c-axis electromagnetic responses, especially in 2 1 4 or other one-layer materials. If (VI) is right, the conductivity which is missing in the normal state should partially reappear in the superconducting response $1/\lambda^2$. (This has been nicely done in 2 1 4 by

Uchida.) Infrared can confirm the absence of an appropriate $\sigma_c(\omega)$ in the whole relevant range in the normal state.

An experiment which can be quite interesting would be to search for the "$\sigma - \pi$" interband absorption in 2-layer materials in the c direction. There is, in every band calculation, a splitting of the Fermi surface, due to $t_\perp$, of ~ 0.1ev, essentially between bands even and odd in the Y or other ion between the close CuO_2 planes. This should, in Fermi liquid theory, cause a very strong c-active infrared absorption, which should be easily visible. In fact, it should also have a very strong tail toward low frequencies because of the vanishing of this splitting in symmetry directions. If it is not there our diagnosis of "confinement" is confirmed very strongly. The absence of ρ_c can be thought of, naively, as some kind of localization effect, but the strong reduction of electronic conductivity in a wider frequency range would be very telling.

As with the other experiments we will discuss, these electromagnetic anomalies will manifest themselves as an apparent *failure of a sum rule*: the superconducting response $1/\lambda^2$ will not be equal to the integral up to the gap frequency of the normal conductivity.

$$\left(\frac{1}{\lambda^2}\right)_c \neq \int_0^\Delta \sigma_{norm}^c(\omega)\, d\omega.$$

This is a striking prediction which may already have been verified.

(**2**) Careful experiments on tunneling along the c axis. "HIH" junctions consisting either of single crystals clamped together, relying on a surface inactive layer, or true tunnel junctions, for instance created by MBE, should manifest extremely interesting properties. The same arguments needed to explain $\sigma_c \simeq 0$ seem to lead to weak tunnel currents as well, in this case. We should see a strong contrast between the normal and superconducting states. In the superconducting state Andreev scattering allows quasiparticles to move freely between planes; spinons pick up charge from the condensate and turn back into electron quasiparticles. Thus tunneling, at least in the energy gap region, will be relatively normal. (Of course, energies above the gap are not affected.) Thus we can expect to see a rather clean superconducting tunneling spectrum complete with gap and Josephson supercurrent in junctions which appear more or less insulating in the normal state. This is a tricky but easy experiment. I suspect that such junctions have already been observed and discarded by a number of experimentalists.

It also may be relevant to reexamine really clean c-oriented normal to superconductor contacts—why are they so bad? Do they also show changes below T_c?

(3) $G_1(k, \omega)$ as measured by ARPES, like the electromagnetic response, will probably not obey the usual sum rules. When the sample is normal, the quasi-particles are composite and G_1 is spread over a broad spectrum. We have shown elsewhere that when the system becomes superconducting the quasiparticle amplitude becomes finite and peaks reappear in the Green's function. Again, the effect is an apparent failure of the sum rule for quasiparticle amplitude: the peaks appear to arise from nowhere, because they are borrowed from a wide range of energies. There is a difference, however, in that in this case the original sum rule is still satisfied if all frequencies are summed over.

In the above we have given a logically consistent overall picture of the physics of high T_c. At the very last stage we find that the picture is susceptible to several clear experimental checks, involving phenomena inexplicable on other grounds. Some other vital areas of experimental study can also be suggested: clearly high-resolution, careful ARPES will continue to be of great value. A less obvious but important area is doping the planes with spin impurities, which by smearing the spinon Fermi surface will rapidly destroy the unique Luttinger liquid properties, reintroducing residual resistance and metallic transport along the c axis as well as reducing T_c. In subsequent chapters we will flesh out these ideas. We will focus on transport, experiment, and theory; on spectroscopy, Raman, IR, and PES; and then on theoretical concepts: the 1d Hubbard model, Luttinger liquids in higher dimensions, and finally the interlayer mechanism for superconductivity.

Epilogue to Chapter 2

Well, How Did the Dogmas Turn Out?

At this point it is possible to look back over the route from the dogmas to the relative certainty we now enjoy and give a synthetic view of the mechanism and the underlying microscopic physics of the cuprate superconductors. Since this is a book, not a review article, and because the detailed arguments and the detailed confrontation with experiment will be the subject of the rest of the book, I will not give literature references or attempt rigorous argument here. The situation is so complex experimentally and theoretically that such a simplified road map is essential. It will also help if I avoid, as far as possible, polemics, in favor of simple description.

THE ELECTRONIC PARAMETERS: HUBBARD MODEL AND $t_\perp$

In chapter 1, we have already shown that the essential physics is contained in the two-dimensional, one-band, Hubbard-like model for the CuO_2 planes, which is essential to the understanding of the insulating, undoped, antiferromagnetic phases which almost all of the materials show. The invariant association of Mott insulating antiferromagnetism with high-T_c superconductivity and of both with the striking, chemically stable planar structure cannot be a coincidence.

It helps if we realize that there are two widely separated energy scales, which elsewhere I have called the "A" and "B" scales. The A region from $\sim 200°$ up to ev, has as its controlling scale the parameters of the 2D Hubbard model $\sim$ several ev. "A" does not vary much from substance to substance; for example, the antiferromagnetic J in the plane is always around .1–.15 ev. One varies the planar properties in this regime only by doping.

The second, B scale is the scale of T_c, and of spin-gap phenomena, and is very variable, extending in temperature downward from $200°$ (or in energy a bit higher), to near zero. We are certain that this scale is set by the hopping interaction between planes, which is the major term left out of the A scale.

For practical purposes we leave phonons out of account, an omission which is simply driven by the absence of substantial evidence that they influence the electronic behavior (no polemics!) except peripherally—essentially, they seem to enter only to mess things up, as in the well-known anomalies for $\delta = 1/8$ in "2 1 4." This must be a general property of non-Fermi liquids: the phonon-electron coupling *is* renormalized by "Z," if you like, and has little effect.

The B scale is very complex. The reliably high T_c's tend to come with 2 or 3 tightly coupled planes, as in 2- and 3-layer BISCO, YBCO, and the Tl, Pb, and Hg compounds. More layers can be persuaded to form but not to take doping at optimum percentage. Between these groups of tightly coupled layers there can be very weak couplings as in BISCO or deoxygenated YBCO, or quite strong as in the Hg compounds, and probably in $YBCO_7$. In these tightly coupled cases one must presume that there are bands in the intervening layers which are at (in the case of chain bands in YBCO) or near (as in $TlCCO$) the Fermi level, and which transmit electrons efficiently between the CuO_2 layers. But, except somewhat for YBCO, we are far from really understanding the electronic structure of these complex systems. In any case, we observe that the couplings are always *weaker* than the direct coupling of CuO_2 layers with only a single + ion (Y, Ba, Ca) between them.

The band structure in the planes is quite simple if the examples of BISCO, YBCO, and NdCCO are considered together. It is well described by a $t - t'$ tight binding structure, which follows from band theory, with its maximum energy at X (total bandwidth $\sim$ 2–3 ev). Until it is doped to $\sim$25%, the Fermi surface is a simple hole surface around X, but at that point it goes through the van Hove singularity at M and for higher doping may be closed around Γ. At $\delta = 0$ the Fermi surface would not be square nor visit the M point.

The same band calculations which give this agreement suggest that the hopping integral between close planes, $t_\perp$, is strongly peaked near "M" and along $\Gamma - M$, and weak or nonexistent along $\Gamma - X$. Its maximum magnitude is 3–500 mev. There is no reason, nor are there good calculations, suggesting that $t_\perp$ in 214, $NdCCO$, or in 1-layer Hg or Tl compounds necessarily has the same structure. This peaking means, we shall see, that $\Delta(k)$ is highly anisotropic independently of other interaction mechanisms.

THE NON-FERMI (LUTTINGER) LIQUID

Experimentally, the electronic state, when the planes are doped to be metallic, is a non-Fermi liquid on the "A" scale with charge-spin separation, an anomalous Fermi-surface smearing exponent α, and a fortiori in that Z, the wave-function renormalization constant, $\equiv 0$. It retains a *finite* Fermi velocity and hence, by a simple argument, $Z \equiv 0$ implies an *incompressible* liquid, if compressibility is defined as if it were a Fermi liquid, as $(\partial E(k)/\partial k)_{kF}|^{-1}$. This property intuitively leads to insensitivity to phonons, to impurity scattering, and to the hopping integral $t_\perp$: the Fermi surface is very rigid and does not respond to one-electron perturbing potentials in the usual way.

Theoretically, it came as rather a surprise that this non-Fermi liquid was a simple consequence of two-dimensionality and repulsive interaction. I had proposed this earlier using an argument which depended on the bounded spectrum of the Hubbard model, but it seems to be general that in 2D the on-energy shell forward scattering phase shift for two opposite-spin particles near the Fermi surface is finite. This makes the addition of a particle at the Fermi surface into what Ludwig has called a "boundary-condition-changing-operator," which causes the orthogonality catastrophe and reduces Z to zero. A recent paper with Khveshchenko (reprinted in this book) shows how to cope with this formally, by recognizing that such interactions modify the commutation relations which control the Fermi surface dynamics. An electron of spin σ carries with it a co-moving partial hole in the $-\sigma$ Fermi gas, hence its operators do not anti-commute with those of opposite spin. The eigenexcitation which results must have backflow added to it, to make it equivalent to an electron, backflow which involves an infinite number of soft quanta of density fluctuations. Also, since up and down spins are not dynamically independent, the eigenexcitations carrying spin and charge move at different velocities. This "Luttinger liquid" behavior is an organic whole and the two phenomena of anomalous exponents—equivalent to $Z \equiv 0$—and spin-charge separation are not distinct.

The formal way of handling this uses an idea due to Luther and Haldane. Because of the Luttinger-Ward identities the Fermi surface is incompressible in momentum space, while particles at each point of the Fermi surface are separately and independently conserved. By using these ideas the dynamics of electrons may be converted into a dynamics of the Fermi surface, each "patch" of the surface obeying its own conservation law. The conservation law couples the number and phase (or gauge) variables separately for each patch, which leads to a simple bosonic dynamics, in harmonic order in the Fermi surface fluctuations.

One recognizes that this means that there is no single, universal gauge field coupling to all electrons; rather, the constraints have become local to the partic-ular patch of Fermi surface. Thus gauge techniques which are meaningful for free particles are irrelevant in the presence of a Fermi surface. One aspect is that the transverse gauge field is an irrelevant relabeling of states. The gauge ideas we originally proposed in the early days of "RVB" are just wrong, regrettably, or at least unusable.

The transport properties of this "Luttinger Liquid" are quite complex. It is possible that many of the existing calculations done on 1-dimensional systems are not correct because of the failure to take into account the existence of two independent fluids of excitations. In the most common mode ("holon nondrag") the charged excitations are nearly stationary and the current is carried by the backflow of charge which accompanies a flow of spin excitations or "spinons." But the spinons feel acceleration only during the brief period before the electron decays into its two components. This is the regime which shows $\sigma \propto \omega^{-1+2\alpha}$ and the Hall angle which is inversely proportional to T^2 or ω^2. Other regimes, depending on temperature and purity, are also seen but less well understood. The recent microwave Hall angle measurements by the Maryland group demonstrate this compound behavior of the current very well. Stafford as well as Coleman and Tsvelik also point out the strange thermopower which results from the two-fluid structure of transport.

The broad angle-resolved photoemission peaks which are seen in these materials are not background but give the true density of states $Im G_1(k, \omega)$. Along the unit cell diagonal, one can see a sharp cusp feature which must be the spinon energy, and in this direction the linear dispersion and power-law decay of $G(\omega)$ are clear. The EDCs are much broader along M-X or $\Gamma - M$, which is a consequence of interlayer coupling: the incoherent interlayer hopping should broaden the spinon peak by $\sim t_\perp^2/t_{||}$ (see later). But none of the curves show a splitting per se which is identifiable as the $\epsilon_k \pm t_\perp$ splitting of odd vs. even bands, especially when we realize that $t_\perp$ is 3–500 mev. My impression is that the proximity or not of the Van Hove singularity is simply another irrelevance; T_c is certainly optimum in BISCO when it is .2ev from the Fermi surface, not when it is at E_F. Nuclear magnetic resonance relaxation in the normal state shows a strong T-dependent enhancement for the Cu ions in the plane, but behaves like a Fermi liquid for most other ions. Complex, delicate fits to the enhancement using "antiferromagnetic spin fluctuations" can in most cases be replaced by the power law enhancement of χ'' appropriate to a Luttinger liquid, except where this enhancement is reduced by the spin-gap phenomenon. I am as baffled as anyone by the complete absence of enhancement for the O nuclei, I am only sure that it cannot be explained by antiferromagnetic structure, which neutron scattering does not see.

INTERLAYER HOPPING: THE SUPERCONDUCTIVITY MECHANISM

Let me now descend to the "B" scales. What does the interlayer hopping $t_\perp$ do to the Luttinger liquid?

Experimentally and theoretically, what it does not do is move the Fermi surface: split it for 2 layers, curve it for many equivalent ones. This calculation is actually trivial, though widely denigrated. The parameter which tells us the Fermi surface shape is $t_\perp/t_{||}$. The renormalization exponent for this may be trivially calculated, and for *all* Luttinger liquids it is negative: $t_\perp/t_{||}$ renormalizes to zero as $(b)^{-2\alpha}$: the band renormalizes to 2 dimensions from 3, 1 dimension from 2, if $t_\perp/t_{||}$ starts out $\ll 1$.

This leaves a dilemma, since the exponent for $t_\perp$ itself is positive. Its effects are *not* irrelevant. Chakravarty, and Strong and Clarke, found a fascinating resolution to this dilemma: hopping between Luttinger liquids is incoherent rather than coherent. An only slightly schematic way of describing what happens for a pair of chains or planes coupled by $t_\perp$ is to say that in the Fermi liquid, the two energies for a given k split along the real axis: $\epsilon_k \to \epsilon_k \pm t_\perp$. But in the incoherent case, they stay identical but become complex: $\epsilon_k \to \epsilon_k + it_\perp^2/t_\parallel$: they move off into the complex plane. This is a familiar type of behavior of a splitting interrupted by some perturbation—the earliest example I know of being the ammonia inversion splitting under pressure broadening. In the Luttinger liquid case, the Lorentzian broadening implied by a complex pole is slightly different, but its effect is the same: the Green's function is no longer sharp. It is noticeable that the photoemission features are sharp near the $\Gamma - X$ line, but broadened near M where $t_\perp$ is larger: we expect a broadening of the same order as the energy gap when superconductivity occurs. The immediate consequence of this incoherence is that transport along the c-axis is always diffusive. When $T > t_\perp$ there is no possibility of coherence since the mean velocity is zero even in a Fermi liquid, and a simple argument due to Kumar gives us $\rho_c \propto T$ (possibly $T^{1-2\alpha}$), actually $\sigma_c \simeq t_\perp^2/t_\parallel T$ which for $T > t_\perp$ is always well below the Mott limit. Below $t_\perp$, Clarke and Strong have derived a diffusion constant $D \propto \omega^{2\alpha}$ so that σ at low temperature or any frequency is

$$\sigma \propto \omega^{2\alpha}, \ T^{2\alpha}.$$

There is quite good experimental evidence for this kind of power law behavior.

The low-frequency matrix elements of $t_\perp$ cause the incoherent single-particle interlayer hopping. The matrix elements to higher-energy states in the spectrum lead to *second*-order perturbative effects: superexchange, and pair hopping. Both of these have the capability of becoming coherent interactions, that is, of providing ground-state to ground-state matrix elements and thereby providing an energy which is *first*-order in the second-order perturbation term. It is this remarkable and counterintuitive quantum-mechanical effect which is at the heart of the mechanism of high-T_c superconductivity. It is essentially Josephson's discovery, that one could obtain coherent supercurrents from incoherent hopping processes. But there is also a parallel with the "superexchange" mechanism of magnetic coupling between spins in the Mott insulator.

The way to describe the process in words is as follows. The energy $t_\perp(c_{1k}^+ c_{2k} + c_{2k}^+ c_{1k})$ gives repeated virtual hops of the electrons of momentum k from layer (1) to layer (2). But in the majority of such hops, it ends up in a high-energy state of spinon and holons, and because of the high energy E_n it stays there only for a time $\tau_n = 1/E_n$ (the few low-frequency states into which it can hop cause the incoherent tunneling). Normally it will simply "bounce," returning to the original state after a time τ_n. But if *during* this hop another electron of the opposite spin hops, the net result can be a coherent process and the original hop becomes quasi-permanent. The second hop can be of two kinds.

(1) In a superexchange process, the second hop is from 2 to 1, and the net result is the exchange of a spin between chains 1 and 2:

$$\mathcal{H}_{eff} = t_\perp^2 \sum_{k,k'} c_{k\uparrow}^+ (1)\, c_{k\uparrow} (1)\, c_{k'\downarrow}^+ (2)\, c_{k'\downarrow} (1) \times \left(\frac{1}{E_n}\right) + \mathcal{H}.\mathcal{C}.$$

(2) in a pair-hopping process, the second hop is in the same direction,

$$\mathcal{H}_{eff} = t_\perp^2 \left(\frac{1}{E_n}\right) \sum_{k,k'} c_{k\uparrow}^+(1)\, c_{k'\downarrow}^+(1)\, c_{k'\downarrow}(2)\, c_{k\uparrow}(2) + \mathcal{H}.\mathcal{C}.$$

In the first case, the process can be coherent because the final state has no net charge exchanged between the chains, so can represent only a spinon exchange; in the other there is no net spin exchanged, so that only holons are transferred. Both can serve as effective pair interactions and cause superconductivity. But if they serve as pair interactions, they both have a unique character: the interacting pairs are of the *same* transverse momentum because $t_\perp$ conserves momentum. If we set $k' = -k$, (1) and (2) represent two effective pairing Hamiltonians

$$\mathcal{H}_{SE} = \sum_k \lambda_{SE}^{(k)}\, c_{k\uparrow}^+(1)\, c_{-k\downarrow}^+(2)\, c_{-k\downarrow}(1) c_{k\uparrow}^+(2)$$

$$\mathcal{H}_{PT} = \sum_k \lambda_{PT}^{(k)}\, c_{k\uparrow}^+(1)\, c_{-k\downarrow}^+(1)\, c_{-k\downarrow}(2)\, c_{k\uparrow}(2). \qquad (3)$$

These are the pairing Hamiltonians which lead to the high T_c's of the cuprates, and also are responsible for the many anomalous observations on cuprate superconductivity.

The other counterintuitive feature of these terms is that they act only in the paired state (where the operator $c_{k\uparrow}^+\, c_{-k\downarrow}^+$ has matrix elements within the ground-state manifold) while there is no corresponding energy in the normal state. As I said, the distortion of the Fermi surface caused by $t_\perp$ renormalizes away, and this distortion is the way in which $t_\perp$ modifies the ground-state energy. This unexpected absence of the compensating normal state term is the key to the physics of the cuprates.

SUPERCONDUCTING PROPERTIES

The peculiar nature of the main part of the pairing interaction leads to many unique features of the superconducting state, most of which have no alternative explanation.

One of these is not at first sight a superconducting property: it is the "spin-gap," the effects of which appear $100°$ or so above T_c, primarily in bilayer materials in which there is weak coupling between the bilayers. Strong and I have proposed that this is a consequence of the "k-diagonal" form of (3), in which pairing can occur between a pair of planes for each k-value separately without phase-correlation between the different pairs. For a given quartet of electronic states $k \uparrow, -k \downarrow$ in layers (1) and (2), the only occupancies which are coupled by (3) are those with exactly 2 opposite-spin electrons (only 4 out of 16, at most). The splitting among these is larger than the gap which could be obtained by conventional BCS splitting, but there is no cooperative interaction and no phase transition. One may show that generally, at low enough temperatures, a transition into conventional pairing will occur, because of the coupling to other planes and among different k-values. But it is predicted—and observed—that the nature of the superconducting transition in spin-gap systems will be very much altered from simple BCS: the specific heat shows large fluctuation effects above T_c, for instance, showing that much of the binding energy has already appeared above T_c.

Solution of the actual superconducting gap equation (see chap. 7) requires assumptions about the more conventional particle-particle scattering. The first, sound assumption one can make is that this will be dominated by strong local repulsion, so that the pair amplitude at the origin in real space should be zero, which is the average pair *amplitude* (not gap) in k-space. There are two ways to average to zero: averaging in angle (a "d-wave") or averaging in energy (dynamic screening). It seems possible that dynamic screening is less efficient in these materials than in Fermi liquids, possibly because the Luttinger liquid pair susceptibility may be slightly sublinear, so that the linear perturbation theory may not work (this idea was suggested by Balatsky). If so, the energy gap will tend to be "d-wave," as observed so far (except in $NdCCO$, which does not have a spectacularly high T_c.) For the multilayer, canonical high T_c materials, BISCO, YBCO, and TlCCO, this now seems to be the case. Even the apparently isotropic gap in $NdCCO$ could be a complex d-wave rather than an isotropic state.

One question which strikes one is: to what extent is "d-wave"caused by an attractive interaction, to what extent is it merely the least repulsive state? Since almost any scattering process will disfavor higher harmonics and there may be coupling between nearby k-values caused by deviations from the strict tomographic picture, we cannot be sure that *any* specifically d-wave interaction is necessary to cause a d-wave gap. In actual calculations, it is probably best to use an infinitesimal attractive d-wave interaction for modeling purposes.

If we do so, and also schematize the form of the underlying pair susceptibility for a particular k (not very accurately, but for simplicity)

$$\chi_{pr}(k) \simeq \left(\sqrt{\epsilon_k^2 + \Delta_k^2 + T^2} \right)^{-1}$$

we get a remarkably simple gap equation which demonstrates the various special features of the interlayer tunneling phenomenon. It is

$$\Delta(\vec{k}) \simeq \frac{\lambda(\vec{k})\Delta(\vec{k})}{\sqrt{\Delta_k^2 + \epsilon_k^2 + T^2}} + v(\hat{k})\delta(T).$$

Here $v(\hat{k})$ is the d-wave eigenfunction of the conventional pairing interaction $V_{kk'}$ and $\delta(T)$ is

$$\sum_{k'} v(k')\, x(k')$$

which is assumed small (x is the pair amplitude).

The symmetry of the gap is determined by the small term $v\delta$ connecting different directions of $\hat{k}$, but its magnitude and shape are determined entirely by λ_k. In fact,

$$\Delta_k^2 + \epsilon_k^2 + T^2 = E_k^2 + T^2 \simeq \lambda_k^2 \quad \text{for} \quad E_k^2 + T^2 < \lambda_k^2$$

($\Delta \ll \lambda$ outside this limit).

Thus we see two striking phenomena quite different from conventional BCS:

(1) The dispersion of quasiparticle energy E_k is zero up to $\sim \lambda$ where Δ drops to small values; thus the density of states peaks are much sharper than in BCS.

(2) The *shape* of the gap is strongly T-dependent. At the Fermi surface, with

$$\epsilon = 0, \quad \Delta^2(\hat{k}) + T^2 = \lambda^2(k)$$

$$\text{for } T < \lambda(k)$$

$$\Delta = 0 \quad T > \lambda$$

so that the gap spreads out along the Fermi surface rather than rising everywhere proportionally to $v(k)$.

There are experimental hints of both of these strange phenomena in the photoemission measurements. Figure 2.1 shows schematically how the gap varies with angle around the Fermi surface using the schematic model $\lambda(\hat{k}) \simeq \lambda_0 \sin^2 2\theta_k$ which I have used in several preprints to fit the observations of Campuzano et al.

These considerations about the structure of the gap equation use a highly schematized, heuristic version of the gap equation. The primary justification for it is that its parameters can be fitted to perturbation theory for small x_k while

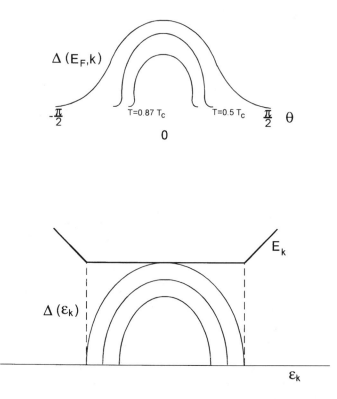

Figure 2.1

for large Δ x_k must saturate at 1. The Luttinger liquid character of the states enters only through the novel interaction term (3).

The equivalent of BCS theory for a fully spin-charge separated Luttinger liquid has not been formally worked out. But there are at least two—and perhaps three—ways in which it can be made clear that the lowest quasiparticle excitation is an isolated pole of the Green's function, not a cut as are the excitations in the normal fluid. Perhaps the simplest way to see this is to realize that a gap for electrons must necessarily imply that there is a gap for both the holon (charge) and spinon excitations. The spinon condensate carries no quantum numbers—as we shall see, spinons are Majorana (real) Fermions with no underlying Fermi sea and in order for spin to flow there must be real spinons—but the holon condensate is simply the charged electron condensate and of course can carry arbitrary amounts of charge. Thus this lowest excitation with the quantum numbers of an electron is a spinon with the appropriate backflow of charged condensate to give it charge e. There is, then, a further gap of Δ before one can create *both* a holon and a spinon. The amplitude which, in the normal state, is found in the continuous holon-spinon spectrum must become concentrated in the one-spinon state. This is then the extraordinarily sharp quasiparticle peak seen in photoemission, and it is the disappearance of the electron $\rightarrow$ spinon + holon decay mode which leads to the long quasiparticle mean free path seen in many kinds of experiments. There is no other explanation of this effect, and in fact it is more mysterious than it seems. The mean free paths (estimated by Ong to be as long as 2500 Å in YBCO) are incompatible with the magnitude of impurity scattering which a conventional quasiparticle would encounter, and the explanation for this is that the current is carried by superflow, only the spin is a true excitation, and spin scattering is weak. It would be interesting to confirm this by deliberately inserting magnetic scatterers.

A final interesting point can be made about the superconducting gap parameter's symmetry. We expect the two terms of (3) to have approximately equal coefficients. The two forms of pairing to which they lead are not mutually exclusive but complementary. Since the wave functions on the separate planes are not necessarily orthogonal (compare the case of superexchange between magnetic ions in a Mott insulator) but somewhat mixed, pairing within planes contains a component of pairing between planes and vice versa. For a pair of isolated planes, we can transform to a basis of even and odd linear combinations of the two planes, $c_{ek}^+ = c_{1k}^+ + c_{2k}^+$ and $c_{0k}^+ = c_{1k}^+ - c_{2k}^+$. It then turns out that both terms of (3) favor pairing of the even linear combination, while they have opposite signs for the odd one; so we may assume the pairing is dominated by the even combination

$$x_k^+ = \langle c_{e\uparrow k}^+ \, c_{e\downarrow -k}^+ \rangle.$$

The observation by Keimer of quasiparticle pair creation by neutron scattering in YBCO exhibits selection rules which confirm that the pairing is predominantly in the even subspace.

This, then, is pretty much the extent to which observations on the cuprate have been, or in some cases could be, checked against the theory. Undoubtedly there are many more checks to come, and some of the above may turn out to be misguided, but I don't, at this moment, think so. Taken all together they seem to form a consistent picture. As you see, the outline of the theory follows the "dogmas," but a great many complications ensued. From this précis of the present, up-to-date picture, we now proceed to detailed discussion of the observations and of the theory.

December 23, 1995

NOTES

*a The number of superconducting compounds continues to grow, and most of the new ones have no chains; but "2 4 8" $YBa_2Cu_4O_8$ has two sets of chains; so the text statement is right in import but legally incorrect.

*b Chain Fermi surfaces are claimed to have been seen by Z.-X. Shen in photoemission. The data are consistent with perfect 1-dimensionality, as we would expect (see later). When the chains are depleted of oxygen, it appears that the pairs of planes in YBCO become decoupled from each other.

*c Recent measurements, theoretically discussed in the 1995 preprint, suggest that our estimates of ρ (residual) were slightly large, and that ρ_{rs} is comparable with the lowest $\rho(T > T_c)$ but not much larger. The argument does not depend on factors of ~ 2 in ρ.

*d The large variation of splitting with k_x, k_y is correct, but the reason given here is not. S. Chakravarty and O. K. Andersen have traced the variation to another pathway, the $Cu\,4s$ orbital (see reprint section).

*e Here I was remarkably and regrettably optimistic. Interchain hopping does not renormalize to irrelevance by the standard reckoning. (D. H. Lee did not join me in stating otherwise, incidentally.) The subject was not closed by my article in JETP Letters, and was further exhaustively studied by my students Clarke and Strong, who ended up confirming much of my statement here, although some—still mostly semantic—confusion remains in the literature. The Fermi surface splitting does renormalize away, but higher-order effects of interchain hopping such as incoherent hopping conduction and superconductivity do not. And, finally, the real results indeed depend on α, and the one-dimensional model is the correct template.

*f The Anderson-Khveshchenko reprint takes the point of view that such relationships are at best tenuous: it suggests that, on broad symmetry grounds, the Luttinger liquid cannot be approached validly from different viewpoints.

*g The situation is more complex and more interesting than this—see reprint section. Only some of the band energy is restored, and this gives us a handle on c-axis charge dynamics.

*h The spin gap phenomenon appears to be a piece of the puzzle which has fallen easily into place: see preprint, '95, by Strong and Anderson. It is singlet-pairing, but not a CDW. It can be thought of as spinon pairing.

3

Normal State Properties in the High-T_c Superconductors: Evidence for Non-Fermi Liquid States

INTRODUCTION

In the previous chapter, on the Central Dogma, I relied heavily on certain assertions about the properties of the normal cuprate materials; in chapter 6 in which we discuss the theory of the normal state, I shall attempt to explain these properties theoretically. In this chapter I shall give an overview of the reasons I believe these properties so clearly indicate a unique, anomalous state of the "normal" metal. Some, but not all, of this material was reviewed in an article in Science.[1] The most important of the data refers to transport, and an increasing role is played by infrared spectroscopy, which tells us essentially the same information. Also highly important is photoemission and related spectroscopies; and, finally, NMR relaxation.

I shall mention neutron scattering spectroscopy in passing, but after early promise this has been the most disappointing of the probes, for several reasons. As we became surer that the normal state, whatever else may be true of it, has a true Fermi surface of "Luttinger" volume (i.e., containing as many momentum states as there are electrons above a filled band) it became clear that the low-energy excitation spectrum, by very general arguments, must have its support on that Fermi surface (aside from long-wavelength collective modes). This is as true of a Luttinger liquid as it is of a Fermi liquid, where the Landau formalism tells us that all excitations may be thought of as multiples of quasiparticle states. But both because of the poor resolution of neutron scattering, and perhaps for a less acceptable reason, namely, custom and habit, neutron scattering results, unlike photoemission, have not been referred to the Fermi surface structure, but fitted empirically to relaxation functions suitable for paramagnets. Great areas of the Fermi surface are mysteriously absent from the neutron scattering data. One can only assume that the technique has not been sensitive enough to reveal actual Fermi surface structure. Fortunately, the recent work of Aeppli et al.[2] on $(La - Sr)_2CuO_4$ (and of others) has remedied this situation to a great extent in one system.

*a

A glance at any other review of properties of these materials (for instance those in the Ginsberg series of books) will find a very different emphasis and, often, rather different data. The latter is not as serious as it appears to be, for four reasons:

(a) I emphasize data on the "optimally doped" metallic crystals. Underdoped phases are often unstable and always subject to localization and "spin gap" phenomena; overdoped phases become 3-dimensional.

(b) For reasons too various to catalogue, much of the data is simply wrong; I have exerted judgment, taken advice from good experimentalists, and done a lot of ad personam judgments to select data I trust. For instance, I don't quote most film transport data because Montgomery method techniques can't be used on films.

(c) A not uncommon fallacy is reinterpretation of the same data in new terms. Tanner and Timusk for instance[3] describe featureless curves in terms of two different components while showing curve after curve where the infrared reflectivity is perfectly smooth and unitary. (Particularly striking is fig. [1] of the cited article.) I pick the simplest, "Bayesian" interpretation. Another example occurs in NMR relaxation; where a single fit works, why introduce two processes? A third is shown in the interpretation of c-axis infrared data: in order to fit a hypothecated sum rule, conductivity is simply added in an unobservable range, and then the data is claimed to "confirm" the sum rule. Again, the rule must be to use Ockham's razor and not introduce unnecessary hypotheses. Thermal conductivity is another example: completely implausible parameters are used to fit it to a phonon model.

(4) Most important, I have a (possibly the only extant) theoretical framework into which to fit the observations; they make so much more sense in this framework that I tend to stop there. Much of the data is invalidated by attempts to squeeze it into conventional theoretical ideas, and I ignore many experiments where such an interpretation is part of the experimental paper, as intrinsically doubtful.

RESISTIVITY AND OTHER TRANSPORT PROPERTIES

Almost the first striking measurements of high-T_c superconductivity revealed a structure of resistivity vs. temperature which showed unusual features in the normal state in addition to the high transition temperature. I well remember Laura Greene and John Rowell calling my attention to this in March 1987. Well-made ceramics characteristically showed a "metallic" normal state with ρ linear in T with a very small $T = 0$ intercept; while poorly compacted samples, showing poorly metallic or insulating properties in the normal state, still showed

MULLER DATA

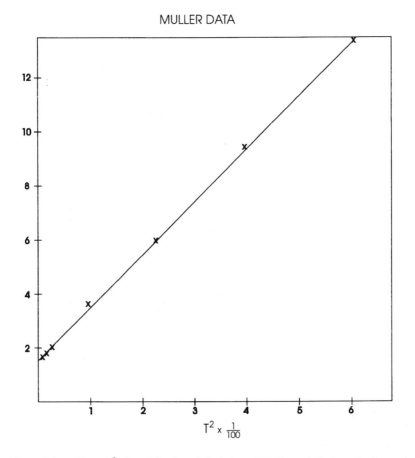

Figure 3.1a. ρT vs. T^2 plotted for the original data of Muller and Bednorz to show ρ is a series combination of increasing and decreasing terms.

superconductivity easily. With the advent of good single crystals and films, we now know that both kinds of behavior are characteristic. Anyone who attended those early sessions will remember the striking similarity of resistivity plots for sample after sample from all kinds of diverse provenance. The first, and perhaps the most successful, heuristic description of data on these materials was the plot, for these ceramics as well as for many single crystals, of ρT vs T^2, which for most materials is fairly accurately linear. See fig. 3.1. This indicates that the resistivity is the series combination of a linear $\rho_{ab} = AT$ (which is usually the planar component) and of a term $\rho_c \simeq \frac{B}{T}$ which is characteristic both of intergranular contacts and of the c-axis resistivity at low temperature. Chapter 6 gives a more careful treatment of ρ_c showing that that alone may

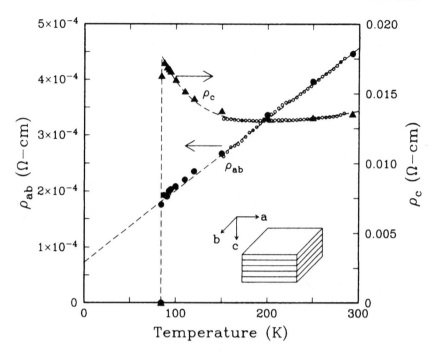

Figure 3.1b. First anisotropic measurements of resistivity by IBM group.

have a rough $\frac{A}{T} + BT$ shape. Although a useful heuristic, this one has met with little acceptance. It tells us a great deal merely by its form: it means that the most universal term in metallic resistivity in ordinary metals, the residual resistance (independent of T), is quite absent in many, many samples. From the very start, then, it was clear that we did not have to do with conventional poor metals.

As single crystal measurements became available (see fig. 3.1b), it became clear that there is an underlying characteristic behavior of the anisotropic resistivity. The resistivity along the planes is quite low, and, in good pure single crystals and films of many materials, measured carefully, extrapolates as $T \to 0$ nearly to zero. Qualitatively we may say

$$\rho_{res}^{ab} \ll \rho(T > T_c) \qquad (III-1)$$

and

$$\rho_N^{ab}(T) \simeq BT \qquad (III-2)$$

where the conductivity per layer per degree, B, is remarkably constant for good high-T_c material. The numerical value is about $\rho = (10 \pm 5)T$ ohms per layer.

We discuss this intralayer conductivity first, going on later to the much smaller, if equally anomalous, interlayer conductivity.

$(\rho_{ab})^{-1}$ from (III-2) is a remarkably high conductivity. It is important to express it in terms of the Mott-Thouless-Yoffe "localization" parameter $k_F l$. The expression for conductivity in two dimensions may be written

$$\sigma = \frac{ne^2\tau}{m} = \frac{e^2}{2\pi\hbar}k_F l \qquad (III - 3)$$

where the universal "Mott" constant (sometimes called the von Klitzing conductance) is

$$\frac{e^2}{2\pi\hbar} = \frac{1}{25,000}\Omega^{-1} \ .$$

With $\rho^{-1} = 10^{-3}$ at $100°$K, we find $k_F l \sim 25$, very far on the metallic side of the Mott-Yoffe-Regel limit $k_F l \sim 1$. These materials are "good metals" in the ab plane, with l's comparable to pure copper at room temperature.

It is not widely enough recognized that numerical values of conductivity are, in Fermi liquid theory, very independent of complicated interaction effects. The Mott conductivity at the metal-insulator transition, as Mott himself emphasized, does not renormalize, and is quite universal experimentally and theoretically. It represents simply a coherence criterion as to how well and whether electrons can propagate by wave motion.

Its universality can be understood most easily in the context of quantum tunneling, and the relevant theorem was proved (I believe by Schrieffer)[4] in the early years of quantum tunneling theory. It is easily shown using the tunneling Hamiltonian technique that the tunneling conductance into or out of a metal is proportional to the one-particle density of states per channel times the velocity of approach to the junction, summed over all the possible channels (for tunneling, read channel = momentum transverse to the junction)

$$\frac{dI}{dV} \propto \sum_{k_\perp^a} \sum_{k_\perp'^b} \left| T_{k_\perp k_\perp'}^{ab} \right|^2$$

$$\times \int_0^V dV' \left(\int dk_\parallel \, v_\parallel \, Im \, G_1(k, \omega = V) \right)_a \qquad (III - 4)$$

$$\times \left(\int dk_\parallel' \, v_\parallel \, Im \, G_1(k', \omega = V - V') \right)_b \ .$$

The integral is known as the tunneling density of states.

The essence of Fermi liquid theory is the idea that quasiparticle occupancies are in one-to-one correspondence with Hilbert space, and that low-energy properties may be calculated in terms of quasiparticles. Within this assumption, it is easily seen that the tunneling DOS is unrenormalized by dynamical effects. We insert $v_{\parallel} = \frac{d\epsilon}{dk}$ and obtain for the tunneling DOS at ω

$$Im \int_{-\infty}^{\infty} \frac{d\epsilon_k}{\omega - \epsilon_k - \Sigma(k,\omega)}. \qquad (III - 5)$$

If $\sum$ vanishes for large $\omega - \epsilon_k$, as it does (or at least goes to a constant), the contour may be closed and we simply pick up $2\pi i$ at the quasiparticle pole (the singularities of $\sum$ are on the other side of the real axis) if ω is on the correct side of the Fermi surface, zero otherwise. Thus as far as transport is concerned, a quasiparticle acts as though it carries exactly e of charge and the density of states and velocity corrections exactly cancel out, even in many-body corrections.

The theory of quantum conductivity may be recast as a "tunneling theory" by generalizing the Landauer theory to many channels[5]: conductivity is a sum of transmission through channels, with the $|T_{nn'}|^2$ factors representing the essentially geometrical scattering of the electrons by impurities or phonons which determines "$k_F l$." In this theory, it has always been noteworthy that the conductivity never participates in many-body renormalizations, due to the above theorem. The essential content of the theorem can be seen to be unitarity: the fact that particles cannot get lost, and if they enter the system in some channel they may be scattered but not lost.

One important exercise is to estimate the residual resistivity that a normal Fermi liquid would experience in the cuprate layers. In most of these materials, with $YBa_2Cu_3O_7$ the *rare* exception, there is very large nonstoichiometry. Let us take 20% doping as a typical, if low, value, realizing that in $(La-Sr)_2CuO_4$ and in the Bi and some of the Tl compounds doping has been shown to be a consequence of nonstoichiometry.

The doping ions in general reside in the neighboring perovskite planes about 4 Å from the CuO_2 layers. It is an overestimate to put the screening dielectric constant at 4 for the essentially insulating oxide material in the intervening nonconducting layer. The screening length in the layers due to the carriers themselves cannot be shorter than the distance to the nearest dopant, since the number of carriers = the number of dopants. We overestimate their screening by assuming that a dopant gives no potential if it is more than one lattice constant away from a given site and otherwise gives the full potential. A good estimate of the mean potential fluctuation from site to site is then

$$\overline{V^2} = .2 \times \left(\frac{e^2}{\epsilon_0}\right)^2 \times (1\text{Å})^{-2} \times \left(4^{-2} + \frac{4.4^{-2}}{2} + 0\right)$$

$$\sqrt{\overline{V^2}} \simeq .69ev. \qquad (III - 6)$$

This is the same order or somewhat larger than t as estimated by Schluter and Hybertson from electronic calculations.[6] Using, nonetheless, Born approximation, we would get

$$k_F l = k_F \, V_F \tau = 2t \, (k_F a)^2 \tau$$

$$\tau^{-1} = \frac{\langle V^2 \rangle}{4t}$$

so that $k_F l \simeq 8(k_F a)^2 \frac{t^2}{\langle V^2 \rangle}$. t^2 is somewhat less than $\langle V^2 \rangle$ so a good estimate is 5–10: about equal to the mean free path deduced from the resistivity as measured at room temperature.

A second approach is to calculate phase shifts for potential scattering from the Friedel Sum Rule. This is done in the Science article[1] with the same result as (III-6).

It hardly needs to be emphasized that in good, pure, single crystals the residual resistance is at most 10^{-1}, and often 10^{-2}, of this estimated value. We approach values like this only when doping with Ni, Zn, or other ions which substitute directly into the CuO_2 plane positions. The argument is only strengthened when we look at estimates of possible resistivity due to phonons and to the rather large static positional disorder often encountered in these materials. These, too, are at least as large as the observed linear T resistivity and in spite of the considerable effort which has gone into fudging them to make them much more linear in T than is remotely plausible, one still is unable to reproduce such resistivity curves as those of Martin et al. on Bi one-layer materials,[7] showing linearity from 10–700°K. The $Cu - O$ layers have strong optical phonons which are known, from other data, to interact reasonably strongly with electrons (specifically, they shift when the energy gap opens up). There is no hint of irregularity in the T-dependence of resistivity near the appropriate temperatures. We show, in figures 3.2a, b, and c, some of the best data on ab plane resistivity, especially that of Ong et al. on YBCO,[8] Martin on Bi materials, and Ginsberg et al. on YBCO[9] to illustrate these points. A preliminary curve on $(La - Sr)_2 CuO_4$[10] is also shown in fig. 3.2.

We will discuss infrared results more thoroughly in the spectroscopy section, but it is worth saying here that the best evidence is, in the first place, that $\sigma(\omega \to 0)$ extrapolates well to the observed d.c. conductivity values, and second, that consistent with $\sigma \propto \frac{1}{T}$, for $\omega > T$ we find, out to $\sim$1000 cm^{-1}, that $\sigma \propto \frac{1}{\omega}$, quite accurately, with no observable features in this region.

A recent measurement shown in fig. 3.3 by Ginsberg et al.[9] has given us the ab anisotropy in YBCO. This is interpreted as implying that there is an appreciable conductivity due to the chains in the normal state but that *both* chains and planes satisfy

$$\rho_{\parallel} \propto T$$

*b

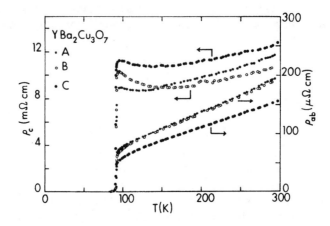

Figure 3.2a. Planar resistivity vs. T for a variety of materials. (a) Anisotropic resistivity of YBCO from Ong et al.

with residual resistivity $< .05 \times \rho(T = 100°)$. (However, recent measurements on substances without chains show similar anisotropy; it is possible this is a band structure effect.)

*c

The simplest and most direct conclusion from these data was drawn by Varma et al.[11] (following Anderson)[12]. The infrared data, as we shall see, support strongly the interpretation that this resistivity implies an *inelastic* scattering rate

$$\frac{\hbar}{\tau_{in}} \simeq \max\left(\frac{kT}{\hbar\omega}\right) \qquad (III-7)$$

with no sign of conventional elastic scattering. But such an inelastic scattering rate for transport seems inevitably to imply a similar inelastic single-particle scattering rate, i.e., the Green's function must decay similarly, and

$$Im\Sigma \sim \frac{\hbar}{\tau} \sim \hbar\omega \, ,$$

for $\omega > T$. But Kramers-Kronig transformation implies

$$Re\Sigma \sim \omega \ln \omega$$

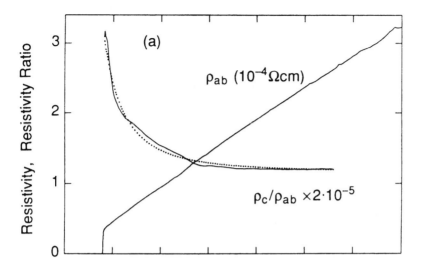

Figure 3.2b. Planar resistivity vs. T for a variety of materials. (b) Anisotropic resistivity of BISCO.

or

$$Z = \frac{1}{1 - \frac{\partial \Sigma}{\partial \omega}} \to 0 \text{ as } \omega \to 0 . \qquad (III - 8)$$

(Note that *elastic* impurity scattering is quite different, in that it can be thought of as simply redefining the single-particle states for the quasiparticles.) More careful study of the data suggests that the dependence is on a low negative power of ω rather than $\ln \omega$. This argument is quite well supported, as we shall see, by direct photoemission measurements of $Im\,G(\omega)$. These arguments are so simple and so consistent among several completely independent sources of data that it hardly seems that the conclusion can be faulted. $Z \to 0$ in turn implies that quasiparticles do not exist.

The planar metallic conductivity in all its strange aspects leaves us convinced that in some sense the *normal* state of the planar layers has superconducting aspects. The conductivity is not affected by impurity or phonon scattering, empirically seeming to be immune to T-invariant scattering as is BCS super-conductivity. It would be a valid description of the empirical facts to say that the conductivity behaves like fluctuation conductivity from superconducting fluc-tuations above a superconducting state at $T_c = 0$, with an exponent -1 as in conventional 2d fluctuation conductivity, $\sigma \propto \frac{1}{T-T_c}$. The coefficient, however, bears no resemblance to that in conventional fluctuation theory. Nonetheless, these facts are powerful arguments that whatever the nature of the supercon-ductivity, it leads to T-reversal invariant—i.e., BCS—pairs.

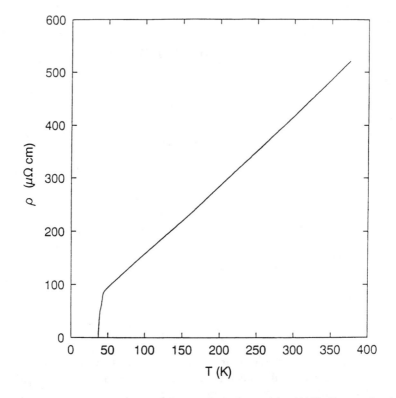

Figure 3.2c. Planar resistivity vs. T for a variety of materials. (c) Single-crystal resistivity of $(La - Sr)_2 CuO_4$.

A second set of data are equally persuasive on this point. Many measurements, for example those of Ong shown in fig. 3.4,[13] demonstrate that certain impurities do succeed in strongly modifying the planar conductivity in such a way as to produce a residual resistivity, while, as we know, many other dopants have an essentially negligible effect. Such effective dopants are Ni, Zn, Co, and Ga in $YBCO$, Pr in $BISCO$, among other cases. These are also the impurities which strongly reduce T_c, in much the same proportion. There seems to be a monotonic relationship between ρ_{res} and T_c in all these cases.

Recently several measurements have shown that the dopants which reduce T_c and increase ρ_{res} have the following two properties:

(a) they are substitutional for Cu in the planar sites.

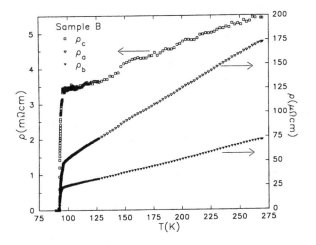

Figure 3.3. Resistivity data of Ginsberg et al. on an untwinned single crystal of YBCO.

(b) they induce a free spin moment in the planes, as measured by magnetic
susceptibility and, recently, by electron paramagnetic resonance (Cieplak
et al.[14] and Finkelstein, et al.[15]). The susceptibility and resistivity measurements are quite quantitative.

There are two points to these measurements, one straightforward and the
other not quite so. Straightforwardly, it is clear that spin scattering is in some
sense "pair-breaking": it does cause resistivity, as well as lowering T_c. We may
describe the effect straightforwardly in our "$T_c = 0$" superconductor language
by saying that "T_c" $\propto -n_s$, i.e., T_c is reduced by the number of pair-breaking
scatterers, and then σ is still $\propto \frac{1}{T-T_c}$. Again, this is precisely the conventional
behavior of BCS superconductivity, except for the large magnitude of the ef-
fects. This effect, almost more than anything else, proves that the conductivity
is not conventional, in that there is a complete contrast between T-invariant
and T noninvariant scattering, which is not compatible in any reasonable way
with the hypothesis of a conventional Fermi liquid. A somewhat more subtle
question is why Zn, a "nonmagnetic" impurity, gives the same spin that Ni, a
nominally magnetic impurity, does. Here it is a straightforward experimental
case of "hole-particle symmetry": adding one relatively $+$ or $-$ charged ion in
the Cu^{++} lattice has the symmetric effect of adding a localized spin. The real
meaning of this observation will become clear later when we come to discuss
"confinement." But we can see intuitively that the spin manifold, which we can
think of as spinons moving about on the manifold of singly occupied (Cu^{++})
ions, is behaving to a great extent like a Fermi gas, and that a local potential
creates a singly occupied bound state for a spinon and its accompanying holon

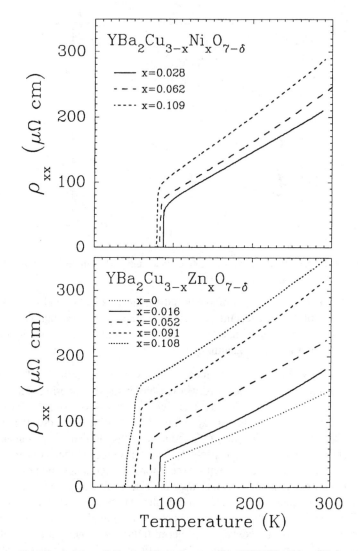

Figure 3.4. Resistivity of Zn-doped single crystals of YBCO. (Also Ni-doped, for comparison; the scattering is somewhat smaller, as is the effect on T_c.)

or antiholon, the two forming effectively a localized electron or hole, carrying both spin and charge. It is clear that this Zn impurity is not a simple Anderson model, which should scatter weakly (phase shift $\ll \frac{\pi}{2}$). Rather, one has to think of it as a bound holon and a bound spinon, each giving $\pi/2$ scattering for its respective entity.

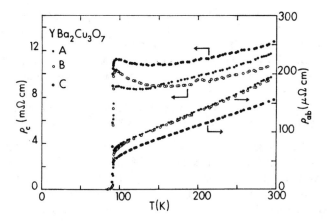

Figure 3.5. Samples of YBCO with varying c-axis conductivity.

The ab plane conductivity is, as described above, bizarre; but it is nowhere near so puzzling as the c-axis and intergranular contact resistivity. As I re-marked, especially the high resistance of intergranular contacts in the normal state was noticed very early, but with the advent of single crystals it became clear that even the bulk conductivity has a truly gigantic anisotropy. The con-ductivities along the c-axis of respectable single crystals of $(La - Sr)_2CuO_4$, $YBa_2Cu_3O_{7-\delta}$, $Bi\ 2\ 2\ 1\ 2$, and some newer materials all are 100 to 10^4 times as low as in the ab plane. Since, as we showed, the ab plane resistivity is only a factor 10–20 below the Mott limit even at 100°K, this means that the c-axis resistivity is nonmetallic in all cases.

More striking still is its T-dependence, especially in the case of the best single crystals measured carefully, emphasizing this parameter. In general, a fair fit to most data is the same empirical formula of fig. 3.1,

$$\rho_c = \frac{A}{T} + BT \qquad (III - 9)$$

with the lowest temperatures dominated by the A/T term. Figure 3.5 is a sequence of samples of YBCO measured by Ong showing that the BT term is sample-dependent, while the A/T term is reasonably constant. One possible explanation for such an expression is that actually the interlayer resistivity is even higher than that which is measured but much of it is bypassed by devious transverse paths. This may indeed be the case in some samples, with spiral dislocations or antiphase layer boundaries. But in fact N. Kumar[16] has shown that a momentum-conserving tunnel junction has a Giaever conductivity linear

*d

in $1/T$, which will appear when $t_\perp < \frac{\hbar}{\tau} \simeq kT$, i.e., at high temperature (see chap. 6). This may be the most common cause of the characteristic change in sign of $d\rho/dT$. Intergranular contacts have the same $1/T$ resistivity characteristic, at low T, as is seen from a ρT *vs* T^2 plot (fig. 3.1) for the original Müller-Bednorz data on poorly compacted ceramics, and the source must be similar.

These data are often rationalized away on the basis of a number of suggestions which are unrealistic. Let us discuss each possible alternative in turn.

(1) The electrons are "localized" in the c direction, extended in the ab plane. This represents a misunderstanding of the nature of localization. If a wave-function is extended in any direction it has matrix elements to a continuous spectrum of final states in all other directions and is hence extended in all dimensions. "Extended" and "localized" states cannot coexist for a Fermi liquid.

(2) The relevant one-electron matrix elements are just too small to give coherent motion in the c-direction, and all motion is phonon-assisted. Entirely aside from the fact that the resistivity often fits a smooth plot indicating no Debye or Einstein phonon bumps (which always appear in phonon-assisted tunneling), the interplanar hopping can be quantified in several ways.

(a) Band theory gives us values for the effective masses in the c-direction which are not negligible either in La_2CuO_4 or in $YBCO$. In the former the hopping matrix element is predicted to be about .05 ev, in the latter, that between the close planes is $> .1$ ev, between chains and planes about .02–.05 ev. Only in BISCO is the hopping via the Bi layers truly small.

(b) In La_2CuO_4 and $YBCO$ the values given by band theory are confirmed experimentally to be correct in two ways: values of superexchange integrals t^2/U which using reasonable U's give the above values for t; and superconducting penetration depths for c-oriented currents. Malozemoff's data, for instance, give a mass anisotropy of at most 10 in YBCO. This is perhaps the strongest and most significant single experimental anomaly: the c-axis, one-electron matrix elements are clearly present below T_c, but fail to cause coherent transport in the normal state. This failure of the conventional sum rule is the key clue to the mechanism of superconductivity.

(3) A third suggestion, which we have already, in effect, dismissed, is that there would be large many-body ("Z") corrections to the coherent transport in the c-direction. This is one of the reasons we emphasized the generality of the proof of nonrenormalization of transport processes in Fermi liquid theory.

(4) A fourth response is to attempt to argue that the c-axis resistivity is *sometimes* "metallic" (by which is usually meant its T-dependence, not

its magnitude). This is a logical fallacy. If insulating behavior ever coexists with metallic, quasiparticle theory has failed.

In general, velocity and density of states corrections cancel, so that conductivity is usually more isotropic than effective masses. What then happens in a true tunnel junction? In fact,the relevant scale of energy is the inelastic scattering time: if $t_\perp < \hbar/\tau$, we get tunnel junction behavior. This is temperature-independent if momentum is not conserved, T-dependent with σ decreasing with T if it is; but can lead to a resistivity greater than the Mott limit. In several of these materials we know that $t_\perp > \hbar/\tau$, and in none of them is the behavior like that of a stack of tunnel junctions, but instead σ grows with T in that region.

This kind of highly anisotropic conductivity is not unknown, but in most cases the cause is essentially the same as here: there is, in the low-dimensional material, a highly correlated state, SDW or CDW, and the charge carriers are not the tunneling entities—electrons—but some form of soliton as in polyacetylene. As we shall see, it is the exceptional 2- or 1-dimensional material which is a true Fermi liquid, and this case is not that exception.

The only reasonable explanation for the peculiar c-axis transport is to abandon the hypothesis of the Fermi liquid. We are already happy to do that to understand ρ_{ab}, we *must* do it to understand ρ_c, because we must abandon the structure in which the quasiparticle states arise by adiabatic continuation from real particle states; i.e., we must abandon the definition of the incoming particle channels.

It is perhaps a somewhat high-flown way of describing what happens but it is a valid analogy to call the process "confinement." In particle physics quarks can only live in the"real," bare vacuum which is not the "physical" vacuum we have all around us; this latter is a complex many-body state containing large amplitudes of gluon fields, etc., and in it only conventional hadrons can move freely. Quarks build themselves a "bag" of "real" vacuum inside the nucleon. Yet there is no true symmetry breaking or order parameter in the physical vacuum. In our case, it appears that electrons can only live in the bare vacuum which we have all around us, and which also inhabits ordinary metals and the intervening layers of insulator. They are the fundamental particles, the "quarks" of the theory. Inside the CuO_2 planes is a "physical vacuum," a correlated state in which electrons are not elementary excitations, but completely unstable: they break up into something else. This is not unfamiliar to us—it is exactly what happens in BCS superconductors, and in charge-density wave systems as well— but it is unfamiliar that it happens in the absence of a symmetry-breaking order parameter such as the BCS gap. Fortunately we have one clear example of this process to provide a model—and very possibly *the* model—for what happens, namely, the exactly soluble 1-dimensional Hubbard model and its near relatives such as the "$t - J$" model and some of the one-dimensional "g-ology" models solved in the '70s by renormalization techniques.

In these models a process called "spin-charge separation" takes place, a process in which the elementary excitations become collective, soliton-like motions of the entire electron fluid, and in which the motions of the charge and spin degrees of freedom become decoupled, in that they have different velocities.

It is not valid to say, as is often implied, that they are "completely" decoupled: clearly the two fluids represent the same set of electrons, and they must, on the average, flow together as emphasized by Larkin.[17] This appears in the dynamics as a "backflow" coupling of the two sets of excitations, a very long-range but strong condition on the overall flow. This can often be well described in terms of a gauge field.

Thus we have two separate sets of soliton-like excitations, which we may call "spinons" and "holons," and which are independent degrees of freedom in the layers. These must coalesce into an electron in order to tunnel through into the adjacent layer, and will then break up again, and we have no reason to expect the conventional transport theory to hold for such a process.

Until recently this picture was our only guide. Now that we have a reasonably clear picture of the one-dimensional model, one may say a bit more. (Those arguments are given in reprinted articles by Anderson[18] and by Clarke, Strong, and Anderson.[19]) In the first place, it is important to take the correct limit, since the wrong conclusion is ensconced in the literature because of calculations using perturbation theory naively. In the usual, incorrect calculations, it has been customary to turn on the interchain or interlayer hopping

$$t_\perp \sum_i (c_{i\sigma}^+)_a \cdot (c_{i\sigma})_b + HC \qquad\qquad (III-10)$$

first, and then to do perturbation theory in U, which of course must then be resummed by renormalization group theory or otherwise, a process which is difficult to control formally and does not necessarily lead to a valid large U limit. Naturally, we then are doing perturbation theory on a state in which the bands have been split by $t_\perp$ and the Fermi surface modified. For instance, the two-chain problem will have been split into two one-chain problems with *different* k_F's which do not interact except at long wavelengths. Thus in this order turning on U leads to two relatively decoupled interacting Fermi gases with two different Fermi surfaces.

The correct order is to turn on U and only then to introduce $t_\perp$. Only in this order is it possible to test in a rigorous fashion whether a small $t_\perp$ can modify the band structure or not. In fact, the problem has been remarkably complicated by the necessity to keep straight *three* separate limits. If $T \sim \hbar/\tau > t_\perp$, the Fermi surface is not sharp to the accuracy of $t_\perp$ and the effect is invisible; while if $t_\perp > U$, as we have explained, it is appropriate to diagonalize $K.E. + t_\perp$ first. So the appropriate limit is

$$U \gg t_\perp \gg kT,$$

Figure 3.6. Lowest-order diagram for the interlayer tunneling process.

i.e.,

$$\lim_{t_\perp \to 0} \ \lim_{T \to 0} \Big|_{U \text{ finite}}.$$

This shows that in doing perturbation theory we use methods appropriate at $T = 0$, and that the "temperature Green's functions" of AGD are not appropriate. What we must do is to examine the "anomalous diagrams" in the sense of Kohn-Luttinger and de Dominicis,[20] and in particular just the simple lowest-order diagram for the tunneling process (see fig. 3.6).

$$t_\perp^2 \sum_k \int d\omega \, n_k \, (1 - n_k) \, G_k (\omega) \, G_k^* (\omega).$$

Here G is the exact Green's function with finite U. For a conventional Fermi sea, this is $0 \times \infty$: state "k" is either empty or full, while the combination of Green's functions is

$$\frac{1}{\omega - \epsilon_k + i\delta} \ \frac{1}{\omega - \epsilon_k - i\delta} = \frac{1}{\delta} \delta (\omega - \epsilon_k)$$

and hence diverges in the limit $\delta \to 0$. This divergence, as pointed out in my original paper, implies a Fermi surface shift. It indicates that there is an energy shift of linear order in $t_\perp$, which requires reoccupying states in the Fermi sea. Thus simple second-order perturbation theory in $t_\perp^2$ formally diverges, though the original ground state is in unstable equilibrium. This is not at all complicated, and is dealt with formally by going to finite T, which is perfectly all right in the conventional case. The breadth of n_k and the breadth of G match precisely and give the correct energy shift. The peculiar thing about the interacting case is that the integral of the product of Green's functions does *not* diverge: the unperturbed ground state of two interacting chains is *stable* to lowest order.

This does not mean that $t_\perp$ is formally irrelevant, since in conventional renormalization procedures it grows with scale; what it does mean is that the Fermi surface is not displaced by it. Since the Fermi momentum $\propto t_{\|}$ which increases faster than $t_\perp$, the system becomes increasingly *one*-dimensional. The effects of $t_\perp$ appear in higher order: superconductivity or magnetism, as we shall see.

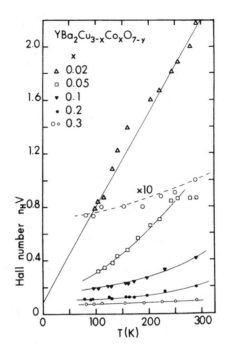

Figure 3.7. Earliest data on the Hall effect, showing confusing picture in terms of R_H.

This is the process we describe as "confinement" and it is implicit in any treatment in which charge and spin separate and, in fact, takes its origin in the gauge couplings of charge and spin solitons. But it is, in effect, very simple; for infinite-time, coherent processes, the electron—which is the only particle which can tunnel through the vacuum—separates into its constituent particles and these eventually get infinitely far apart and cannot recohere to tunnel back.

We must assume that spin-charge separation occurs also in the 2d case and is responsible for the insulating c-axis behavior and for the poor intergranular contacts in the normal state. In a later chapter we will give considerable theoretical justification for this. This argument is also crucial, as we shall see, to the mechanism for superconductivity, which involves coherent pair ("Josephson") tunneling, which *is* possible at zero frequency, since it can take place in the singlet channel without generation of any real spinons.

A third puzzling experimental fact is the temperature-dependent Hall effect. It was evident from early ceramic measurements by Ong's group that the Hall effect had striking and unusual properties (see fig. 3.7). But we had to await the availability of single crystals before the physics became reasonably clear. The first pure single crystals of YBCO, produced by Z. Z. Wang, exhibited the surprising result that the Hall effect $R_H \propto \frac{1}{T}$, with $n_H = \frac{ec}{R_H}$ extrapolating quite accurately to zero at $T = 0$. The group's Ni- and, later, Zn-doped samples showed a progressively smaller temperature dependence.

This data remained a mystery until 1991, when we realized that the key quantity is not R_H but θ_H, the Hall angle. This is given in simple theory by $\omega_c \tau$, and it is found that in a wide variety of substances and samples one finds

$$\theta_H = \omega_c \tau_H = \frac{1}{A + BT^2} \qquad (III - 11)$$

($A = 0$ for crystals without "magnetic" impurities). Thus the relaxation rate which determines θ_H is of conventional form, $\propto T^2$ rather than T. The theory of this phenomenon is given in chap. 6: it involves the idea that τ_{Hall} is the spinon mean free time since the orbital magnetic field part of the Hamiltonian commutes with the anomalous interaction which causes spin-charge separation. Another way to put it is that resistivity is not caused by scattering of the quasiparticles carrying the charge, however else one wishes to calculate it—either scattering from fluctuations of the gauge field, decay of accelerated electrons, or otherwise. The actual Fermion "quasiparticles" are the spinons, which are in general much less scattered than the observed resistivity would indicate, because they are independent of the charged, singlet channel; but it is these Fermions which determine the widths of states near the Fermi surface, and thus determine the τ_H which should be used in the Hall angle $\omega_c \tau_H$. That is, the momentum states from which the collective state is composed are rotated under the action of the magnetic field by the equation of motion

$$\hbar \dot{\mathbf{k}} = \frac{v}{c} \times B \qquad (III - 12)$$

and the dynamic memory of the last transverse acceleration lasts a time τ given by the energy width of *eigenstates* near the Fermi surface, not the energy width of the *electron* states which determines ρ_{ab}. $1/\tau$ for the spinons contains two types of scattering, both involving only spin scattering since T-invariant scattering is ineffective. One type is spinon-spinon scattering which is identical to conventional Fermi liquid electron-electron scattering and has T^2 temperature dependence; the second is whatever magnetic scattering is responsible for the residual resistance if any. Thus we have

$$\frac{1}{\tau_H} = \frac{1}{\tau_{res}} + AT^2 \qquad (III - 13)$$

and

$$\theta_H = \omega_c \tau = \frac{\omega_c \tau_{res}}{1 + AT^2 \tau_{res}}. \qquad (III - 14)$$

This gives a Hall effect of the form

$$R_H = \frac{\rho_{ab}\theta_H}{H}$$

$$= \frac{\omega_c \tau_{res}}{H} \cdot \frac{\rho_{ab}}{1 + \tau_{res}AT^2}. \tag{III-15}$$

Now

$$\rho_{ab} = \frac{m}{ne^2}\left(\frac{1}{\tau_{res}} + BT\right) \tag{III - 16}$$

so that

$$R_H = \frac{1}{nec} \times \frac{1 + BT\tau_{res}}{1 + AT^2\tau_{res}}. \tag{III - 17}$$

Clearly, if

$$\frac{1}{\tau_{res}} \to 0 \qquad \text{this gives} \qquad R_H \propto \frac{1}{T}.$$

If $\frac{1}{\tau_{res}}$ is finite, we find a characteristic temperature dependence which had previously seemed quite puzzling. Figure 3.8 shows a fit of this temperature dependence to a sequence of Zn- and Ni-doped YBCO samples measured by Clayhold et al. The $\Theta_H^{-1} \propto T^2$ behavior is characteristic of almost all good, single crystal specimens (see fig. 3.9). This prejudice needs to be tested against many more samples, especially of good single crystals. Another significant measurement is the Hall effect in the c-axis direction. This is constant and electron-like, according to one measurement by Ong.[21] This is what one expects if transport in this direction is purely inelastic tunneling of electrons, and no holons or spinons can move in this direction.

So far, the measurements of other transport effects are too fragmentary to be very helpful. Heat conductivity and thermopower are so anisotropic and so difficult to interpret for ceramics that we choose to ignore all ceramic measurements. Heat conductivity seems to obey a Wiedemann-Franz law in the ab plane, with some hint that the coefficient is a bit bigger than 1. There is a jump in heat conductivity below T_c. The jump in heat conductivity is consistent with the increase in quasiparticle mean free time which is observed in microwave conductivity and in photoemission. But we must not pretend we completely understand these observations. In the c-direction, the heat conductivity is T-dependent and of a magnitude quite compatible with phonon conduction, and no anomaly occurs at T_c. The observations on YBCO are shown in figs. 3.10a, b, and c.

*e

Thermopower measurements on single crystals are even more fragmentary. In the early days of high T_c one was reassured that the carriers were holes by the thermopower sign, which is quite immune to the kind of band structure problems that the Hall effect has. Otherwise we have seen no clear, reproducible

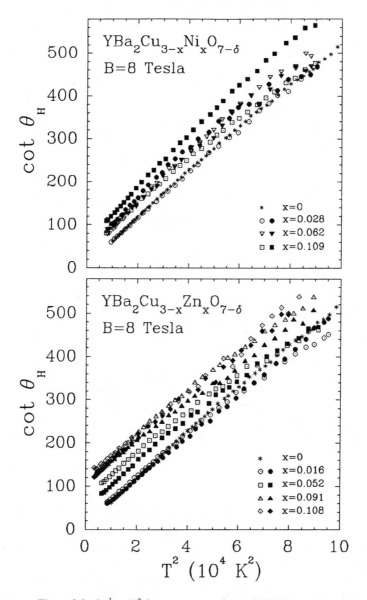

Figure 3.8. Θ_H^{-1} vs T^2 for a sequence of doped YBCO samples.

trends, partly because of the absence of generally agreed data. This is a pity because in many other cases thermopower is the easiest measurement to interpret. The transport theory in chapter 6 seems to make a number of suggestions about thermopower in the ab plane, which are not firm, since a careful theory

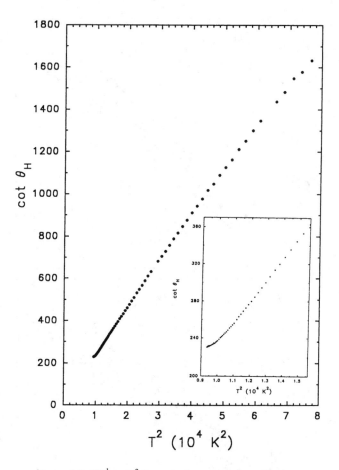

Figure 3.9. Θ_H^{-1} vs T^2 for a single crystal, from Ginsberg.

has not been completed, but more in the nature of a scenario. These are borne out in some of the data. One way to think of that theory is that electrons are accelerated by the field, which then decay into spinons, which are not quickly scattered but no longer carry current, and holons, which are rapidly scattered. In this "two-fluid" type model, there is "spinon drag" in that the momentum of the spinons persists for a time $\tau_H \sim T^{-2}$, and they can transport heat even though they are not carrying current. This leads to a T-independent thermopower, i.e., an intercept at $T = 0$. There will also be a conventional thermopower, possibly of electron sign. c-axis should be interesting: because high-energy particles hop more easily than low, proportionally to ω, we expect a T^2 behavior. Both predictions are not as yet firm, and both seem to be borne out more or less by what experiments there are (see fig. 3.11). Both of these effects can be stud-

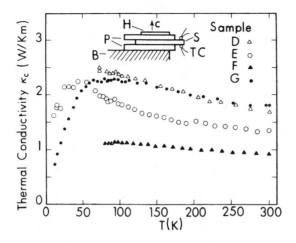

Figure 3.10a. Thermal conductivity in c-direction showing no hint of an electronic peak below T_c.

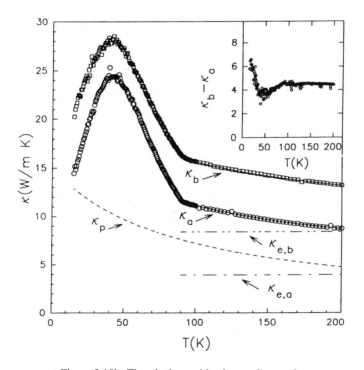

Figure 3.10b. The ab plane with a large, clear peak.

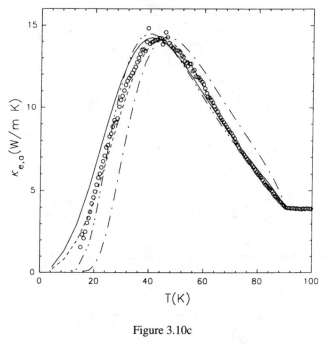

Figure 3.10c

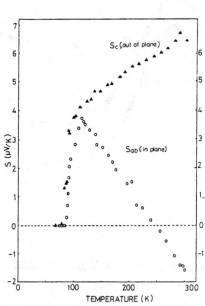

Figure 3.11. Anisotropic thermopower measurements.

ied on the one-dimensional model (unlike the Hall effect) but have not been; Stafford is beginning this investigation.[22]

Tunneling is a transport process. It was, of course, the first hint that non-Fermi liquid behavior was present, but to this day tunneling measurements are ambiguous and controversial. There is one remark we should make: many of the great difficulties with tunneling measurements could well fit under the general rubric of "confinement" effects. We still have no clear evidence that even normal metals can tunnel into high-T_c materials in a normal, coherent way along the c-axis. There is a mysterious "dead layer" on the surface of most samples which could be related to the confinement effect. Certainly intergranular contacts are sources of resistivity, especially in the normal state. This will be discussed again under one-electron spectroscopies.

OPTICAL AND INFRARED SPECTROSCOPY OF HIGH-T_c SUPERCONDUCTORS

I have not chosen to emphasize the results from optical spectroscopy very heavily. Optically, the insulating antiferromagnets exhibit a fairly broad gap (~2.3 ev) which is clearly of Mott-Hubbard origin. Millis and Stafford[23] have shown that the filling in of this gap[24] and the replacement of high-frequency amplitude with Drude intensity follows rather surprisingly well the calculation from the one-dimensional Hubbard model (see fig. 3.12); in Stafford's thesis it is shown that this figure is well fit by 1D Hubbard. There is nothing in this data which is at all unexpected given the picture of chapter 1.

The infrared spectrum of conductivity in the ab plane, as deduced from reflectivity (as well as transmission, which does not, in spite of some claims to the contrary, differ appreciably) is rather strange. It has been controversial in the literature, but not at all so on the basis of any appreciable variability in the measurements; rather in the emphasis of the interpretation. Good, pure YBCO single crystals exhibit a conductivity which is best described as roughly linear in $1/\omega$ from $\omega \sim kT$ to about $\frac{1}{2}$–1 ev(5,000-10,000 cm^{-1}). At least six groups (Schlesinger and Collins, Thomas and Orenstein, Tanner et al., Mihaly, Bozovic, as well as Klein) seem to agree on the rough magnitude as well as variation of the conductivity. Bozovic plots it as inverse dielectric constant, and Tanner divides the smooth curve into two pieces arbitrarily and calls one "Mid-Infrared," but the measurements do not differ.

If we write

$$\sigma \simeq \frac{ne^2}{m\left(i\omega + \frac{1}{\tau}\right)}$$

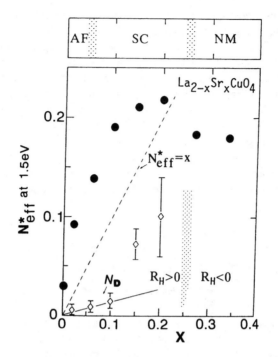

Figure 3.12. Anomalous rise of plasma density with doping; Stafford and Millis showed this is characteristic of the Hubbard model near half-filling, even in 1D.

the results are equivalent to

$$\frac{1}{\tau} \simeq \max\left(\frac{\omega}{\frac{kT}{\hbar}}\right) .$$

There is a tendency for m to be also slowly variable with frequency. Figs. 3.13a and b shows Schlesinger's interpretation of his data and a similar one from a different group.[25]) Most of this data is on YBCO, of which there are excellent single crystals suitable for reflectivity measurements. For this reason it is heartening that Bosovic[26] finds the same general behavior using ellipsometry on films of several other materials.

The "underdoped" YBCOs measured at length by Orenstein et al. show considerable deviation both in $\rho(T)$ and $\sigma(\omega)$. In particular, there seems to be a feature below $\sim$150° at around 500 cm^{-1}. This temperature corresponds to the appearance of the "spin gap" in this material, i.e., a relative reduction in susceptibility and nuclear magnetic relaxation as the temperature is lowered below 150° K. One possible speculation relates this to a tendency toward large fluctuations in a CDW coupled to a $\sim 2k_F$ phonon: an SDW would not be

*f

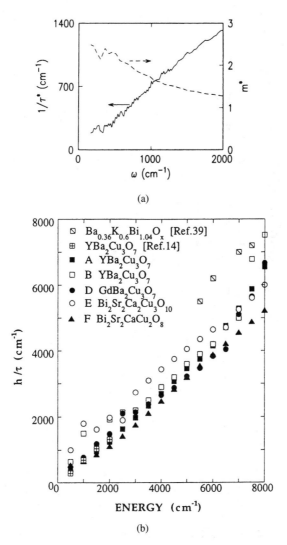

Figure 3.13. (a) Schlesinger's Kramers-Kronig transform of IR conductivity on a good O_7 YBCO sample; (b) The same for a number of samples as measured by N. Bontemps's group, to show the range and universality of the data.

likely to lead to *reduced* magnetic fluctuations. I have called this phenomenon "the Fermi Surface Eaters." (Another possibility is singlet pairing of spinons between planes.) Phenomenologically, it is as though a portion of Fermi surface were gradually disappearing. It seems to have amazingly little relevance to superconductivity, and the spinon relaxation rate remains $\propto T^2$.

*g

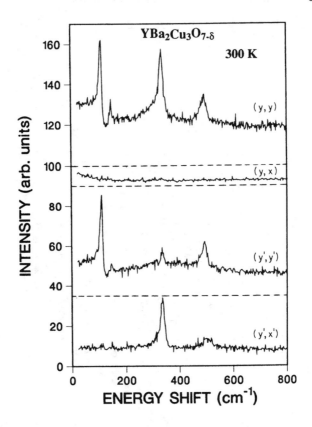

Figure 3.14. Raman Scattering "Background" confirming by indirect arguments linear-ω dependence of $1/\tau$.

A confirmation of the generality of the linear ω scattering appears in observations of the "Raman background" in several of these materials by Klein, Cooper, et al.[27] (fig. 3.14). The magnetic insulators show a clear peaked "2-spin wave" (or 4 spinon) Raman spectrum from which Fleury has been able to deduce values for J. But the doped materials show a much broader, flatter spectrum, which shows Fano resonances with Raman-active phonons, and definitely is gapped in the superconducting state. As Shastry has pointed out, particularly[28] (and also Abrahams and Varma) this is exactly what one expects from $\frac{\hbar}{\tau} \simeq \omega$, since the Raman scattering from electrons is proportional to the polarizability tensor with a factor ω^2:

$$R\left(\omega\right) \propto \omega^2 \alpha\left(\omega\right) \sim \frac{\omega_p^2 \omega}{(\omega + \frac{i}{\tau})}.$$

For normal free electrons, R is not constant but grows linearly with ω. Of course, above the Debye temperature phonon scattering can show the same behavior, but the constant Raman background does not seem to be T-dependent. The constant Raman background is a major experimental fact telling us that $\frac{1}{\tau_{tr}} \propto \omega$ is a generic behavior and not a special property of a few substances.

Infrared spectroscopy polarized along the c-axis is a more difficult matter. There were early measurements on questionable crystals and on stacked crystals, but fortunately excellent measurements have been made on YBCO single crystals by Tanner's group, figs. 3.15a and b, and on 2 1 4 by Uchida's group, fig. 3.16. The essence of these measurements is that the far infrared spectrum for c-axis polarized light is indistinguishable from that of an insulating ionic crystal in the normal state. Uchida's group's measurements[29] are particularly striking in that the superconducting crystal shows an unmistakeable plasma edge, appearing in a region in which the normal state is showing constant, low reflectivity ($\sim .45$). This is also a specially good case because there is only a single type of layer.

The theory of this measurement is very simple and beautiful. The dielectric constant has two components, a constant ϵ_0 (about 20) due to the polarizability of the ions (plus possibly some optical modes) and a contribution due to the superconducting electrons which accelerate freely,

$$\epsilon = \epsilon_0 - \frac{4\pi \, n_s \, e^2}{m_s \, \omega^2} = \epsilon_0 - \frac{\omega_p^2}{\omega^2} .$$

The reflectivity goes through a sharp minimum when $\epsilon \simeq 1$:

$$\epsilon_0 - 1 = \frac{\omega_p^2}{\omega_{min}^2}$$

$$\omega_{min}^2 = \frac{\omega_p^2}{\epsilon_0 - 1} \simeq \frac{\omega_p^2}{4} .$$

Judging by the observed sharpness of the minimum, the residual loss in the material is very small, as is, of course, attested to by the very poor $d\,c$ conductivity along the c-axis. There appears to be no sign of a quasiparticle conductivity consistent with the superconducting response ω_{ps}^2. We will return to this in the next chapter.

Rutter and coworkers measured R at very high frequencies along the c-axis. Like the high-temperature conductivity, the high-frequency region should see simply a low, but not unusual, conductivity, and it does. (See also Klein et al.)[30] Strong and Clarke suggest a $\omega^{4\alpha}$ behavior for $T \ll \omega < t_\perp$. *h

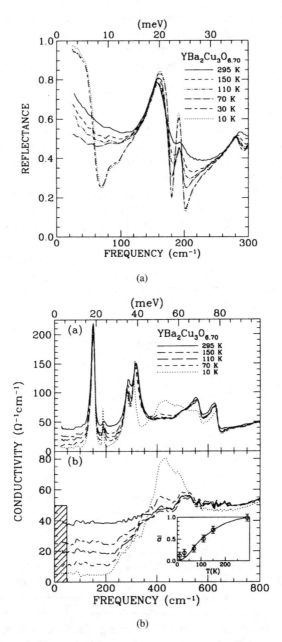

Figure 3.15. *c*-axis IR conductivity; early measurements by Tanner on YBCO. It is strange that this data is characterized as "metallic"!

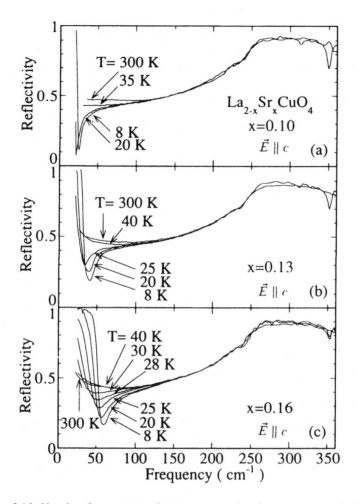

Figure 3.16. Very low frequency c-axis measurement showing appearance of plasma-edge in 214 out of an ionic insulator-like behavior.

PHOTOEMISSION AND TUNNELING SPECTROSCOPY OF THE NORMAL STATE

Photoemission has been the most useful spectroscopy for the high-T_c supercon-ductors, playing the key role which tunneling played for conventional super-conductors. The reasons for this are various and not perfectly understood. Both are methods which nominally probe the most fundamental electronic property, the one-electron density in energy and momentum.

For normal superconductors, tunneling was the method of choice, first because the normal state spectrum was conveniently featureless, due to the fact that for Fermi liquids both the density of states and the interaction effects cancel out against velocity factors. Therefore the spectrum tells us only about superconductivity. Also, in general, momentum information is lost by using polycrystalline films in tunnel junctions, which is actually a convenience rather than otherwise. But for high T_c, from the start it was clear that the real problem is the normal state; and while the tunneling spectrum in the normal state is not featureless, for reasons imperfectly understood, it is sufficiently uninformative to be not much use (and also very variable from one sample preparation means to another).

The main feature of photoemission is that angle-resolved spectroscopy gives one the momentum- and energy-dependent, single-particle density: it is a direct measurement of

$$I_{\text{photoemission}} \propto Im G_1 \, (k, \omega)$$

where G_1 is the one-electron propagator. The contrast with conventional BCS is that momentum-dependence here is vital, in both normal and superconducting states. Unlike BCS quasiparticles, which in many samples, particularly in superconducting films, are strongly scattered and not momentum eigenstates at all, spinons are essentially momentum eigenstates in many samples.

Finally, the superconducting binding energy—the energy gap—is much larger and is accessible to photoemission. This is often cited as the only reason for the renewed interest in photoemission, but is of subsidiary importance. It is true, however, that it is the fortuitous improvement in energy resolution by an order of magnitude (to $\sim$25 mev) which came along in this period which was an essential point. In my opinion, much of the photoemission community cannot believe its good luck and is reluctant to accept that their data are as accurate and detailed as they actually are because they have not in the past had adequate resolution. For these reasons this section will emphasize the photoemission spectra and will touch only briefly on the bewildering phenomenology of tunneling, at the end.

One of the primary reasons why photoemission has been such a useful technique is that it is uniquely suited to measure electrons which are two-dimensional. Several of the high-T_c materials, but especially BISCO, can be prepared with surfaces which are clean ab planes. When such a surface is used, angle-resolved photoemission can achieve its full momentum resolution.

The principle of the measurement is shown in fig. 3.17. An incoming photon of fixed energy $\hbar\omega$ is incident on the surface and excites a single electron from an occupied state with crystal momentum k and energy E_k to an empty state some tens of ev higher in energy at $\hbar\omega + E_k$ and above the vacuum level, but with nearly the same crystal momentum k. (Since $\frac{\hbar\omega}{c} \ll k$.) The momentum and energy can be measured, hence both E_k and k are in principle

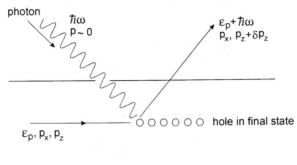

Note: δP_z may be finite; but in
2D this is irrelevant.

Figure 3.17. Principle of angle-resolved photoemission.

available. In general, k_z perpendicular to the surface is considerably less accurately conserved in the process of emission than k_x and k_y, and for conventional 3-dimensional materials this restricts the method's usefulness. In particular, if there is a large uncertainty in k_z, Δk_z, the whole spectrum will be broadened by $\Delta E = v_z \Delta k_z$, even if one has perfect resolution on the final state energy. The perpendicular momentum is, at the very least, broadened by the high-energy scattering processes which restrict the mean free path of the emitted electron within the metal to not many tens of Ångstroms or less. Such processes do not broaden the energy spectrum because they either emit a large energy, removing the electron from the beam, or nothing; but they do cause a Δk_z.

Fortunately for the subject of high T_c, this has no broadening effect in these materials. The absence of coherent motion in the c-axis direction, which is an empirical fact attested to by the low c-axis conductivity, means that $v_z \ll v_x$, v_y, so that $v_z \Delta k_z$ is negligible. This may be quantified. Assuming metallic conduction,

$$\sigma = e^2 \times \text{(Fermi surface area)} \times v_F \, \tau$$

and τ being the same as in the a-direction, $\sim \frac{kT}{\hbar}$, v_F is a factor of several hundred to $\sim 10^4$ slower. If in fact, as this book argues, the bands are literally 2-dimensional, and conductivity in the c-direction is all inelastic, the argument is even stronger. Thus so long as energy *and transverse momentum* are conserved in the emission process, the photoemission amplitude is an accurate measurement of $Im G_1 (k_\perp, E)$, where $k_\perp$ is the k-value in the plane and $E = E$ (emitted electron) $- \hbar\omega_{ph}$. The surfaces are, especially in the case of BISCO, quite perfect and there is no reason to suspect failure of $k_\perp$ conservation.

The earliest angle-resolved measurements which were made with high resolution gave initially surprising results. Measurements made slightly later along

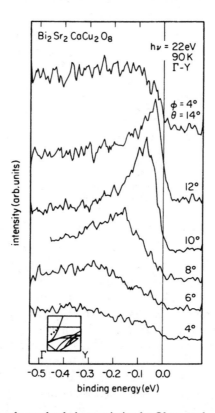

Figure 3.18. Early angle-resolved photoemission by Olson et al., showing broad "tail" and cusp.

a more favorable cut through the Fermi surface are shown in fig. 3.18; these are perhaps the highest resolution results yet achieved. It may be wise to set these in context. Looked at under low resolution in fig. 3.18, one sees a gigantic, broad peak from 1ev below E_F down, which represents the great "spaghetti" of nonbonding and bonding oxygen and copper bands. In the insulating phase, this peak contains also the lower Hubbard band. In the metal, a weaker signal is seen near the Fermi level, corresponding in intensity to a single band, which must be the $Cu - O$ antibonding hybrid or "lower Hubbard band." Because of screening, the Fermi level is well above the original insulating lower Hubbard band, although in fact its electronic character is the same, essentially. It is this band which is magnified in the figure and shown under high resolution. Manipulating the energy of the incoming photon allows one to use resonances with inner-shell levels of the different atoms to enhance features coming from the different layers of the substance, and shows that the band under consideration comes from the Cu-O planes. Part of its weakness is caused by the fact that the outer cleavage layer is almost certainly a $Bi\,O$ layer, through which the emitted electrons must make their way. None of the Fermi level structure seems to come from this layer.

Returning to the original measurements on BISCO[31] of the superconducting composition, fig. 3.18, it was evident from the start that these are presenting a good approximation to the actual spectrum with the claimed resolution. I say evident from the start: I mean evident to one who came to the data with no prejudices based on past experience with lower resolution and with 3-dimensional band structures. To the casual glance, on the other hand, the spectra looked enormously broadened and of very low resolution; in the past, ARPES spectra had always represented only a low-resolution way of following band structure "features" through momentum space, and it had never before been suggested that the entire spectrum could be interpreted quantitatively as literally $Im\,G(k,\omega)$.

The original spectra, however, contained internally sufficient evidence to show that they could be taken literally; (1) There was no intensity above the Fermi level: this region was sharp and clean (and is always so). This shows that there are no stray surface fields and inhomogeneities: there are well-defined vacuum potential levels and a well-defined Fermi level. It is, of course, not an indication that there is no effect of final state scattering, which will be predominantly downward in energy.

That there is very little final state soft energy loss scattering is shown by the spectra taken below T_c, where superconductivity causes a sharp quasiparticle peak to occur. (This will be discussed in the next chapter; here we only use this fact to establish instrumental parameters. The earliest data (Arko et al.)[32] is shown in fig. 3.19. Note that (as we find independently from transport and NMR measurements) the quasiparticle lifetime below T_c is strikingly longer, in fact one may say that below T_c there *is* a quasiparticle, with lifetime well below the nominal instrumental resolution. The key point is that there is no hint of final state scattering of this quasiparticle peak: it *has precisely the instrumental resolution*. The broad "background" on the normal state spectrum is simply not there on this peak. (The same point has been made very ably by J. Allen.)

A final indication is the sharp cusp structure on some of the peaks. The widths of the peaks—if that can be assigned (it has been, using a Lorentzian fitting procedure assuming the nonexistent background, but I do not accept that that fitting has much meaning) is of the same order as the energy of the peak. It is clear that for those peaks near to the Fermi level, the definition of the cusp is very much sharper than this nominal width. That is, the Green's function has a clearly defined and characteristic *shape* which can be estimated quite well from the spectrum. For the region shown in the figure, this shape is as sketched in fig. 3.19 a fairly well-defined line or region from the Fermi level to the cusp, and above the cusp a power-law fall-off with a power between about $\omega^{-1/2}$ and ω^{-1}. This is the kind of Green's function which would result from appropriate

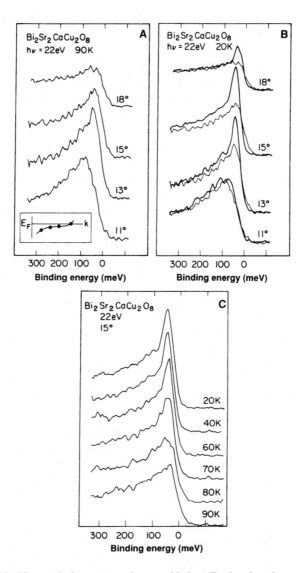

Figure 3.19. Photoemission spectra above and below T_c, showing sharpness of quasiparticle peak in superconductor: a check on resolution.

angular averaging of the one-dimensional Green's function, as we point out in chapter 6.

The scan we have shown is typical of those run from the interior of the Fermi surface, in both BISCO and, with more serious sample preparation problems, in YBCO. The only clearly identifiable Fermi surface in BISCO in this sense is

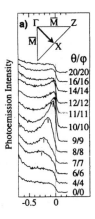

Figure 3.20. A scan showing dispersion through a Fermi surface.

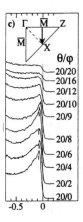

Figure 3.21. A scan through a "ghost" Fermi surface, with no dispersion but only amplitude variation.

the simple tight-binding Fermi surface of the CuO layers, centered on Γ and rather flat in the $\Gamma - S$ direction (see fig. 3.20).

From the start the measurements of Olson et al. and, later, much more detailed scans by Dessau et al.[33] showed other puzzling features. Olson described his observations as broad areas of the Brillouin zone which had amplitude in the photoemission spectrum reaching up to the Fermi surface. Dessau's data, fig. 3.21, show a typical scan through such a region. It is seen here that there is no visible cusp or peak dispersing toward a Fermi surface, merely a more or less rapid dying-out of the Fermi surface amplitude in a certain region along the line M-Y.

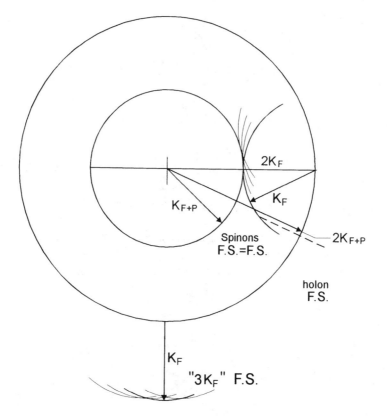

Figure 3.22. How "ghosts" arise in holon-spinon picture.

Dessau et al. have interpreted this region of "dying out" as a second Fermi surface in figures like 3.20. I believe that this interpretation is incorrect and that this phenomenon is what has been called "ghost" Fermi surface amplitude. This occurs because the low-frequency holon and spinon bosons at the Fermi surface can be coupled via the "singular interaction" term which we will discuss in chap. 6 not just to the electron operators at the corresponding point at the Fermi surface but to electrons which are quite far from the Fermi surface. The electron of momentum $\delta k + k_F$ couples to a whole "patch" of Fermi surface of size $\sim \delta k$, and this allows noncolinear holons and spinons to be generated. The corresponding construction is shown in fig. 3.22. A quite reasonable description of these "ghosts" results.

Their ascription to actual electron band dispersion curves is irrational. The band structure postulated by Dessau et al. and shown in fig. 3.23 is unrealistic both theoretically and experimentally. One feature of the curves on BISCO, the pseudo-Fermi surface shown by Dessau et al. in the $\overline{M} - Y$ region, has a possible

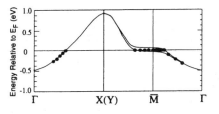

Figure 3.23. Unrealistic "band structures" necessary to interpret ghost Fermi surface structure.

explanation as caused by the prominent peak near $Y = \pi, \pi$ in the two-spinon and holon amplitudes. This is caused by the "nesting" of the original Fermi surface at this vector, which must be reflected in the $2k_F$ amplitudes (and is, in the neutron scattering). This holon combined with the Fermi surface spinon can account for the sharp drop-off in Fermi surface amplitude at their "ghost Fermi surface."   *i

Photoemission has given a much simpler result for the "electron" supercon-ducting cuprate $Nd - Ce\ CuO_2$. Here the Stanford group[34] finds a smaller, almost circular Fermi surface with no marked anomalies or ghosts and with the correct volume, fig. 3.24. The electron Green's functions, though, still have the cusp—power-law behavior of fig. 3.18. It seems likely that this T_c is too high and is caused by the interlayer enhancement mechanism, even though the characteristic resistance linear in T is not observed. (Perhaps, since this is probably a stoichiometric compound, we are in the holon drag regime.) The crucial measurement of ρ_c has not been carried out.

Finally, we mention angle-averaged photoemission. The most noteworthy measurements are those showing the opening of a gap below T_c, but in the normal state the key message is that the band density of states as measured in

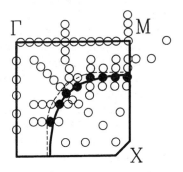

Figure 3.24. "Normal" Fermi surface of $Nd - CeCuO_4$.

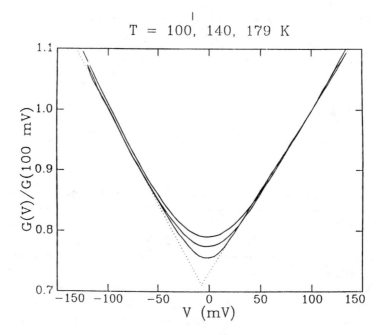

Figure 3.25. Characteristic "V" shape of c-axis tunneling into normal metal.

photoemission is *featureless*. There are no sharp peaks of the sort which would necessarily accompany a flat band of the type postulated to fit the angle-resolved photoemission. In fact, perhaps the strongest argument that the structure is an interaction effect is its disappearance when angle-averaged.

Finally, we comment on tunneling phenomena in the normal state. It has not been possible to obtain clear data on normal state tunneling in STM or in "break junctions" (obtained by breaking a ceramic or crystalline sample, and pressing the pieces together) because of the lack of temperature stability. The best data use films of an ordinary metal evaporated on to either a simple or an oxidized surface. By looking below the superconducting T_c of the ordinary metal (say Pb) one checks the quality of the film by measuring the characteristic tunneling spectrum at and above the Pb superconducting gap. If the Pb gap structure appears with full amplitude, we are justified in assuming the junction current is all carried via single-electron tunneling.

Such junctions show a very characteristic "V" structure of the conductance as a function of applied voltage. (See fig. 3.25.) This is quite unlike conventional tunnel junctions, where the characteristic is normally parabolic near $V = 0$. (Kirtley has pointed out that a few other types of junctions, notoriously those involving the metal Cr, also show this structure. Cr, being a well-known spin density wave structure, can hardly be described as a conventional weakly interacting metal.) Most often other cases involve localization effects and are relatively well understood.

The "$\vee$" structure can be tentatively identified as another marker for non–Fermi liquid states. It is much more widespread than the cuprates alone, unlike other such markers, appearing in the $BaBiO_3$-based superconductors and even in some older "bad actors" in a weak fashion.[35]

A tentative, heuristic "explanation" is that, like the resistivity and c-axis conductivity, it is an indication that $\frac{1}{\tau} \propto \omega \simeq V$, *and* that coherent tunneling is blocked for some reason so that only—or primarily—inelastic processes which lose coherence in the CuO_2 layer are allowed. These then occur at a rate $\propto V$. The background at the apex of the "$\vee$" is more puzzling; this could be thought of as an inelastic amplitude which is not ω-dependent, either caused by inelastic scattering in the junction, or by thermal events. This background seems to diminish as the system becomes less Fermi liquid–like.

Some junctions, especially those not parallel to the c-axis, do not show this phenomenon. We must assume that in these cases an ordinary conductivity process akin to a metallic contact is operative: perhaps a normal-state version of Andreev tunneling direct into the charge degrees of freedom, with momentum conservation. Momentum plays a much more interesting role in the cuprates than in normal metals: the optimal tunneling electrons in the normal metal cannot, in a c-axis junction, match transverse momentum with the large values characteristic of cuprate electrons. Tunneling is, as we said in the introduction, a fascinating study which will provide a lot of physics when the more fundamental issues are better understood.   *j

NMR AND MAGNETIC FLUCTUATIONS IN THE NORMAL STATE

La_2CuO_4, $YBa_2Cu_3O_6$, and various others of the cuprates doped so that the planar coppers are all divalent, are all antiferromagnetic insulators. The "generalized phase diagram" (fig. 3.26) describes the result of even a small doping percentage of holes, which is to destroy antiferromagnetic long-range order. At small doping localized holes create local order defects; at larger doping mobile holes, as described by Shraiman and Siggia,[36] cause chiral magneto-polarons which, in small percentages, combine to create spiral orderings, according to mean field theory. Since the direction of the spiral is undetermined, spiral ordering seems to break down into disorder. All of this happens long before we reach the superconducting phase. As we have elsewhere emphasized, this low doping region is intrinsically not very stable and the true metallic phase, which is stable, appears at $\delta \gtrsim 10\%$. We are not, in this book, concerned with the complex magnetic behavior of these insulating and transitional phases, fascinating as they may be. There is, undoubtedly, much new physics in them but it is obscured by instabilities, trapped holes, and other epiphenomena.

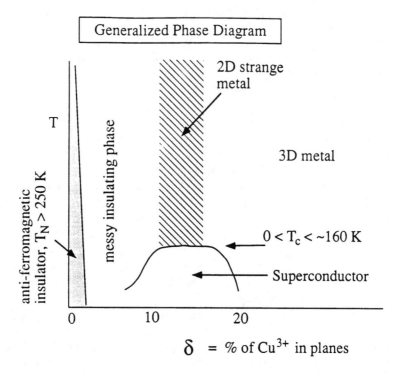

Figure 3.26. "Generalized Phase Diagram" as seen (roughly) in $(La - Sr)CuO_4$.

It is worth emphasizing this point, which is widely misunderstood. Mott transitions such as this are commonly quite first-order and involve a very complete restructuring of the quantum state. The metal develops a Fermi surface of Luttinger size, while the insulator has local moments or a magnetic structure or both. We should have had no expectation of continuity between the two extremes except at very high temperature or energy. Indeed, optical doping experiments show droplets of superconductor coexisting with the insulating phase: a miscibility gap. A "Stripe phase" which is a hybrid is seen in a few cases; also defect phases of various sorts. But attempts to explain the non-Fermi liquid as a hybrid show no promise: once metallic, there are no remnants of the insulator, and the metallic phase is very stable.

In the metallic phase one begins to see an orderly behavior. As we have discussed elsewhere, there is a level of doping above 10% and below ∼50% or less (optimal doping) where T_c is high and the properties tend to fit simple power laws in frequency and temperature. This we describe as the "ideal" normal metallic state. The susceptibility in the ideal metal is Pauli-like, constant with T, and of a magnitude indicating reasonable masses and enhancement factors.[37] In somewhat underdoped compounds one sees the development of the mysterious "spin gap" phenomena, signaled by a diminished susceptibility at low ($<150°$ or so) temperatures, and the susceptibility rising to the Pauli value at higher T. This can't be caused by "antiferromagnetic fluctuations" (spin density waves = SDW) because magnetic systems above their ordering temperature show a Curie-Weiss susceptibility, falling as T rises. We suppose that possibly phonons coupled to charge-density waves ("CDW"s) are responsible. Certainly phonon-like features appear in the infrared spectra, coupled to the electric response. Whatever is responsible, it clearly involves singlet formation, not magnetic order.   *k

Neutron scattering shows marked magnetic fluctuations in underdoped samples, persisting into the optimally doped regime. These appear to peak near a commensurate wave-vector $\pm\pi/a$, $\pm\pi/a$ in YBCO, but to be slightly incommensurate, compatible with $2k_F$ of a reasonable Fermi surface geometry, in   *l
$(La - Sr)_2CuO_4$. We have no reason to believe that peaks are not compatible with fluctuations involving a Fermi surface.

Such peaked fluctuations occur in mean field perturbation theory when the Fermi surfaces "nest," but they are quite different in a "Luttinger Liquid" such as the one-dimensional Hubbard model. In fact, insofar as the Fermi surface "nests" it has become effectively a one-dimensional model (since even in Fermi liquid theory collective excitations at $Q \neq 0$ decouple for different parts of the Fermi surface unless they coincide in velocity and momentum) so that it is inconsistent to use RPA on "nested" Fermi surfaces. There have been extensive attempts, nonetheless, to parameterize the neutron data as RPA-like, which, however, do not really work well, nor should they. The susceptibilities and exchange constants must be very finely tuned to fit the experimental data. As pointed out in the introduction to this chapter, the bulk of the neutron data on YBCO is suspect because of the absence of any feature which is identified with the chain Cu's. There should be a strong one-dimensional peak which is not seen, yet seems likely to be the most easily identifiable feature in the spectrum. One cannot be sure that the observed features are uncontaminated by this missing amplitude. Its observation would be very useful in settling the controversy about the valence states of chain and plane Cu's, as well. In general, we feel that neutron data, involving unique, large, but imperfect single crystals and extensive background corrections, are not very reliable.

The data on NMR in the normal metallic state of the high-T_c superconductors present a remarkably simple and uniform picture, overlain by a sequence of relatively minor complications. The big picture is the following: there are two characteristic behaviors, exhibited by the planar copper ions and by the planar oxygen ions (and by most of the other nuclei in the unit cell as well), respectively. The chain Cu and O in YBCO and the apical oxygen in 2 1 4 have not been as well studied (though they should be) and we shall not emphasize them. The planar Cu nuclei in all of the cuprates exhibit non-Korringa relaxation. The Korringa process in a metal is spin-flip scattering by the electrons at the Fermi surface, and is simply proportional (in rate) to the Fermi surface spread, $\sim kT$, and the density of initial × final states, $(N(0))^2$. $N(0)$ is the spin susceptibility, give or take a small factor, and the Knight shift is proportional to that, so $K^2 T_1 T$ is constant with T, and roughly uniform in magnitude in many metals. The planar Cu relaxation rate, on the other hand, rises more slowly than T, and is much larger (about an order of magnitude) than predicted by TK^2. The conventional value is not bad as $T \rightarrow \sim 1000°$, but the observed rate is much higher at $\sim 100°$K. This anomalous relaxation seems to be one of the common features of the high-T_c state. It is actually relatively more pronounced for somewhat lower T_c materials, according to early measurements, so is not so closely related to superconductivity. (See fig. 3.27.) It seems, however, to be a characteristic deviation from Fermi liquid theory.

As we pointed out, the planar O, Y, and various other nuclei, in general, exhibit Korringa ($T_1 T = $ const) relaxation, of the Korringa order of magnitude.[38] This is strong evidence for a Fermi surface. The striking difference in behavior of the two kinds of nuclei seems to have a simple explanation. These other nuclei are coupled equally to both sublattices of the Cu lattice, so that if the anomalous fluctuations were antiferromagnetic (anywhere in Q-space along the line π, O to π, π) they would not couple to the O nuclei (Shastry, Mila, and Rice).[39] This has been adduced as strong evidence for the RPA-spin fluctuation framework, but of course is equally true for any $2k_F$-based scenario, so long as the most strongly fluctuating $2k_F$ is fairly close to an antiferromagnetic plane in one direction or another, as it is for both the incommensurate 2 1 4 and the apparently commensurate YBCO. The underlying spin structure of the large U Hubbard model is basically antiferromagnetic, as is shown by the "t-J" model, to which it is equivalent, and the physical signal of this antiferromagnetic structure is primarily the singularity in the responses at $2k_F$.

The spin correlation function of the 1d Hubbard model is strikingly different near $Q = 0$ from that at $Q = 2k_F$. Near $Q = 0$, the spin correlation function and the spin susceptibility are of the same form as those of a conventional Fermi liquid: the susceptibility is constant as $T \rightarrow 0$, and is not peaked in Q-space. The corresponding Fourier transform gives a time-dependence as $1/t^2$ as $t \rightarrow \infty$. Contrastingly, the susceptibility shows a marked peak at $Q = 2k_F$. The corresponding correlation function has the asymptotic form

$$\langle S(x, t) S(0) \rangle \propto \frac{\cos 2k_{Fx}}{\sqrt{x^2 - v_s^2 t^2} \, (x^2 - v_c^2 t^2)^p}$$

with $1/2 < p < 1$, and correspondingly the susceptibility as a function of Q also has a peak

$$\chi(Q) \propto \frac{1}{|Q - 2k_F|^{1-p}}$$

$$\chi(T) \propto \frac{1}{T^{1-p}}$$

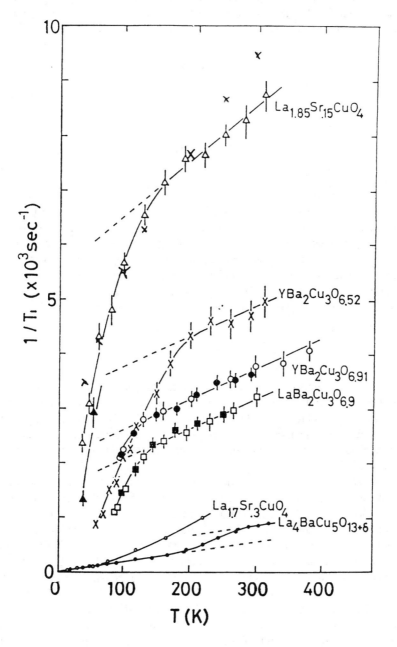

Figure 3.27. An early compendium of non-Korringa Cu relaxation rates.

with the same exponent in Q, ω, and T. Note that this is a stronger singularity than that exhibited by the *free* Fermi gas, which is a BCS-like logarithm, but of course much weaker than the RPA-mean field theory gives.

$$\chi_{\text{free}}^{\text{MF}} \propto N(0) \ln \frac{\wedge}{T}$$

and

$$\chi_{\text{int}}^{\text{MF}} \propto \frac{\chi_f}{1 - \chi_f} \to \infty$$

$$\text{at} \quad T_{\text{mf}} = \Delta e^{-\frac{1}{N(0)J}}.$$

This is the singularity which would be predicted by a mean field theory based on exact nesting such as is envisaged by a number of authors. No singularity necessarily appears without nesting. Nesting, however, is not a consistent description, as we already pointed out, since nesting implies approximate one-dimensionality which in turn implies non-mean-field behavior. That is, in a region where the Fermi surface is flat so that ϵ_k is only a function of one component of k, k_M, we may sum over the other component

$$c_{k_{\parallel},R}^{+} = \sum_{k_{\perp}} e^{ik_{\perp} \cdot R} \, c_{k_{\parallel},k_{\perp}}^{+}$$

to make a sequence of one-dimensional interacting chains at position R. More generally, any tendency for the transverse kinetic energy to be smaller than in a genuine 2D model reinforces the arguments we will make in chapter 6 in favor of independent dynamics of the different parts of the Fermi surfaces, which in turn makes it less rather than more likely that a mean field-RPA dynamic will give a correct description.

Our point of view will be that the fluctuations in these materials are correctly dealt with by ignoring the conventional mean field or "spin fluctuation" corrections and assuming that the great bulk of the interaction is taken into account in the rediagonalization of the bosonic fluctuations, which redefines the elementary excitation spectrum. Thus, we do exactly what is done in Fermi liquid theory, which assumes that the basic fluctuations are describable by redefined quasiparticles, which give us the basic power laws for physical quantities, with high-order corrections (in the case of F.L.T., $\sim\omega^3 \ln \omega$, for example) due to collective excitations, treated by RPA. However, in the present case the basic power laws are modified because the excitations are no longer quasiparticles and have separate spin and charge dynamics, as well as singular behavior at $Q = 2k_F$ which is not a trivial reflection of that near $Q = 0$. This is the natural way to arrive at "antiferromagnetic spin fluctuations"; RPA superimposed on

Fermi liquid theory is not, since the "spin fluctuations" do not naturally modify the single-particle properties in that theory. In this view, antiferromagnetism arises when the $Q = 2k_F$ singularities reinforce each other by the umklapp phenomenon as is possible in the cuprates.

We shall not, for these reasons, dip into the morass of spin fluctuation "theory," since that is fundamentally flawed. Instead we shall see whether we can reinterpret the phenomena in our own terms. The outline of this reinterpretation is due to Yong Ren. We borrow from Shastry et al. the idea that the O nuclei do not couple strongly to the $2k_F$ singularities because of their equal coupling to the two sublattices, while the Cu nuclei are strongly affected by the singularity and are primarily relaxed by it.

In fig. 3.28 we show the relaxation rate $1/T_1\,(T)$ for the planar Cu in $YBCO_7$ fitted to a power law in temperature. This fit requires no sensitive special choices of several parameters, but only a single, very insensitive one, the exponent. (Since at high T it should reduce to simple Korringa, there is no free amplitude parameter.)

The Cu nucleus's spin is known to have considerable coupling both to the $d(x^2 - y^2)$ component of the wave function and to an isotropic, s-wave component. The latter has been ascribed to a transferred hyperfine coupling to the Wannier function on the neighbor, but such a coupling would have considerable cancellations at $2k_F$ and would not be strongly affected by the singular fluctuations, so we suspect that the isotropic coupling is core polarization of the 2s and 3s electrons on the original site, as is observed in many transition metals. We assume, then, that the Cu nucleus couples to the local spin and is therefore relaxed by the *origin* to *origin* correlation function

$$\langle S\,(0, t = 0)\,S\,(0, t)\rangle.$$

This is an "$l = 0$" quantity, coupling symmetrically to every point in k-space and particularly to every $\vec{k}_F(\Delta)$ on the Fermi surface. Thus it couples only to the single $l = 0$ channel and is in that sense essentially a one-dimensional response function. We presume, given the arguments of chapter 6, that this response, at least in 2D with strong interactions, is the same as that of a 1-dimensional Hubbard model. A somewhat less schematic way to derive this fact is simply to sum the $2k_F$ responses over all k-vectors as follows: (Note that Luther has proved that all correlation functions—at least in the Fermi liquid case—are correctly given, asymptotically, by such sums)

$$\chi\,(0, t) = \int ds \cos 2k_F^{(s)} \cdot (0) \frac{1}{i v_s^{(s)} t} \frac{1}{[i v_c^{(s)} t]^p}$$

$$\propto \frac{1}{(it)^{1+p}} \;\; ; \;\; \frac{1}{2} < p < 1.$$

(The correlation functions for each $k_F^{(s)}$ are identical in form at $x_s = \vec{r} \cdot \vec{v}_s \equiv 0$).

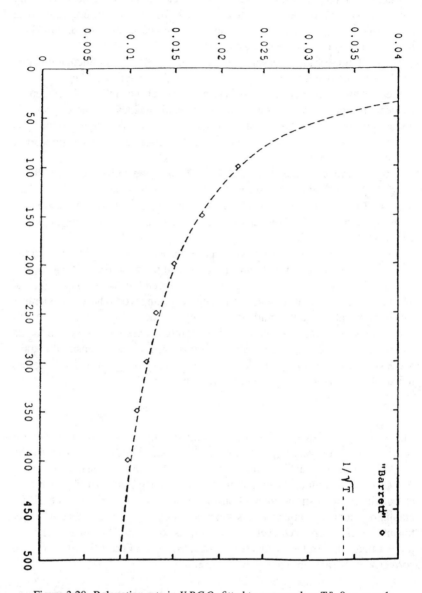

Figure 3.28. Relaxation rate in $YBCO_7$ fitted to a power law T^p, $0 < p < 1$.

The relaxation rate $\frac{1}{T_1}$ is given by

$$\gamma^2 \int_{T^{-1}}^{\infty} \langle S(0,0)\, S(0,t) \rangle \, dt = \gamma^2 [N(0)]^{1+p}\, T^p$$

to within factors of order unity, and where γ is the strength of the hyperfine coupling. In this manner we find very directly and simply the enhanced relaxation, without any need for delicate parameter adjustment. (See fig. 3.24.) Where there is some need for adjustment is at lower doping where the "spin gap" phenomenon intervenes. As we go to lower doping, we are approaching the half-filled case and several things are happening at once. At high temperatures our Luttinger liquid approach is still correct, presumably with a smaller p (greater enhancement); as the temperature is lowered the umklapp terms begin to enhance the magnetic responses. On the other hand, the "spin gap" due to CDW formation (equivalent to valence bond formation) seems to be developing at the same time, which causes a simultaneous *decrease* in all responses, both spin and charge; the conductivity, susceptibility, Knight shift, and superconducting condensation energy all fall off with similar T-dependence. Correspondingly, $1/T_1$, both for oxygen and Cu sites, is decreased. This "spin-gap" phenomenon is fascinating but not central to the problem of superconductivity. Our philosophy is that the Luttinger liquid of the CuO_2 planes is essentially a high-temperature, large energy-scale state whose properties are quite universal, and that superconductivity as well as the spin-gap phenomenon are the result of interactions outside of the planar Hubbard Hamiltonian which controls the high-temperature state. In the case of superconductivity, this interaction is interlayer coupling. The spin gap may be caused either by interlayer magnetic coupling or by phonons, but its nature is clearly to "eat away" large chunks of the Fermi surface, leaving all Fermi surface phenomena proportionally weaker. Its study is good physics but would not contribute much to our understanding of superconductivity. *m

Finally, we come to the problem of the oxygen nucleus—and, for that matter, all of the remaining non-$Cu(2)$ nuclei. So far no measurement has revealed any $\frac{1}{T_1}$ enhancement in any of these. This is a problem for any theory, since it is implausible that the anomalous fluctuations should center entirely at the π, π point which is the only place where all oxygens have form-factor exactly zero, even in the spin-fluctuation scenario. We offer no particularly brilliant solution to this dilemma. The oxygen nucleus couples to, presumably,

$$S(R_i, t) + S(R_i + 1, t)$$

and so the $2k_F$ part of its relaxation rate will be multiplied by

$$\int ds \, \gamma^2(s) \left| 1 + e^{2ik_F^{(s)} \cdot (R_{i+1} - R_i)} \right|^2$$

while the $Q = 0$ term is multiplied by $\int ds \, \gamma^2(s) \cdot 4$. The ratio of these two magnitudes cannot be zero but may be quite small, we suppose. An additional factor may be that in fact the oxygen $p\pi$ wave functions may be admixed into Wannier functions belonging to even further neighbors, giving an even closer approximation to an extended source which would not scatter strongly at $Q = 2k_F$. This will be even more true of the other nuclei in the unit cell, which cannot be expected to couple to the $2k_F$ singularities. Nonetheless, until experimentalists quantify the amount of anomalous relaxation of the planar oxygen nuclei exactly, this kind of explanation remains open.

NOTES

*a New observations by Keimer et al. (see also note *l) in the superconducting state hold the promise of great usefulness, as well.

*b The actual *absence* of ρ_{res} is brought out clearly in modern measurements by Ando and Boebinger discussed in a preprint in section II. Where the "holon non-drag" regime of spinon only transport kicks out, there is an immediate sharp rise in resistivity which is the restoration of ρ_{res}.

*c Materials continue to improve and the latest data (12/95) show an isotropic *ab* plane conductivity, with a chain contribution not obeying $\rho \propto T$.

*d Low temperature c-axis resistivity has a number of complications. In "spin-gap" samples there is an Arrhenius component $e^{+E_g/T}$. In the best data so far on $(La - Sr)_2CuO_4$, as well as on 1-layer BISCO, it seems to be T^{-p} with p small and negative. These are discussed in some recent papers reprinted in sec. II.

*e More recent Righi-LeDuc (Thermal Hall Effect) measurements by Ong's group very much clarify and confirm these conclusions. They will play more of a role in the superconducting state.

*f More recently, the data shown in fig. 3.13 have been interpreted in detail in terms of $\sigma \propto \frac{1}{\omega^{1-2\alpha}}$, which is straightforward to derive by the methods of chapter 6. See preprint submitted Oct. '95.

*g See the note above on spin gaps (chap. 2, p. 17) and the Strong-Anderson paper.

*h Clarke finds that Uchida's high-frequency data confirm this; see also note (chap. 2, p. 11).

*i I believe that this entire discussion is flawed. The true Fermi surface must lie somewhere between the two surfaces in fig. 3.21 (by hole count) and rather close to the van Hove singularity at M. Theoretically, we must realize that virtual admixture of the two layers by $t_\perp$ must exist over a broad range of k's, as well as the noncollinearity effect mentioned, so in this region the Fermi surface may appear quite fuzzy for a number of reasons. Later data at other doping levels does show what has been interpreted as an unsplit Fermi surface. The theory is not yet capable of predicting actual EDCs with any precision.

*j Recently tunneling from a point in vacuum to BISCO surfaces has been measured by O. Fischer's group. The curves in the superconducting state are reasonably compatible with the photoemission data at the same temperature: it appears that this type of tunneling measures $Im G_1$ simply.

*k See, again, note chap. 3, p. 22.

*l Within the past year more accurate neutron measurements by Keimer et al. (reproduced among the reprints) have shown that the π, π peak in the normal state of YBCO is an experimental artifact—a phonon. My doubts about neutron data turn out to have been well justified. I have not fully corrected the text.

*m (both) The CDW possibility for the spin gap seems less likely than the Strong-Anderson theory that it is a kind of superconducting fluctuation. This may make the sentence flagged by the second note incorrect. The effects on NMR are predicted to be simply proportional, and proportional to the drop in χ. (As is observed, apparently.)

REFERENCES

1. P. W. Anderson, Science, **256**, 673 (1992).
2. T. E. Mason, G. Aeppli, S. M. Hayden, A. P. Ramirez, and H. A. Mook, Phys. Rev. Lett. **71**, 919 (1993).
3. D. B. Tanner and T. Timusk, in *Physical Properties of High T_c Superconductors III*, D. M. Ginsberg, ed., p. 363. (World Scientific, Singapore, 1992).
4. J. R. Schrieffer, Rev. Mod. Phys. **36**, 200 (1964); also in *Phonons and Phonon Ints.*, T. Bak, ed., Benjamin, N.Y., 1984.
5. P. W. Anderson, Phys. Rev. B, **23**, 4828 (1991).
6. M. A. Schluter and M. S. Hybertson, Physica C **162**, 583 (1989).
7. S. Martin, A. T. Fiory, M. Fleming, L. Schneemeyer, and J. V. Woozcak, Phys. Rev. B **41**, 846 (1990).
8. S. J. Hagen, T. W. Jing, Z. Z. Wang, J. Horvath, and N. P. Ong, Phys. Rev. B **37**, 7928 (1988).
9. T. A. Friedmann, D. M. Ginsberg, M. W. Rabin, J. Giapintzapakis, and J. P. Rice, Phys. Rev. B **42**, 6217 (1990).
10. S. Kambe, K. Kitazawa, M. Naito, A. Fukuoka, I. Tanaka, and H. Kojima, Physica C **160**, 35 (1989).
11. C. M. Varma, P. B. Littlewood, S. Schmitt-Rink, E. Abrahams, and A. E. Ruckenstein, Phys. Rev. Lett., **63**, 1996 (1989).
12. P. W. Anderson, in Proceedings of Fermi International School of Physics, *Frontiers and Borderlines in Many-particle Physics* (North-Holland, Amsterdam, 1987).
13. T. R. Chien, Z. Z. Wang, and N. P. Ong, Phys. Rev. Lett. **67**, 2088 (1991).
14. M. Z. Cieplak, G. Xiao, A. Bakshai, C. Z. Chien, Phys. Rev. B **39**, 4222 (1989).
15. A. M. Finkelstein, V. E. Kataev, E. F. Kukovitskii, and G. B. Teitelbaum, Physica C **168**, 370 (1990); see also Alloul et al. (A. V. Mahajan, Alloul, G. Collin, and J. F. Marucco), submitted to Phys. Rev. Lett (1994), for nuclear resonance data reinforcing these observations.
16. N. Kumar and A. M. Jayannavar, Phys. Rev. B **45**, 5001 (1992).
17. A. I. Larkin and L. B. Joffe, Intl. J. Mod. Phys. B **3**, 2065 (1989).
18. P. W. Anderson, Phys. Rev. Lett **67**, 3844 (1991).
19. D. G. Clarke, S. Strong, and P. W. Anderson, Phys. Rev. Lett., to be published.
20. W. Kohn and J. M. Luttinger, Phys. Rev. **118**, 41 (1960); C. Bloch and C. de Dominicis, Nucl. Phys. **7**, 459 (1958).
21. J. M. Harris, Y. F. Yan, and N. P. Ong, Phys. Rev. B **46**, 14293 (1992).
22. C. Stafford, Phys. Rev. B **48**, 8430 (1993).
23. C. Stafford and A. J. Millis, Phys. Rev. B **48**, 1409 (1993).
24. S. L. Cooper, G. H. Thomas, J. Orenstein, D. H. Rapkine, A. J. Millis, S.-W. Cheong, A. S. Cooper, and Z. Fisk, Phys. Rev. B **41**, 11605 (1990).
25. A. El Azrak, R. Nahoon, A. C. Boccara, N. Bontemps, M. Guilloux-Viry, C. Thiret, A. Perrin, Z. Z. Li, and H. Raffy, Journal of Alloys and Compounds, **195**, 663 (1993); also preprint.
26. J. Kim, I. Bosovic, J. S. Harris Jr., N. Y. Lee, C.-B. Eom, and T. H. Geballe, preprint (1993). Also reported in Ref. 3.
27. S. L. Cooper, M. V. Klein, B. G. Pazol, J. P. Rice, and D. M. Ginsberg, Phys. Rev. B **37**, 9860 (1988).
28. B. S. Shastry and B. Shraiman, Phys. Rev. Lett., **65**, 1068 (1990).

29. K. Tanasaki, Y. Nakamura, and S. Uchida, Phys. Rev. Lett. **69**, 1455 (1992). These measurements have recently been confirmed and extended by Timusk and Van der Marel.

30. M. V. Klein et al., preprint.

31. C. Olsen, R. Liu, A. Yang, D. W. Lynch, A. Arko, R. S. List, B. Veal, Y. Chang, P. Jiang, and A. Paulikas, Science **245**, 731 (1989).

32. J.-M. Imer, F. Patthey, B. Dardel, W.-D. Schneider, Y. Baer, Y. Petroff, and A. Zettl, Phys. Rev. Lett. **B62**, 336 (1989).

33. D. S. Dessau, Z.-X. Shen, D. M. King, D. S. Marshall, L. W. Lambordo, P. H. Dickinson, A. G. Lovesco, J. di Carlo, C.-H. Park, A. Kapitulnik, and W. E. Spicer, Phys. Rev. Lett. **71**, 2781 (1993).

34. D. S. Dessau, B. O. Wells, W. E. Spicer, A. J. Arko, D. S. Marshall, E. R. Ramer, J. L. Peng, Z. Li, and R. L. Greene, submitted to Phys. Rev. Lett. (1993).

35. F. Sharafi, A. Parrellis, and R. C. Dynes, Phys. Rev. Lett. **67**, 509 (1991); M. Garsitch, J. M. Valles, A. M. Cucciola, R. C. Dynes, J. P. Garno, L. F. Schneemeyer, and J. Wasczak, Phys. Rev. Lett. **63**, 1008 (1989).

36. B. Shraiman and E. Siggia, Phys. Rev. Lett **62**, 1564 (1989).

37. D. C. Johnston, S. K. Sinha, A. J. Jacobson, and J. M. Newson, Physica C **153-55**, 572 (1988).

38. M. Takigawa, P. C. Hammel, R. H. Hoffener, Z. Fisk, K. C. Ott, and J. D. Thompson, Phys. Rev. Lett. **63**, 1865 (1989).

39. B. S. Shastry, Phys. Rev. Lett. **63**, 1288 (1989); F. Mila and T. M. Rice, Physica C **157**, 561 (1989).

4

The Superconducting State

This chapter will be much less complete than the last. For one thing, data on the superconducting state are still the subject of deep experimental controversy. On many of the controversies, I have little to say and my intervention into them would be pointless, since most of them seem not to be relevant, as far as we can see, to the decision as to mechanism.

The theory we have presented is primarily a theory of the normal metal and is not easily extended to the superconducting state in any precise form, so that direct calculation is in many instances not possible. The primary obstacle is the absence of a theory of the superconducting Luttinger liquid, so that all we can do is to glue BCS superconductivity to the Luttinger liquid, hoping that the joint between these incompatible concepts will not be too disruptive. Attempts to do so are given in the reprinted articles by the group of Chakravarty, Sudbo, Strong, and myself in part II of the book.

I will only, in this chapter, comment briefly on some experimental anomalies which are relevant. These are generally and reliably observed but not necessarily accepted as to their meaning and significance. I think they represent striking differences from the conventional superconducting state. Added together, they represent a picture which is very hard to fit into the normal BCS framework of quasiparticles binding together in pairs by an interaction of some local sort (hence broad in momentum space) and thereupon opening up a gap. On the other hand, all of these phenomena fit naturally into the theory of this book, and are listed under four headings.

THE SPIN-GAP PHENOMENON

In chapter 3 we described some of the phenomena associated with the "spin gap" observed in oxygen-poor YBCO, and less clearly in other cases. It is, as I pointed out, not clear what is causing the spin gap; a likely suggestion is fluctuating CDW ordering, caused perhaps by interlayer spin coupling or by umklapp effects. What is clear is that its effect is to wipe out some fraction of the Fermi surface, possibly by opening a gap or pseudogap *not* of superconducting type. Normal state measurements which indicate this are drops in $1/T_1$, Knight shift, susceptibility (see fig. 4.1), and—relative to the extrapolated $\frac{1}{T^{1-\epsilon}}$ law—

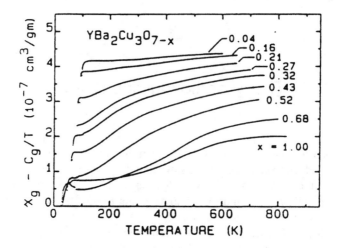

Figure 4.1. The "spin-gap phenomenon" in susceptibility.

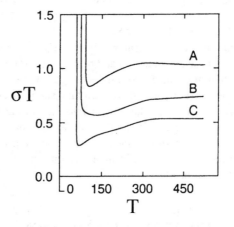

Figure 4.2. Same in electrical conductivity.

conductivity (see fig. 4.2). Nonetheless some Fermi surface is left intact, as is indicated by the fact that the spin gap does not reduce the Hall angle below the $1/T^2$ law extrapolated from high temperature[1] (see fig. 4.3). Thus what Fermi surface is left seems to be unaffected by whatever is happening elsewhere.

This independence of different pieces of Fermi surface seems also to be the case for superconductivity. Loram[2] has pointed out that the total specific heat

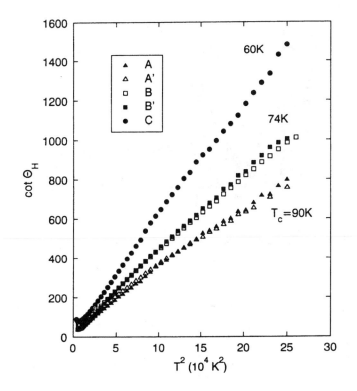

Figure 4.3. Θ_H vs T^2 for samples exhibiting spin gaps, showing that the spin gap does not affect spinon life time.

peak and hence the total superconducting binding energy drops rapidly as the spin gap develops, *even before T_c has been appreciably affected* (see fig. 4.4) and by much more than T_c drops. This is also indicated by the Knight shift,[3] where the singularity at T_c decreases in the same spectacular way (see fig. 4.5).

The fact that T_c holds up (decreasing by only 40%) while the total binding energy decreases by an order of magnitude or more is incompatible with BCS theory as it is normally used, with a local attractive interaction. In BCS, T_c is an exponential function of density of states, with fixed interactions; the situation in real superconductors is more complex than that but it is a good rule of thumb that high T_c and high density of states are strongly correlated, *more* strongly than linear. The only interaction which can leave T_c large while the density of states is decreased by cutting down the area of the Fermi surface is one which is very short-range in momentum space, i.e., extremely long-range in real space, a property which is unique to the tunneling mechanism. In conventional BCS the

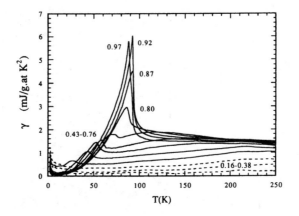

Figure 4.4. Spin-gap effects on specific heat are large, but not those on T_c.

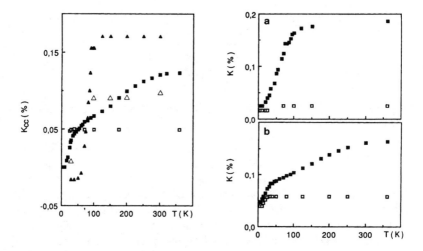

Figure 4.5. The Knight shift (spin susceptibility) is like specific heat.

gap at k is the sum of contributions from all other k', whereas this phenomenon requires independence of different k'''s.

A possible theory of the spin-gap phenomenon has been developed by Strong and myself, using the interlayer mechanism expressed in systems with independent bilayers. A preprint is appended.

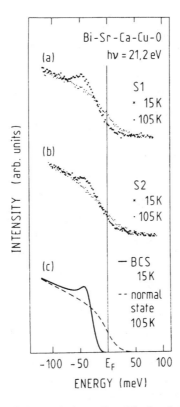

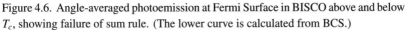

Figure 4.6. Angle-averaged photoemission at Fermi Surface in BISCO above and below T_c, showing failure of sum rule. (The lower curve is calculated from BCS.)

PHOTOEMISSION AND OTHER ANOMALIES:
"VIOLATED" SUM RULES

Both angle-averaged and angle-resolved photoemission studies show striking anomalies in comparisons of normal state and superconducting state spectra.

Angle-averaged photoemission: the early high-resolution spectra of Baer and Petroff[4] (fig. 4.6) immediately presented us with a dilemma. They triumphantly confirmed that much of the intensity at the Fermi level was indeed moved out of a gap of approximately the right size, estimated to be $\sim$ 15–20 mv. However, what was observed that was unexpected was not what was missing but what was present: *more* intensity appeared "piled up" above the gap than disappeared from the gap region (see fig. 4.6). If the angle-averaging is done properly, there is an *exact* sum rule which states that the *total* intensity averaged over all energies is constant. (The BCS density of states at energy ω is $N(\omega) = \dfrac{\omega}{\sqrt{\omega^2 - \Delta^2}}$, which, when integrated to $\omega = \infty$, gives the same total intensity.) But the extra intensity is very slowly decaying (as $1/\omega$) so that integrating over any finite energy aperture should give *less* total intensity; in fact, Baer and Petroff's numerical estimate gave about 1/2 to 2/3.

Experimentally, on the other hand, one could see that the intensity missing from the gap reappeared *magnified* by a factor 3, i.e., the sum rule was *over*satisfied by a factor 6! (On the figure we have attempted to show the missing area in the gap and the enhancement in the peak. The papers reporting the experimental results have downplayed this discrepancy, though it is obvious in the published data.)

Since the sum rule is very fundamentally based, the conclusion is that BCS theory does not describe the intensity changes properly. In BCS theory one-electron intensity moves up in energy from the gap region, indicating that the net electronic kinetic energy is increased in the superconducting state (a change which is compensated by the gain in interaction energy due to pair binding and overcompensated by the decrease in zero-point kinetic energy of the phonons). What this experiment shows is that in the cuprates there is a considerable shift of one-electron energy downward in the superconducting state, from quite high energies (of order U or bandwidth, presumably). This is a very direct and rigorous demonstration that the primary mechanism for superconductivity is electronic (thus excluding, aside from conventional BCS, any "bipolaronic" possibilities).

Very similar, but much less unequivocal, results have often been indicated in tunneling from the cuprates to normal metals, where superconducting structure often is quite a bit stronger than the "normal" conductance would allow,[5] conventionally (fig. 4.7).

ANGLE-RESOLVED PHOTOEMISSION

From the angle-resolved spectra taken above and below T_c one can see explicitly the process which can only be inferred from the angle-averaged data. (The latter, however, can be reasonably well compared quantitatively above and below T_c, while the angle-resolved data suffer from an arbitrary experimental normalization factor.) In the very earliest data (fig. 4.8) it is seen that the power-law and cusp structure of the normal state[6] which we interpret as the composite holon-spinon spectrum, develops a very sharp and strong peak near the original spinon energy (presumably shifted by the energy gap). This peak has never been observed to be broader than the experimental resolution. This sharp peak is not unexpected in any theory in which the anomalous breadth of the spectrum is due to electron-electron interaction—i.e., spin-fluctuation[7] or marginal Fermi liquid[8] as well as Luttinger liquid theories. The threshold in an electron-electron theory for any real decay process is 3Δ, so that $Im \sum(\omega) \equiv 0$ for $\Delta < \omega <$

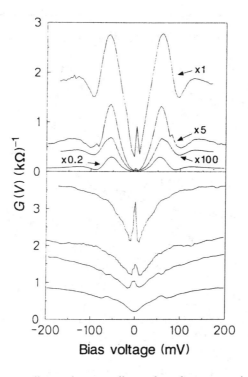

Figure 4.7. Large tunneling peaks on small tunnel conductance, typical of break junctions.

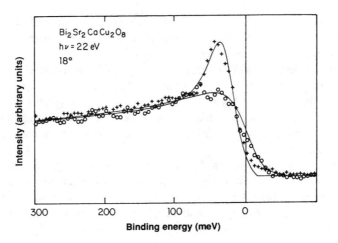

Figure 4.8. Broad cusp of photoemission develops a strong and sharp peak below T_c.

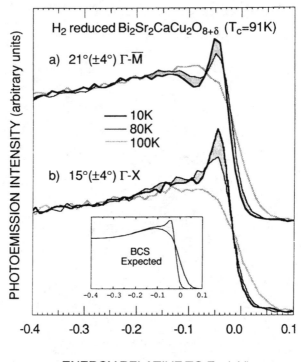

Figure 4.9. Anomalous shift of photoemission intensity at T_c: may be 3Δ effect.

3Δ. This is quite a contrast to conventional BCS, and even if phonon or bipolaron explanations for the T-dependent resistivity and superconductivity were otherwise viable, these phenomena would eliminate them.

Any of the electron-electron theories is compatible with a possible anomaly in the spectrum at $\omega = 3\Delta$, which may have been observed in some spectra by Shen et al.[9] (fig. 4.9). My own interpretation of this singularity suggested at the time[10] does not seem as likely. It is unfortunate that no detailed theory of the superconducting Luttinger liquid exists because the spectrum of fig. 4.9 would be a crucial test of it.

The main physics on which the three theories seem to differ is the sum-rule anomaly revealed best by angle-averaged photoemission. In the Luttinger liquid, the one-electron Green's function is homogeneous of order $-1 + 2\alpha$ in k and ω, so that the integral of its imaginary part—the total weight of the one-electron Green's function—is not convergent without an upper cutoff Λ. Therefore, the peak of finite amplitude which appears in the superconducting state must bring the bulk of its amplitude from energies near the cutoff, which accounts for the downward shift of one-electron energy. In the other two alternative theories, it is assumed that the Green's function has either a simple pole (in the case of spin fluctuations) or is quasiconvergent, in the marginal case, and the amplitude which goes into the pole at $\omega = \Delta$ comes from a region of energy at the scale Δ, not $\wedge$.

The article by Chakravarty and myself which is reprinted, as well as other previous work by me, shows how the quasiparticle pole develops within perturbation theory. But this is only a specific example of a very general theorem about Green's functions and self-energies. In a normal metal with a Fermi surface the inverse Green's function must vanish at $\omega = 0$ and $k = k_F$:

$$G^{-1}(\omega = 0, \, k_F) = 0 \, .$$

This is a consequence of the unbroken symmetry of the system. One can define an unperturbed system where

$$G_0^{-1}(\omega = 0, k_F) = 0 \; \text{also} \quad ; \quad \text{hence} \quad \sum_{im}(k_F) \equiv 0.$$

The Luttinger (and marginal) liquids obey this identity as well as does the Landau liquid; the difference is that $\sum_{Landau}$ is regular while $\sum_{Lutt}$ has branch cuts.

When gauge symmetry is broken we must introduce the Nambu generalization of the Green's function, as a 2×2 tensor. In this case it is only at special k's, if at all, that both the anomalous and normal components of $\sum_{im}$ can vanish, so that $\underset{\sim}{G}^{-1}$ will always exhibit poles or pole-like singularities in the region of the Fermi surface, signifying quasiparticles. The branch cuts become irrelevant since $\sum$ never vanishes.

ANOMALOUS TRANSPORT PROPERTIES

Various anomalous transports all confirm the striking observation of sharp, resolution-limited quasiparticle peaks in photoemission. This in turn indicates that free quasiparticles are quite strikingly present in the superconducting state, where they did not exist in the normal state—or at least, were extremely heavily scattered. The "fundamental process" which gives $\hbar/\tau \sim \hbar\omega$ or kT for the electrical conductivity drops out below T_c. As a result, there is an anomalous peak in quasiparticle conductivity as the temperature falls below T_c. This peak is seen in two types of experiment: microwave conductivity and thermal conductivity. Both of these phenomena contrast with the NMR relaxation rate $1/T_1$, in showing anomalous rises ("Hebel-Slichter peaks") below T_c. There

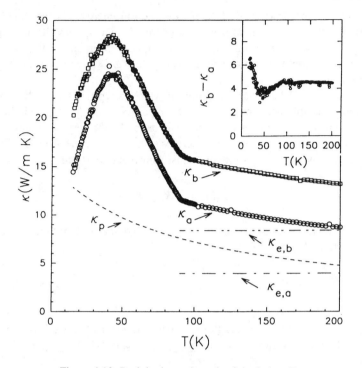

Figure 4.10. Peak in thermal conductivity below T_c.

is some controversy over the interpretation of both phenomena as well as over experimental data in the case of microwave conductivity, but it is now clear that the two sets of data have settled down to a Wiedemann-Franz relation to each other, and demonstrate that the heat conductivity is almost all electronic both in normal and superconducting states. The review article by Uher[11] is one of the more interesting examples of papering over inconsistencies; for instance, he cites without comment experimental data (see fig. 4.10) giving a 17 : 1 anisotropy in thermal conductivity, which is of course inconceivable for a phonon mechanism; yet continues to ascribe the peak to phonon scattering by electrons. The subject has been adequately covered in a Letter.[12] (See figs. 4.10, 4.11.) With the photoemission experiment, a consistent and convincing picture has emerged, one which requires a theoretical structure of the sort we have described. One thing which is completely decided by these data is that the normal state resistivity is caused by an *electronic* mechanism which is gapped below T_c. A striking confirmation is given by Ong et al.'s measurements of a magnetic "Hall effect" in thermal conductivity, which shows enormous enhancement of

*a   mean free path below T_c.

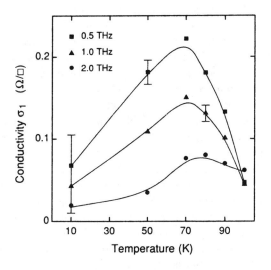

Figure 4.11. Peak in μ-wave conductivity below T_c.

POSTSCRIPT: JOSEPHSON INTERFERENCE EXPERIMENTS AND THE SYMMETRY OF THE ORDER PARAMETER

There has been a great deal of discussion of the question of the symmetry of the superconducting order parameter ever since the first observations on cuprate materials. In my opinion, emphasis on this aspect of the phenomenon is misplaced in view of the very many anomalies of the cuprates, both in the superconducting and in the normal state, which can have little to do with the relatively minor modifications of BCS theory which result from a different order parameter symmetry. In actual fact, for instance, the most widely proposed alternative order parameter, "d-wave" $\Delta(k_x, k_y) \propto k_x^2 - k_y^2$, belongs to the same representation Γ_1 of the orthorhombic crystal group of YBCO as does "s-wave" $\Delta \propto k_x^2 + k_y^2$, so that the two should mix at will, and no very profound differences can be expected. In fact, the whole debate on this point has been characterized by oversimplification.

Since the early days of BCS the possibility of anisotropic order parameters belonging to symmetries other than the BCS Γ_1 has been understood (the first fully articulated discussion being that of Morel and Anderson[13]). Anderson incidentally pointed out that in metals such order parameters would be very sensitive to impurity and phonon scattering which would always be pair-breaking; this was formally confirmed by Balian and Werthamer.[14] Nonetheless some heavy-electron metals are thought to have such order parameters, and indeed only very pure samples of these are superconducting (although even here some very mysterious phenomena occur). Measurements of thermal and transport properties on these heavy electron metals to very low values of T/T_c show power laws in T rather than exponentials, indicating that there is no complete gap and that Δ must vanish along some manifold of directions in k-space.

Such measurements for cuprates are very difficult to disentangle from various possible defect and impurity effects: two-level centers, surface effects, etc. Firm statements on both sides of issues such as the temperature-dependence of penetration depth, specific heat, and NMR properties exist in the literature. The recent consensus has been that large anisotropy of $\Delta(k)$ seems proven by photoemission and NMR measurements, but vanishing of the gap is not proven.

The most convincing recent evidence has involved Josephson interference phenomena. Very early, D. Wohlleben remarked that some granular samples often exhibited paramagnetism at low fields, suggesting that some intergranular loops had finite magnetic flux in equilibrium. Several recent Josephson interference experiments[15–17] show that junctions on "a" and "b" faces of the same crystal with a conventional superconductor are in equilibrium when the phase of the latter's order parameter is opposite in sign, as would be expected if (a) the order parameter is of "d"-like symmetry; (b) the k-vectors which have the larger Josephson matrix elements are those perpendicular to the face. These experiments are not absolutely conclusive, and contradictory experiments exist, but they are carefully carried out and the real possibility exists that the sign of the order parameter does change over the Fermi surface.

At first sight this observation seems impossible in view of the observation that impurities have little effect on T_c, except for those (such as Ni and Zn) which have magnetic moments and are in the plane. These are clearly pair-breaking; we have discussed them in chapter 3. Such impurities also cause a high residual resistance, having unitarity limit scattering cross-sections.

Most impurities do not cause residual resistance, for the reasons discussed in chapters 3 and 6, but calculations based on Friedel's theorem show that their scattering cross-section should be easily large enough to do so and therefore, to cause pair-breaking and T_c reduction for a d-wave superconductor, at a level which is not even marginally observed. I consider that the general problem of impurity scattering in the superconducting state—and in fact in the Luttinger liquid in general—is very much still an open question. It needs careful quantitative study. The observation of the rapid lengthening of τ below T_c shows that impurity scattering by the doping impurities is still surprisingly ineffectual even in the superconducting state, so that the argument that a non-Γ_1 order parameter would be impurity-sensitive needs quantitative discussion.

Another related puzzle is the isotope effect. YBCO, where well oxygenated, shows none, in agreement with a hypothetical d-like order parameter. But other cuprates—notably 124—show isotope effect of normal sign and small magnitude (when not underdoped). However, isotope studies are notoriously unreliable and other data do not necessarily confirm this. The d-wave state relies on repulsive interactions for its motivation. The repulsive interaction scatters predominantly between the two regions where Δ is of opposite sign, and attractive interactions due to phonons will tend to *reduce* T_c. This means that any isotope effect should be of the opposite sign, if it exists at all, for d-wave. This is not observed. I feel it is crucial to redo the interference experiments on single crystals of 2 1 4 superconductors, as well as those showing anisotropy. It would be fascinating if both d- and s-symmetry gaps occur, in different substances. This would certainly confirm the outlines of our theory.

There is one final point to be made. In conventional BCS, the repulsive Coulomb interaction of electrons is removed, mostly, by the "dynamic screening" mechanism of Anderson and Morel.

$$U \to \frac{U}{1 + U \ln \frac{B}{\omega_{ph}}}$$

where B is the bandwidth or upper cutoff. It is not necessarily true that this mechanism survives in the Luttinger liquid. At high energies the response scales as $\frac{1}{E^{1-\alpha}}$ not as $\frac{1}{E}$, and the logarithmic divergence changes to a small positive power. Numerically it may not matter much but this effect may be big enough to favor a $\Delta(k)$ which vanishes on average. Again, we are at a loss, in the absence of a precise "BCS" formalism for the Luttinger liquid. It is even possible that higher energy states do not respond to the presence of a gap and do not screen at all, enforcing a zero-averaging gap a fortiori.

Two more recent experiments further support a sign-changing gap. ARPES by Campuzano et al.[18] shows a gap with very large peaks in the 0, π and π, 0 areas and deep nodes, possibly complex, between. The shape is not at all d-wave (more like the structure given by Chakravarty et al.) but it seems likely there are real nodes. Second, neutron scattering at 2Δ is observed by Keimer et al.[19] and coherence factors vanish unless the gap changes sign.

ADDENDUM TO CHAPTER 4

Since writing chapter 4, I have realized that the c-axis superconducting dynamics of the cuprates lends itself to a more detailed analysis than had previously been carried out. A Letter[20] has been submitted analyzing, in terms of the interlayer tunneling theory, the infrared data of Uchida, Timusk, and van der Marel, and will be reprinted with this chapter, but I feel that it can bear a little expansion because it has, in combination with the T_c considerations of chapter 7, considerable predictive power.

The infrared measurements of Uchida, Timusk, and van der Marel demonstrate clearly that normal c-axis conductivity is negligible in the 2 1 4 material. (Though there is some residual conductivity which can also be measured and gives useful information about T-dependence.) At T_c, however, superconductivity appears in the form of a δ-function conductivity peak at $\omega = 0$, which leads to a reflectivity which can be analyzed using the dielectric constant

$$\epsilon = \epsilon_0 - \frac{\omega_p^{02}}{\omega^2} \, (+i\sigma_{\text{res}})$$

which (since σ_{res} is small) gives a sharp plasma edge in the reflectivity where the dielectric constant becomes positive, and R drops from 1:

$$\omega^2 = \omega_p^2 = \frac{\omega_p^{02}}{\epsilon_0}.$$

Using Landau formalism, the penetration depth may be written

$$\frac{1}{\lambda^2} = \frac{\omega_{p0}^2}{c^2}$$

since ω_p^{02} measures the δ-function peak in the complex conductivity; so (with a knowledge of $\epsilon_0 \simeq 20$) measuring ω_p^2 is an accurate way of measuring λ^2. Incidentally, the temperature-dependence of ω_{p0}^2 as measured by van der Marel (see fig. 4.11) is quite incompatible with conventional d-wave theory: it is on the "other" side of BCS, flatter at low T and steeper at high. ω_p^2 can also be interpreted in terms of a Josephson plasma frequency[21] for the junctions formed by the contacts between planes. Since virtually the only electron motion is the supercurrent between planes, this *is* the plasma frequency. In the standard Josephson analysis, the Josephson coupling free energy is

$$-F_J \cos(\varphi - \varphi') = -F_J \cos \Delta\varphi$$

where φ and φ' are the phases of the order parameters on the two planes, and the electrostatic energy of two planes is

$$E_{es} = \frac{Q^2}{2c} = \frac{(2e)^2 (\Delta n)^2 d}{2\epsilon_0}$$

(both are per unit area). Since Δn and $\Delta \varphi$ are conjugate dynamical variables, Hamilton's equation gives us

$$\hbar \frac{\partial n}{\partial t} = \frac{\partial \mathcal{H}}{\partial \varphi}$$

which gives us

$$2e\dot{n} = J = \frac{2e F_J}{\hbar} \sin \varphi$$

identifying F_J and the Josephson critical current. The plasma frequency then comes from

$$\hbar \dot{\varphi} = \frac{\partial \mathcal{H}}{\partial n} = \frac{(2e)^2 d}{\epsilon_0} n$$

$$\omega_p^{02} = \frac{(2e)^2 d}{\epsilon_0 \hbar^2} F_J.$$

F_J is an energy per unit area; to get a Josephson energy per electron $\hbar \omega_J$ we divide F_J by $n = r_s^{-2}$ and we have

$$(\hbar \omega_p)^2 = \left[\frac{(2e)^2 d}{r_s^2} \right] \hbar \omega_J.$$

What we can now do is to estimate $\hbar \omega_J$. We can describe two contrasting cases.

(1) In the conventional superconductor the effective Josephson energy is just the bandwidth in the appropriate direction; the opening of a gap removes the coherent single-particle motion and replaces it by pair tunneling. Correspondingly, the penetration depth is given by Landon's expression $\frac{1}{\lambda^2} = \frac{ne^2}{mc^2}$. This would give an $\hbar \omega_p$ enormously larger than the observation, of order of $t_\perp$ itself, around 500–1000 cm^{-1}.

A second plausible possibility is that the pair binding energy comes almost entirely from within the planes. This is the assumption of all other theories of cuprate superconductivity: that the CuO_2 plane has a strong superconducting interaction. The belief in this idea seems ineradicable in spite of the evidence of fig. 4.12. But if this idea were the case, and if (as we observe) there is no coherent single-particle motion in the c-direction to be replaced by pair

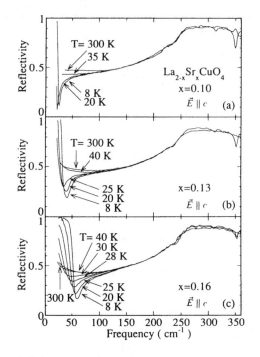

Figure 4.12. *C*-axis infrared spectra showing sharp plasma edge growing from lossy dielectric.

tunneling, the dependence of the superconducting energy on the relative phases of the planes should be weak, much weaker than the pair binding energy itself: $\hbar\omega_J \ll N(0)\Delta^2$ and $\hbar\omega_p \ll \Delta$.

But what the interlayer theory gives is neither of the above. We assume that the bulk of the binding energy comes from the interlayer interaction, but that this binding energy is limited by the fact that the gap must develop self-consistently, and it is only a portion of the matrix elements of $t_\perp$ which are made coherent. We do not restore the full Josephson effect, but only a portion of it of order $\frac{t_\perp^2}{t_{||}/t_\perp} \simeq t_\perp/t_{||}$. What is used to quantify the Josephson effect is the argument that if the full binding energy comes from it, and we can estimate the binding energy as in the BCS theory as

$$\text{B.E.} = -N(0)\Delta^2 \, ,$$

this must also in effect be the Josephson free energy

$$F_J = \text{B.E.} = -N(0)\Delta^2.$$

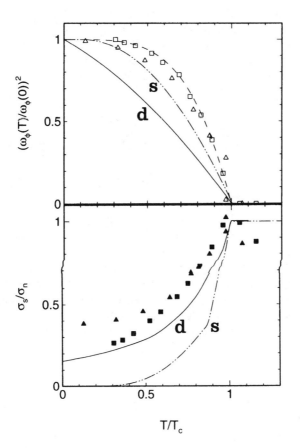

Figure 4.13. $\omega_p^2(T)$ below T_c (open symbols) compared with BCS (dot-dash) and d-wave (solid) predictions.

This means that in the case of the c-axis in cuprates, unlike the conventional BCS case, there are not two independent parameters, ξ and λ, characterizing the electromagnetic and energetic (i.e., thermal) behavior, but only one. As we remark in the Letter, it turns out that this assumption is numerically confirmed in the case of 2 1 4. The coherence length in the Landau-Ginsburg free energy is essentially d, the interlayer spacing:

$$F \simeq -N(0) \left\{ \Delta^2 + d^2 \left[\left(\nabla - \frac{2e}{\hbar c} A \right) \Delta \right]^2 \right\} + \text{nonlinear terms.}$$

In other cuprates this neat numerical agreement will be subject to several modifications. The most important is the effect of complex stacking of layers. The superconducting T_c and Δ will be the sum, essentially, of the various interlayer interactions. In a two-layer system with close layers connected by a coupling $t_\perp^2/t_{||} = \lambda_>$, and distant layers connected by $\lambda_<$,

$$kT_c, \Delta \propto \lambda_> + \lambda_< \ .$$

On the other hand, a phase gradient will divide itself inversely to the magnitude of λ, so that

$$F_J \propto \left(\frac{1}{\lambda_> + \lambda_<}\right)\lambda_>\lambda_<$$

which will mean that $\xi^2 \propto d^2\lambda_>\lambda_</(\lambda_> + \lambda_<)^2$.

Thus in YBCO, and particularly in BISCO, the c-axis coherence length may be a tiny fraction of the lattice constant.

A second factor which may reduce ξ and lengthen λ is the fact that in some materials the gap may appear only over a smaller fraction of the Fermi surface (as, we believe, in $Bi\ 2\,2\,1\,2$). This will mean a reduction in coherence length. $2\,1\,4$'s gap is not subject to large "holes" as in $Bi\ 2\,2\,1\,2$, but numerical fits must wait for detailed knowledge of the gap function.

NOTES

*a Ong's measurements, interpreted using the assumption of skew scattering by vortices, suggest that in fact considerably over 50% of the thermal conductivity in the normal state is phonons, though the peak is entirely electronic. This is a little surprising but not very disturbing, except that it is hard to see why the Wiedemann-Franz ratio is so low. More theoretical study is called for on the effect of magnetic field on heat transport.

REFERENCES

1. J. M. Harris, Y. F. Yan, and N. P. Ong, Phys. Rev. B **46**, 14293 (1992).
2. J. W. Loram, K. A. Mirza, J. R. Cooper, and W. Y. Liang, J. Supercond. **7**, 347 (1993).
3. M. Takigawa, in *High-T$_c$ Superconductivity*, Bedell, Coffey, Meltzer, Pines, and Schrieffer, eds., p. 236 (Los Alamos, 1990).
4. J. M. Imer, F. Patthey, B. Dardel, W.-D. Schneider, Y. Baer, Y. Petroff, and A. Zettl, Phys. Rev. **62**, 336 (1989).
5. D. Mandrus, L. Forro, D. Koller, and L. Mihaly, Nature **351**, 460 (1991) (As commented upon by P. W. Anderson, Phys. Rev. B **42**, 2624 [1990]).
6. C. G. Olson, R. Liu, D. W. Lynch, R. S. List, A. J. Arko, P. Z. Jieng, A. P. Paulikas, B. U. Veal, and Y. C. Chang, Phys. Rev. Lett. B**42**, 381 (1990).
7. T. Moriya, Physica C**185–89**, p. 114 (1991).
8. C. M. Varma, P. B. Littlewood, S. Schmitt-Rink, E. Abrahams, and A. E. Ruckenstein, Phys. Rev. Lett., **63**, 1996 (1989).
9. D. S. Dessau, B. O. Wells, Z.-X. Shen, W. E. Spicer, A. J. Arko, R. S. List, D. B. Mitzi, and A. Kapitulnik, Phys. Rev. Lett. **66**, 2160 (1991).
10. P. W. Anderson, Phys. Rev. Lett. **67**, 660 (1991).
11. C. Uher, in Physical Properties of High-T$_c$ Superconductors, D. M. Ginsburg, ed., Vol. III, p. 159. World Scientific, N.Y. (1992).
12. D. A. Bown et al., Phys. Rev. Lett. **68**, 2390 (1992).
13. P. Morel and P. W. Anderson, Phys. Rev. **123**, 1911 (1961).
14. R. Balian and N. R. Werthamer, Phys. Rev. (1963).
15. D. A. Wollen, D. J. van Harbigen, W. C. Lee, D. M. Ginsberg, and A. J. Leggett, Phys. Rev. Lett. **71**, 2134 (1993).
16. P. Chaudhari and S.-T. Lin, Phys. Rev. Lett. **72**, 1084 (1994).
17. D. A. Brawner and H. R. Ott, Phys. Rev. B**50**, 6350 (1994).
18. H. Ding et al., Phys. Rev. Lett. **76**, 1533 (1996).
19. H. F. Fong, B. Keiner, et al. Phys. Rev. Lett. **75**, 321 (1995).
20. P. W. Anderson, Science **268**, 1154 (1995).
21. P. W. Anderson, in Lectures on the Many-Body Prb. II, E. R. Caianello, ed., p. 113. Academic Press, NY (1964).

5

The One-Dimensional Hubbard Model

P. W. Anderson and Yong Ren

This chapter on the one-dimensional Hubbard Model was written a few years ago by Yong Ren, with some assistance from me. It contains the basic facts and technology, which are available in a number of alternative articles such as those of H. J. Schulz or of C. Bourbonnais. The key results have been known for many years; most notably, the asymptotic Green's functions were derived as early as 1974 by Larkin and Dzialoshinsky, and are available in G. D. Mahan's text; these are the primary formulas used in the rest of the book. But Ren's treatment is particularly helpful in relating these formal results directly to the Lieb-Wu explicit, totally nonperturbative solution, and I have left it as it was written. I have reprinted some of our subsequent work in part II rather than writing it into the chapter. This point of view on the Luttinger liquid solution, as, in a sense, an "incompressible quantum fluid," is not available elsewhere.

Let me add only a remark on our point of view. Throughout the book we keep in mind the picture of the Lieb-Wu solution as a "squeezed Heisenberg model solution" living on the occupied sites of a gas of spinless interacting Fermions. The spinons are the Fermionized (in Landau's old sense) spin fluctuations of this Heisenberg model. The recent work on spinons by Haldane based on the Haldane-Shastry exact solution of the 1D $1/r^2$ Heisenberg model allows us to really understand spinons in this sense in all detail. It also identifies them very closely with our 1987 picture of the spinon as a Gutzwiller-projected quasi-particle excitation ($P_G^A q_{k\sigma}^+$) of a gapless BCS state, which gives us a physical picture for them and allows us to understand their peculiar "Majorana" fermion character, or equivalently, as Baskaran and I discussed, their additional SU(2) symmetry: ($s_{k\sigma}^+ \equiv s_{-k-\sigma}$). The charge excitations, or holons, are quite different in character: in general they involve shifts of the entire spectrum of momenta in k-space, leading to orthogonality catastrophe exponents. They are best understood by bosonization. Ren's chapter is closer to leading the reader to this picture than the alternative articles, and is quite far in spirit from "g-ology" which tends to be nonphysical.

PHYSICAL PICTURE OF PROPERTIES OF 1D HUBBARD MODEL

The one-dimensional Hubbard model has had an exact solution for a long time; it was solved by Lieb and Wu[1] back in 1965 following C. N. Yang's solution[2] of the continuum equivalent. The solution is one of the triumphs of the Bethe Ansatz, which is nothing but a statement of the fact that in one dimension elastic scattering among particles keeps the collection of the individual particle momenta unaltered and leads to an infinite number of conserved quantities. Nowadays, the beauty of the physics in $1 + 1$ dimensions due to this simplification is described by seemingly different and elegant terms such as conformal symmetries or inverse scattering. The Lieb-Wu solution has been explored by various authors to calculate properties such as magnetic susceptibility[3] and the excitation spectrum.[4,5]

The first calculation of the ground state correlation function from this solution was given by Ogata and Shiba.[6] They calculated numerically the momentum distribution and the spin-spin correlations on a finite size 1D lattice and fitted the Fermi surface singularity with a power law. This leads eventually to the analytic calculation of all the asymptotic correlation functions by identifying the low-energy behavior of the Hubbard model with that of the well-known Tomonaga-Luttinger model in one dimension.

As we have emphasized in previous chapters, one of the central dogmas is that the "normal" metallic state is not a Fermi liquid in the sense that $Z = 0$. The one-dimensional Hubbard model serves us as the best available example. The failure of the Fermi liquid theory is not a consequence of any special feature present in one dimension, but of the fact that strong correlation produces a finite phase shift on the Fermi surface. This will allow us to generalize our result to the two-dimensional systems which we are really interested in.

We begin by considering the relationship between the quasiparticle wave function renormalization Z and the momentum shift δ at the Fermi surface. The quasiparticle assumption in Landau's Fermi liquid theory means that when one bare particle is added to the system at the Fermi surface, it behaves like a free particle because the inter-particle interaction effects can be renormalized. Z, by definition, is the overlap between the N particle ground-state wave function before the particle is added, plus the free particle, with the $N + 1$ particle ground-state wave function after the particle is added. This overlap can be calculated using the technique developed by Anderson in the treatment of the Kondo problem.[7] The method is for finding overlaps between wave functions of determinant form in terms of the single-particle phase shifts. For two single-particle plane wave functions that differ in momentum by a phase shift of δ, that is, one particle with momentum $\pi n_k'/L$ and another with $(\pi n_k + \delta_k)/L$, the overlap is

$$(\psi_k, \psi_k') = \frac{1}{2L} \int\limits_{-L}^{L} dx\, e^{-i\pi n_k' x/L} e^{i(\pi n_k + \delta)x/L} = \frac{\sin \delta}{\pi(n_k - n_k') + \delta}$$

thus the overlap integral between the determinants made up of states ψ_k and ψ'_k is

$$\text{Det}|(\psi_k, \psi'_k)|$$

$$= \text{Det}\left|\frac{\sin \delta}{\pi(n_k - n'_k) + \delta_k}\right|$$

$$= \prod_{n}^{N} \frac{\sin \pi \delta_n}{\pi} \prod_{m<n}^{N} (n - m + \delta_n - \delta_m)(m - n) \bigg/ \prod_{m,n}^{N} (n - m + \delta_n)$$

$$\propto \exp\left[\frac{1}{2\pi^2} \sum_{n=1}^{N} \delta_n^2 (n^{-1} + (N - n + 1)^{-1}) + \frac{1}{2} \sum_{n=1}^{N} \sum_{m=1}^{N-n} \frac{(\delta_{n+m} - \delta_n)^2}{m^2}\right]$$

$$= \exp\left[-\frac{1}{2}\left(\frac{\delta}{\pi}\right)^2 \ln N\right].$$

The size dependence can be translated into dependence on t when there is only one velocity in the part of the system we consider. We then have

$$Z \propto \omega^{\frac{1}{2}(\frac{\delta}{\pi})^2}.$$

As can be easily seen, finite phase shift gives $Z = 0$ at the Fermi surface $\omega = 0$ and vice versa.

Now let us find the phase shift in the one-dimensional Hubbard model. The Lieb-Wu solution of the one-dimensional Hubbard model

$$H = -t \sum_{i,j,\sigma} c_{i\sigma}^{\dagger} c_{j\sigma} + U \sum_{i} n_{i\uparrow} n_{i\downarrow}$$

has the Bethe-ansatz form wave function

$$\psi = \sum_{P} [Q, P] \exp\left(i \sum_{j=1}^{N} k_{P_j} x_{Q_j}\right)$$

where $P = (P_1, P_2, \ldots, P_N)$ and $Q = (Q_1, Q_2, \ldots, Q_N)$ are two permutations of the numbers $(1, 2, \ldots, N)$ for a system with N electrons and M down spins. It is supposed that $N \leq N_a$ and $2M \leq N$. $1, 2 \ldots M$ are down spins.

The solution is expressed in equations determining the $N! \times N!$ coefficients $[Q, P]$ which lead to coupled equations for quantum numbers k_j and Λ_α:

$$N_a k_j = 2\pi I_j + \sum_{\beta=1}^{M} \theta(2 \sin k_j - \Lambda_\beta)$$

$$j = 1, 2, \ldots, N$$

$$-\sum_{j=1}^{N} \theta(\Lambda_\alpha - 2\sin k_j) = 2\pi J_\alpha - \sum_{\beta=1}^{M} \theta(\Lambda_\alpha - \Lambda_\beta)$$

$$\alpha = 1, 2, \ldots, M$$

$$\theta(p) = -2\tan^{-1}(2p/U) \qquad -\pi \le \theta \le \pi$$

I_j's are integers (half-odd integers) for M even (odd), and J_α's are integers (half-odd integers) for $N - M$ odd (even). N_a is the length of the system. The momentum and energy of the state is given by

$$p = \sum_{j=1}^{N} k_j \quad ; \quad E = -2\sum_{j=1}^{N} \cos k_j.$$

To get a better understanding of the ground-state wave function given by the Lieb-Wu solution, we rewrite the wave function in the large U limit, after Ogata and Shiba, as

$$\psi(x_1, \ldots, x_N) = (-1)^Q \det\left[\exp(ik_i x_{Q_j})\right]$$

$$\phi(y_1, \ldots, y_M).$$

The $\phi(y_1, \ldots, y_M)$ is the Bethe solution to a 1d Heisenberg model on the "squeezed" lattice x_i (i.e., all hole sites are squeezed out). Here $y_1, \ldots, y_M$ are positions of the down (or up) spins on this squeezed lattice of the spins. The wave function now is an explicit product of a charge part and spin part and this kind of separation in the ground-state wave function only happens in the large U limit. It is, however, a good approximation for finite U. We have shown that a consistent and convergent expansion in $1/U$ exists. The opposite case, $U = 0$, is not continuously connected to finite U, i.e., $U_{crit} = 0+$. The spinon sector is always described by the Heisenberg solution equations but the spatial part is more complex and slightly spin-dependent in a complex way.

The form of the wave function immediately tells us that the charge part is renormalized for large U to a spinless fermion system, and most properties are easily understood in this picture. Naively, the correlation functions are just simple products of the spinless fermion charge part and the Heisenberg spin part, but there is an important constraint to be taken into account. While the quantum numbers Λ_α depend only on J_α's, the k_j's depend on the sum of all J_α's through (we are assuming large U so that $\sin k_j$ may be neglected in the argument of θ.)

$$N_a k_j = 2\pi I_j + \frac{2\pi}{N}\sum_\alpha J_\alpha.$$

When only charge or spin density excitation exists, i.e., the I_j's and J_α's do not differ from their ground-state value simultaneously, the coupling between charge and spin may be neglected. However, when an electron excitation is created, the last term in the last equation generates a phase shift for the charge part. Thus, whether or not there is an explicit charge excitation, I_j, the electron carries a current. One may think of this as spinons carrying current by backflow of charge. This phenomenon plays a central role in the transport theory.   *a

Let's say that we remove an electron of momentum k_F from the system. The way that costs the least energy will be removing the I_j at the "charge Fermi surface" $2k_F$ and removing the J_α at the "spin Fermi surface" k_F. This in turn will shift all the rest of the k_j's, in terms of phase shift, by

$$\delta = N_a \delta k_j = \frac{2\pi}{N} \frac{M}{2} = \frac{M\pi}{N}$$

which is equal to $\pi/2$ for equal numbers of up and down spins. Let us consider the change in n_k when we add one particle to the system at k_F. This is given by

$$\Delta n_k = |\langle \psi_{N+1,k} | c_k^\dagger | \psi_N \rangle|^2.$$

The overlap above, that of the state of $N + 1$ with total momentum k_F with the ground state of N particles plus a free particle of momentum k_F, is just in the form very handy to calculate using the overlap formula given above, since every holon momentum is shifted by a universal amount $\pi/2$. We have

$$\text{Det}|(\psi_k, \psi_k')|$$

$$= \text{Det} \left| \frac{\sin \delta}{\pi(n_k - n_k') + \delta_k} \right|$$

$$\propto \exp\left[-\frac{1}{2} \left[\frac{1}{2\pi^2} \sum_{n=1}^{N} \delta_n^2 (n^{-1} + (N-n+1)^{-1}) + \frac{1}{2} \sum_{n=1}^{N} \sum_{m=1}^{N-n} \frac{(\delta_{n+m} - \delta_n)^2}{m^2} \right] \right]$$

$$= \exp\left[-\frac{1}{4} \left(\frac{\delta}{\pi} \right)^2 \ln N \right].$$

The extra factor $1/2$ in the formula is due to the fact that we should only pair particles with the same spin direction while we have a determinant of both up and down spins.

The overlap actually consists of two parts; the charge wave function overlap and the spin wave function overlap. The wave function of the added particle is also divided into two parts. For the charge part, a particle is added to the $2k_F$ holon Fermi surface, the extra k_F is absorbed into the spin part to shift the momentum of the ground state of the squeezed Heisenberg model by π. The spin wave function overlap can be expressed as

$$\sum_{y_1, y_2, \ldots, y_{M+1}} e^{i\pi y_{M+1}} \Phi_N(y_1, y_2, \ldots, y_M) \Phi_{N+1}(y_1, y_2, \ldots, y_{M+1}).$$

If the Bethe solution were at the exact free fermion fixed point described by Sutherland, Haldane, and Shastry, the overlap used here turns out to be a constant. This is obviously the case when the interaction is spin-rotation invariant and the spin part of the wave function is not modified by the interaction.

$$\langle P_G^{(N+1)}(c_{1\downarrow} c_{2\downarrow} \ldots c_{M\downarrow} c_{M+1\uparrow} \ldots c_{N+1\uparrow}) | c_{k=\pi\downarrow}^{\dagger}$$

$$| P_G^{(N)}(c_{1\downarrow} c_{2\downarrow} \ldots c_{M-1\downarrow} c_{M\uparrow} \ldots c_{N\uparrow}) \rangle \sim \text{const.}$$

As it is, there may be logarithmic corrections to correlation functions due to spin function overlaps. In our case, the annihilation operator for the spin commutes with P_G, thus the overlap between the spin part of the wave functions before and after taking one spin out is a finite constant independent of N. We have

$$\Delta n_k = n^{-1/8} \propto (\Delta k)^{-1/8}$$

this implies near the Fermi surface,

$$n_k = n_{k_F} - \text{const.} |k - k_F|^{1/8} \text{sgn}(k - k_F).$$

We want now to show that by further formalizing the phase shift idea, more correlation functions can be "visualized" in the large U limit. Even though it is not necessary at all to introduce boson operators in our calculation of the correlation function, we briefly mention bosonization here just to show there is one phase shift at each of the two Fermi points in 1D. For a 1D fermion system with linear low energy excitations, we can write for the wave functions of the left and right moving branch

$$\psi_R(x) = e^{i\phi_R(x)}, \qquad \psi_L(x) = e^{i\phi_L(x)}$$

with

$$\phi_R(x) = \sum_{k>0} \left(\frac{2\pi}{|k|L}\right)^{1/2} [b_k^{\dagger} e^{-ikx} + b_k e^{ikx}] e^{-\alpha|k|/2}$$

$$\phi_L(x) = \sum_{k<0} \left(\frac{2\pi}{|k|L}\right)^{1/2} [b_k^{\dagger} e^{-ikx} + b_k e^{ikx}] e^{-\alpha|k|/2}$$

where α is introduced for convergence. The ϕ fields are connected to the density field $\rho(x)$ by

$$\nabla \phi_{R,L}(x) = 2\pi \rho_{R,L}(x).$$

Each time a fermion is added, $\phi(x)$ increases by 2π, so $\phi(x)$ is the properly normalized "phase shift" field. The effect of interaction is to change the phase shift coefficients, i.e., $e^{i\phi_R} \to e^{i\lambda\phi_R}$, for example, by mixing the left and right branch through a Bogoliubov transformation as we usually observe in the Tomonaga-Luttinger model.

$$\phi_R(x) \to \cosh \varphi \phi_R(x) + \sinh \varphi \phi_L(x)$$

$$\phi_L(x) \to \cosh \varphi \phi_L(x) + \sinh \varphi \phi_R(x).$$

For our purpose here it is enough to write down the expression for calculating the correlation function

$$\left\langle e^{i\lambda\phi_R(x,t)} e^{-i\lambda\phi_R(0,0)} \right\rangle = \left[\left\langle e^{i\phi_R(x,t)} e^{-i\phi_R(0,0)} \right\rangle \right]^{\lambda^2}.$$

Let us define the "holon" as the excitation associated with adding or removing one I_j, which are called type-II, type-III excitations in Lieb and Wu. The Green's function for the holon is easy to find. When we add one I_j to the system, the Λ_α's do not change. We are simply adding one k_j to the Slater determinant, which behaves like a noninteracting spinless fermion system, filled up to momentum $2k_F$. So we have

$$G^c(x, t) \sim \frac{e^{2ik_Fx}}{x - v_ct + i\delta t}$$

where v_c is the velocity of the charge excitations in the Hubbard model, which together with the velocity of spin excitations v_s have been calculated by, e.g., Coll.[8]

The $i\delta$ term is put in in such a way to make the analyticity correct.

We next consider the excitation associated with removing one of the J_α's, i.e., type-I in Lieb and Wu. When we remove a J_α at $-k_F$, we get a spinon—a spin half excitation with momentum k_F, basically an electron removed from a Heisenberg system

$$\langle \psi(x, t) \psi^\dagger(0, 0) \rangle \sim \frac{e^{ik_Fx}}{(x - v_st)^{1/2}}.$$

(The $2k_F$ contribution in the spin-spin correlation function of the Heisenberg model consists of one left [right] moving particle and a right [left] moving hole, so

$$\langle \vec{S}(x,t) \cdot \vec{S}(0,0) \rangle \sim \frac{\cos 2k_F x}{(x^2 - v_s^2 t^2)^{1/2}}$$

see, e.g., Luther and Peschel.[9]) Moreover, each $k_j L$ is shifted by $\frac{\pi}{2}$, compared to the interspacing 2π. This leads to $1/4$ for the phase shift coefficient we mentioned earlier, and since there appear a right moving particle and a left moving hole, we have

$$G^s(x,t) \sim \frac{e^{ik_F x}}{(x - v_s t + i\delta t)^{\frac{1}{2}}(x - v_c t + i\delta t)^{\frac{(\frac{1}{4})^2}{}}(x + v_c t - i\delta t)^{(\frac{1}{4})^2}}$$

$$= \frac{e^{ik_F x}}{(x - v_s t + i\delta t)^{\frac{1}{2}}(x - v_c t + i\delta t)^{\frac{1}{16}}(x + v_c t - i\delta t)^{\frac{1}{16}}}.$$

The spin-spin correlation is now given by the joint spinon-(spinon-hole) correlation. For the $2k_F$ contribution, the phase shifts are in the same direction, which determines the relative signs when we perform the calculations in the bosonization formalism. We easily get

$$\langle \vec{S}(x,t) \cdot \vec{S}(0,0) \rangle \sim \frac{\cos 2k_F x}{(x - v_s t)^{\frac{1}{2}}(x + v_s t)^{\frac{1}{2}}(x - v_c t)^{(\frac{1}{4} + \frac{1}{4})^2}(x + v_c t)^{(\frac{1}{4} + \frac{1}{4})^2}}$$

$$= \frac{\cos 2k_F x}{(x - v_s t)^{\frac{1}{2}}(x + v_s t)^{\frac{1}{2}}(x - v_c t)^{\frac{1}{4}}(x + v_c t)^{\frac{1}{4}}}.$$

This gives

$$\langle \vec{S}(x) \cdot \vec{S}(0) \rangle \sim x^{-\frac{3}{2}} \cos 2k_F x.$$

(Except, again, for logarithmic corrections.) On the other hand, the calculation of the susceptibility involves only the zero momentum contribution of the spin-spin correlation function, we will get exactly the same result as in the Heisenberg model which is more or less the same as free particles. The $T = 0$ susceptibility has been calculated by Shiba et al.

Another correlation function of interest is the density-density correlation. The $4k_F$ contribution is rather simple:

$$\langle \rho(x,t)\rho(0,0) \rangle \sim \frac{\cos 4k_F x}{(x - v_c t)(x + v_c t)}$$

also confirmed by Sorella et al.

Finally we calculate the electron Green's function and the momentum distribution. To make an electron hole, we have to remove a I_j at $2k_F$ and a J_α at $-k_F$, but then the shifting of k_j's adds a quarter k_j at $2k_F$ and removes a quarter k_j at $-2k_F$. Thus at $2k_F$ we have two opposing phase shifts. Noting this, we immediately have

$$G^e(x,t) \sim \frac{e^{ik_F x}}{(x - v_c t + i\delta t)^{(1-\frac{1}{4})^2}(x + v_c t - i\delta t)^{(\frac{1}{4})^2}(x - v_s t + i\delta t)^{\frac{1}{2}}}$$

$$= \frac{e^{ik_F x}}{(x - v_c t + i\delta t)^{\frac{9}{16}}(x + v_c t - i\delta t)^{\frac{1}{16}}(x - v_s t + i\delta t)^{\frac{1}{2}}}$$

which also leads to the momentum distribution

$$n(k) = -i \int_{-\infty}^{\infty} dx\, e^{-ikx} G(x, 0^-) = \text{const.} - \text{const.}|k - k_F|^{\frac{1}{8}}\,\text{sgn}(k - k_F).$$

But there are other ways of making an electron excitation: we can remove a I_j at $2k_F$, and remove a J_α at k_F, giving an electron excitation at momentum $3k_F$, with the Green's function

$$G_e^{3k_F} \sim \frac{e^{3ik_F x}}{(x - v_c t + i\delta t)^{(1+\frac{1}{4})^2}(x - v_c t + i\delta t)^{(\frac{1}{4})^2}(x - v_s t + i\delta t)^{\frac{1}{2}}}.$$

The momentum distribution singularity is given by $|k - 3k_F|^{.9/8}$ There are also $5k_F, 7k_F, \ldots$ contributions that consist of different combinations of the holon and spinon excitation.

BOZONIZATION AND THE HALDANE LUTTINGER LIQUID

In one dimension, Green's functions in such forms have been obtained in the Tomonaga-Luttinger model by the bosonization technique.[10] Only recently it has been realized that the low-energy behavior of the 1D Hubbard model can be mapped to a modified Tomonaga-Luttinger model even though such a proposition has long existed, e.g., Haldane's concept of the "Luttinger liquid." The bosonization formalism is very suitable for calculating various correlation functions and for discussion of the renormalization of the backward scattering and umklapp term.

For spinless fermions with the Hamiltonian

$$H = v_F \sum_k |k| a_k^\dagger a_k + \frac{1}{2L} \sum_k V_k \rho(k)\rho(-k)$$

the transformation to the bosonic representations

$$\rho(k) = \rho_1(k) + \rho_2(k)$$

$$\rho_1(k) = \sum_{p>0} a^\dagger_{p-k/2} a_{p+k/2} = b_k \left| \frac{kL}{\pi} \right|^{1/2}$$

$$\rho_2(k) = \sum_{p<0} a^\dagger_{p-k/2} a_{p+k/2} = b_{-k} \left| \frac{kL}{\pi} \right|^{1/2}$$

makes both the kinetic term and the interactions quadratic in boson operators,

$$H = v_F \sum_k |k| b^\dagger_k b_k + \frac{1}{2\pi} \sum_k |k| V_k (b_k + b^\dagger_{-k})(b^\dagger_k + b_{-k})$$

and the Hamiltonian is easily diagonalized. For spin $1/2$ fermions, it is convenient to introduce the charge and spin operators

$$c^\dagger_p = (b^\dagger_{p\uparrow} + b^\dagger_{p\downarrow})/\sqrt{2}$$

$$s^\dagger_p = (b^\dagger_{p\uparrow} - b^\dagger_{p\downarrow})/\sqrt{2}.$$

If the original Hamiltonian is invariant under spin reversal, there will be no terms in the product of one charge and one spin creation (annihilation) operators and we may say the charge and spin degrees of freedom are separated. This is a consequence of the fact that in one dimension the only possible excitations are density fluctuations.

$$H = H_c + H_s$$

$$= v_c \sum_k |k| c^\dagger_k c_k$$

$$+ \frac{1}{2\pi} \sum_{k>0} (g_1 - 2g_2) |k| (c^\dagger_k c^\dagger_{-k} + c_{-k} c_k)$$

$$+ v_s \sum_k |k| s^\dagger_k s_k + \frac{1}{2\pi} \sum_{k>0} g_1 |k| (s^\dagger_k s^\dagger_{-k} + s_{-k} s_k).$$

However, the existence of the spin degrees of freedom allow more possible terms to appear. There are the backward scattering

$$H_{bs} = g_1 \int dx \, \psi^\dagger_{R\uparrow} \psi^\dagger_{L\downarrow} \psi_{R\downarrow} \psi_{L\uparrow} + \text{H.c.}$$

in which electrons of opposite spin cross the Fermi surface in opposite directions, and the umklapp scattering

$$H_{um} = g_3 \int dx e^{i(4k_F - G)x} \psi_{L\uparrow}^\dagger \psi_{L\downarrow}^\dagger \psi_{R\downarrow} \psi_{R\uparrow} + \text{H.c.}$$

The Hamiltonian including the backward and umklapp scattering cannot be diagonalized exactly. As Luther and Emery[11] showed, at particular values of g's,
it is possible to reduce H_c and H_s to a free fermion Hamiltonian by a canonical transformation, and the resulting Hamiltonian can again be diagonalized.
The result combined with the renormalization group analysis determines the
ground-state properties in different regions of interaction parameters. In the
case of the Hubbard model, for repulsive U the backward scattering Hamiltonian has the wrong sign for producing a gap in the spin wave spectrum while the
umklapp scattering responsible for a gap in the charge density wave spectrum
is only important when the momentum transfer to the lattice is $4k_F$, i.e., at
half-filling. This agrees with the Lieb-Wu result.

The renormalization group analysis for the Tomonaga model has been done
by Solyom et al.[12] In the case of the Hubbard model, we expect that the charge
renormalization is only relevant at half-filling. The spin renormalization flow
has a fixed point at $g_{1\parallel} = g_{1\perp} = 0$.

A better framework for formulating the low-energy structure of one-dimensional models has been developed by Haldane in a series of papers.[13] The basic
idea is that the low-energy effective Hamiltonian of 1D quantum models could
be mapped onto the spectrum of the Tomonaga-Luttinger model. In particular,
Haldane demonstrated the applicability of the Luttinger liquid formalism to
several Bethe Ansatz type systems.[14]

The "Luttinger liquid" is a universality class of 1D fermion systems with a
gapless linear low-energy spectrum. The structure of such a model is determined
by a single parameter $e^{2\varphi}$. In the long wavelength approximation, a fermion
field can be represented as

$$\psi_F(x) \sim |\rho(x)|^{1/2} e^{i\theta(x)} e^{i\phi(x)}.$$

Here we define $\phi(x)$ and $\theta(x)$ as

$$[\rho(x), \phi(x')] = i\delta(x - x')$$

$$\nabla\theta(x) = \pi\rho(x).$$

$\rho(x)$ is the density so $\phi(x)$ is the phase field and $\theta(x)$ is the "phase shift"
field. The factor $e^{i\theta(x)}$ comes from the "Jordan-Wigner" transformation for 1D

fermions. The Fourier transforms are

$$\rho(x) = \rho + \frac{1}{\sqrt{L}} \sum_k e^{-ikx} \left| \frac{k}{2\pi} \right|^{1/2} (b_k + b_{-k}^\dagger)$$

$$\theta(x) = \bar{\theta} + \frac{\pi N x}{L} + i \sum_{k \neq 0} \frac{1}{k} \left| \frac{\pi k}{2L} \right|^{1/2} e^{-ikx} (b_k + b_{-k}^\dagger)$$

$$\phi(x) = \bar{\phi} + \frac{\pi J x}{L} + i \sum_{k \neq 0} \left| \frac{\pi}{2kL} \right|^{1/2} e^{ikx} (b_k^\dagger - b_{-k}).$$

The Hamiltonian

$$H = \frac{\hbar^2}{2m} \int dx |\nabla \psi|^2 + \frac{1}{2} \int \int dx\, dy\, V(x - y) \rho(x) \rho(y)$$

can be rewritten in the low-energy regime as

$$H = \frac{\hbar}{2\pi} \int dx \left[v_J (\nabla \phi)^2 + v_N (\nabla \theta - \pi \rho_0)^2 \right]$$

where $v_J = \pi \hbar \rho_0 / m$ and $v_N = \kappa / \pi \hbar \rho_0^2$ are the current and number velocity, and κ is the compressibility per unit length. The Hamiltonian can be diagonalized by introducing a Bogoliubov transformation parameter e^φ

$$\theta(x) = \bar{\theta} + \frac{\pi N x}{L} + i \sum_{k \neq 0} \frac{1}{k} \left| \frac{\pi k}{2L} \right|^{1/2} e^{-ikx} e^\varphi (b_k + b_{-k}^\dagger)$$

$$\phi(x) = \bar{\phi} + \frac{\pi J x}{L} + i \sum_{k \neq 0} \left| \frac{\pi}{2kL} \right|^{1/2} e^{ikx} e^{-\varphi} (b_k^\dagger - b_{-k})$$

and choosing

$$v_J e^{-2\varphi} = v_N e^{2\varphi} = v_S$$

so

$$H = \frac{\pi}{2L} \left[v_J J^2 + v_N (N - N_0)^2 \right] + \hbar v_S \sum_k |k| b_k^\dagger b_k$$

$$P = \frac{\pi N J}{L} + \sum_k k b_k^\dagger b_k$$

with the selection rule

$$(-1)^{\Delta J} = (-1)^{\Delta N}.$$

$v_S = (\kappa/m\rho_0)^{1/2}$ is the velocity of the density oscillation, or "sound velocity." For free fermions $e^{-2\varphi} = 1$ and for free bosons $e^{-2\varphi} = 0$.

For the Hubbard model, we have

$$H = \frac{\hbar}{2\pi} \int dx \left\{ v_c \left[e^{2\varphi}(\nabla\phi_c)^2 + e^{-2\varphi}(\nabla\theta_c - \pi N_0/L)^2 \right] \right.$$
$$\left. + v_\sigma \left[2(\nabla\phi_\sigma)^2 + \frac{1}{2}(\nabla\theta_\sigma)^2 \right] \right\}.$$

Here, we have taken into account the fact that the charge part does not get renormalized from the umklapp term when the band is not half-filled, and that the marginal backward scattering term for the spin part renormalizes logarithmically to zero. In the limit of large U, the charge density velocity v_c and spin wave velocity v_σ are given by

$$v_c = 2t \left(\sin \frac{\pi N}{N_a} \right) \frac{L}{N_a}$$

$$v_\sigma = \frac{1}{2}\pi \left(\frac{4t^2}{U} \right) \left(1 - \frac{\sin 2\pi N/N_a}{2\pi N/N_a} \right) \frac{L}{N_a}.$$

The parameter e^φ is decided by the ratio of the two velocities v_N and v_c. It is easy to determine e^φ in both the large and small U limits. For large U,

$$v_N = \frac{L}{\pi} \frac{\partial\mu}{\partial N}$$

$$= 2t \frac{L}{N_a} \sin Q$$

$$= v_c$$

i.e., $e^{-2\varphi} = 1$. This is expected since in the large U limit the charge part renormalizes to free fermions. In the small U limit,

$$v_N = \frac{L}{\pi} \frac{\partial\mu}{\partial N}$$

$$= 2t \frac{L}{2N_a} \sin Q$$

$$= \frac{1}{2} v_c$$

so $e^{-2\varphi} = 1/2$. For general U, one can obtain e^φ using the method similar to that given by, e.g., Haldane for a Bethe ansatz system without internal degrees of freedom. The results of Bethe-ansatz-soluble models are given in terms of a linear inhomogeneous Fredholm integral equation of the second kind:

$$2\pi \rho(k) = 1 + \int_{-Q}^{Q} dk' \frac{\partial \vartheta(k, k')}{\partial k} \rho(k').$$

If we define $\sigma(k)$ by

$$2\pi \sigma(k) = \vartheta(k, Q) - \int_{-Q}^{Q} dk' \frac{\partial \vartheta(k, k')}{\partial k'} \sigma(k')$$

then

$$e^{-\varphi} = 1 - \sigma(Q) + \sigma(-Q).$$

In the large U limit,

$$\frac{\partial \vartheta(k, k')}{\partial k} = \cos k \frac{\ln 2}{U}$$

or

$$\vartheta(k, k') = \frac{\ln 2}{U} (\sin k - \sin k')$$

this gives

$$\sigma(Q) - \sigma(-Q) = \frac{\ln 2}{\pi U} \sin Q$$

so we have

$$e^{-\varphi} \to 1 - \frac{\ln 2}{\pi U} \sin Q.$$

For convenience in the following formulas, we redefine

$$e^{-2\varphi} = 2[e^{-2\varphi}]_{\text{old}}.$$

Now we can write down all kinds of correlation functions by using the bosonization technique for the Tomonaga-Luttinger model. We have

$$\theta_c(x) = \bar{\theta}_c + \frac{\pi N_c x}{L} + i \sum_{k \neq 0} \frac{1}{k} \left| \frac{\pi k}{L} \right|^{1/2} e^{-ikx} e^{\varphi} (c_k + c_{-k}^{\dagger})$$

$$\phi_c(x) = \bar{\phi}_c + \frac{\pi J_c x}{L} + i \sum_{k \neq 0} \left| \frac{\pi}{4kL} \right|^{1/2} e^{ikx} e^{-\varphi} (c_k^{\dagger} - c_{-k})$$

$$\theta_s(x) = \bar{\theta}_s + \frac{\pi N_s x}{L} + i \sum_{k \neq 0} \frac{1}{k} \left| \frac{\pi k}{L} \right|^{1/2} e^{-ikx} (s_k + s_{-k}^{\dagger})$$

$$\phi_s(x) = \bar{\phi}_s + \frac{\pi J_s x}{L} + i \sum_{k \neq 0} \left| \frac{\pi}{4kL} \right|^{1/2} e^{ikx} (s_k^{\dagger} - s_{-k})$$

and

$$\theta_c = \theta_\uparrow + \theta_\downarrow \qquad N_c = N_\uparrow + N_\downarrow$$
$$\theta_s = \theta_\uparrow - \theta_\downarrow \qquad N_s = N_\uparrow - N_\downarrow$$
$$\phi_c = (\phi_\uparrow + \phi_\downarrow)/2 \qquad J_c = (J_\uparrow + J_\downarrow)/2$$
$$\phi_s = (\phi_\uparrow - \phi_\downarrow)/2 \qquad J_s = (J_\uparrow - J_\downarrow)/2.$$

Note that according to this definition, at large U $e^{2\varphi} = 1/2$, and for $U = 0$, $e^{2\varphi} = 1$.

In calculating the correlation functions, we have to follow the selection rules

$$(-1)^{\Delta N_\uparrow} = (-1)^{\Delta J_\uparrow}$$
$$(-1)^{\Delta N_\downarrow} = (-1)^{\Delta J_\downarrow}$$

and the term to enter will be

$$\psi_c(x) = e^{i\Delta N_c \phi_c(x)} e^{\Delta J_c \theta_c(x)}$$
$$\psi_s(x) = e^{i\Delta N_s \phi_s(x)} e^{\Delta J_s \theta_s(x)}$$

which will have $\Delta J_c \times 2k_F$ oscillations. The leading term in the electron Green's function is given by $\Delta N_\uparrow = 1$, $\Delta J_\uparrow = 1$, $\Delta N_\downarrow = 0$, $\Delta J_\downarrow = 0$ (k_F) piece and $\Delta N_\uparrow = 1$, $\Delta J_\uparrow = 3$, $\Delta N_\downarrow = 0$, $\Delta J_\downarrow = 0$ $(3k_F)$ piece

$$G(x, t) = \langle \psi_\uparrow(x, t) \psi_\uparrow^{\dagger}(0, 0) \rangle$$
$$\sim e^{ik_F x} (x - v_c t)^{-1/2} (x - v_\sigma t)^{-1/2} |(x - v_c t)(x + v_c t)|^{-\alpha}$$
$$+ e^{3ik_F x} (x - v_c t)^{-(3e^{\varphi} + e^{-\varphi})^2/8} (x + v_c t)^{-(3e^{\varphi} - e^{-\varphi})^2/8}$$
$$\times (x - v_\sigma t)^{-1/2}$$

with

$$\alpha = (e^{2\varphi} + e^{-2\phi} - 2).$$

This gives

$$n_k \sim -\text{sgn}(k - k_F)|k - k_F|^{(e^{-2\varphi} + e^{2\varphi} - 2)/4}$$

$$\sim -\text{sgn}(k - 3k_F)|k - 3k_F|^{(9e^{2\varphi} + e^{-2\varphi} - 2)/4}.$$

The first few terms in the density-density correlation function are

$$n(x, t) = \psi_\uparrow^\dagger \psi_\uparrow + \psi_\downarrow^\dagger \psi_\downarrow$$

$$\langle n(x, t)n(0, 0)\rangle = n_0^2 + \frac{e^{2\varphi}}{x^2 - v_c^2 t^2}$$

$$+ C_1 \frac{\cos 2k_F x}{(x^2 - v_c^2 t^2)e^{2\varphi}/2(x^2 - v_\sigma^2 t^2)^{1/2}} + C_2 \frac{\cos 4k_F x}{(x^2 - v_c^2 t^2)^{2e^{2\varphi}}}.$$

The first two terms come from the $k = 0$ piece in : $\nabla\theta_c(x)\nabla\theta_c(0)$:. The last two terms correspond to $\Delta J_c = 1$, $\Delta J_s = 1$ and $\Delta J_c = 2$, $\Delta J_s = 0$, respectively. For large U, we expect the $4k_F$ term to be dominant. The leading terms in the spin-spin correlation are

$$\langle S_z(x, t)S_z(0, 0)\rangle = \text{const.}$$

$$+ D_1 \frac{1}{x^2 - v_\sigma^2 t^2} + D_2 \frac{\cos 2k_F x}{(x^2 - v_c^2 t^2)e^{2\varphi}/2(x^2 - v_\sigma^2 t^2)^{1/2}}.$$

*b   The singlet pairing $\Delta N_\uparrow = \Delta N_\downarrow = 1$ has leading terms $\Delta J_\uparrow = \pm 1$, $\Delta J_\downarrow = \pm 1$.

$$\hat{O}_{SP}(x) = \psi_\sigma^\dagger(x)\psi_{-\sigma}^\dagger(x)$$

$$\langle \hat{O}_{SP}^\dagger \hat{O}_{SP}\rangle \propto \frac{1}{|x^2 - v_c^2 t^2|^{e^{-2\varphi}/2}|x^2 - v_\sigma^2 t^2|^{1/2}}$$

$$+ E_0 \frac{e^{i2k_F x}}{(x - v_c t)^{(e^\varphi + e^{-\varphi})^2/2}(x + v_c t)^{(e^\varphi - e^{-\varphi})^2/2}}$$

$$+ E_0 \frac{e^{-i2k_F x}}{(x + v_c t)^{(e^\varphi + e^{-\varphi})^2/2}(x - v_c t)^{(e^\varphi - e^{-\varphi})^2/2}}.$$

The triplet pairing $\Delta N_\uparrow = 2$, $\Delta N_\downarrow = 0$ has leading terms $\Delta J_c = \Delta J_s = 0$

$$\hat{O}_{TP}(x) = \psi_\sigma^\dagger(x)\psi_\sigma^\dagger(x)$$

$$\langle \hat{O}_{TP}^\dagger \hat{O}_{TP}\rangle \propto \frac{1}{|x^2 - v_c^2 t^2|^{e^{-2\varphi}/2}|x^2 - v_\sigma^2 t^2|^{1/2}}$$

$$+ \text{ terms} \propto e^{i2k_F x}/x^{3+8\alpha} + \text{etc.}$$

In the large U limit, we recover our result obtained earlier. In fact, one can show that $e^{-\varphi}$ is proportional to the phase shift on the Fermi surface when we add an equal number of electrons on both sides of the Fermi surface; similarly, e^{φ} is proportional to the phase shift when we add some number of electrons on one side and the same number of holes on the other. Thus, it is obvious why the exponents in the electron Green's function have such forms as $\frac{1}{2}(e^{\varphi} \pm e^{-\varphi})$.

It is important to remember that the mapping to the Tomonaga-Luttinger model only provides us with the exponents but puts no restriction on the coefficients of each possible term. We have to go back to the exact wave function for the coefficients. At $U = \infty$, it can be easily seen from the Ogata-Shiba wave function that the $2k_F$ density correlation vanishes. We can expect that the coefficient of the $2k_F$ CDW instability is at least proportional to $1/U$. On the other hand, the $4k_F$ CDW is an irrelevant term. This implies that in the limit of large U, the competition for the ground state is between $2k_F$ SDW and superconducting instabilities. An interesting possibility is that when the SDW order is disrupted by the presence of holes, superconducting instabilities take over. (But see Note.)

The $t - J$ model is usually investigated in the place of the Hubbard model. This is what one will get in the $1/U$ expansion of the Hubbard model to the second order and restrict the states to the lower Hubbard band.

$$H = -t \sum_{\langle i,j \rangle} c_{i\sigma}^{\dagger} c_{j\sigma} + J \sum_{\langle i,j \rangle} \vec{S}_i \cdot \vec{S}_j$$

where $J = 4t^2/U$ and the states are restricted to singly occupied lattice sites. The $t - J$ model has to be handled with care since it assumes a Hilbert space with no double occupancies. The importance of this model is that it shows that the $U = \infty$ fixed point in the model, i.e., $U_{\text{eff}} = \infty$ and the corrections due to finite U are proportional to J. Finite U will manifest itself only by corrections to the large U fixed point. This will lead us directly to the appearance of a $2k_F$ holon Fermi surface, which does not analytically continue to $U = 0$.

Now, let us look at the analytic behavior of the 1D Green's functions. Following Dzyaloshinskii,[15] we write the k_F Green's function in the form

$$G(x,t) = \frac{1}{2\pi} \frac{1}{x - t + i\delta \text{sgn}(t)} \left[\frac{x - t + i/\Lambda \text{sgn}(t)}{x - vt + i/\Lambda \text{sgn}(t)} \right]^{1/2}$$
$$\times \left[\Lambda^2 (x - vt + i/\Lambda \text{sgn}(t))(x + vt - i/\Lambda \text{sgn}(t)) \right]^{-\alpha}$$

where $v_{\sigma} = 1$, $v_c = v$, and Λ is a cutoff. The signs of the δ and $1/\Lambda$ term are so chosen to satisfy the analytic continuation condition. For example, to see that it has the correct analyticity, let us calculate the momentum distribution

$$n(k) = -i \int_{-\infty}^{\infty} dx\, e^{-i(k-k_F)x} G(x, 0^-)$$

$$= -\frac{i}{2\pi} \int_{-\infty}^{\infty} dx\, e^{-i(k-k_F)x} \frac{1}{x - i\delta} (1 + \Lambda^2 x^2)^{-\alpha}$$

$$= -\frac{i}{2\pi} \int_{-\infty}^{\infty} dx\, e^{-i(k-k_F)x} i\pi\delta(x)(1 + \Lambda^2 x^2)^{-\alpha}$$

$$- \frac{1}{2\pi} \int_{-\infty}^{\infty} dx\, \frac{\sin(k - k_F)x}{x} (1 + \Lambda^2 x^2)^{-\alpha}$$

$$= \frac{1}{2} - \text{const.} |k - k_F|^{2\alpha} \text{sgn}(k - k_F).$$

For Green's function in momentum space, naive Fourier transform using $v_c = v_s$, i.e., $v = 1$, gives

$$G(p, \omega) = \int_{-\infty}^{\infty} dx \int_{-\infty}^{\infty} dt\, e^{-ipx} e^{i\omega t} G(x, t)$$

$$= \frac{1}{\Lambda^{2\alpha}(\omega + p)^{-\alpha}(\omega - p)^{1-\alpha}}.$$

(The cutoff terms which would appear in a direct integration are converged by spin-charge separation.)

A second and very important point is the homogeneity property of G for the asymptotic region x, t large or $p - p_F, \omega \to 0$. Note that in this region

$$G = \frac{1}{t^{1+2\alpha}} F\left(\frac{x}{vt}\right)$$

which implies that near the Fermi points

$$G = \frac{1}{\omega^{1-2\alpha}} F\left(\frac{v(k - k_F)}{\omega}\right).$$

For many results this is all that is necessary.

To find the behavior of $G(p, \omega)$ for various regions, we write $G(x, t)$ in two parts

$$G(x,t) = \frac{1}{2\pi} \frac{1}{x-t} \left[\frac{x - t + i\,\mathrm{sgn}(t)/\Lambda}{x - vt + i\,\mathrm{sgn}(t)/\Lambda} \right]^{1/2}$$
$$\times \left[\Lambda^2(x - t + i\,\mathrm{sgn}(t)/\Lambda)(x + t - i\,\mathrm{sgn}(t)/\Lambda) \right]^{-\alpha}$$
$$+ \frac{1}{2\pi}(-i\pi)\mathrm{sgn}(t)\delta(x - t) \left[\frac{x - t + i\,\mathrm{sgn}(t)/\Lambda}{x - vt + i\,\mathrm{sgn}(t)/\Lambda} \right]^{1/2}$$
$$\times \left[\Lambda^2(x - t + i\,\mathrm{sgn}(t)/\Lambda)(x + t - i\,\mathrm{sgn}(t)/\Lambda) \right]^{-\alpha}.$$

The second term is

$$G_2(p, \omega) = -\frac{i}{2} \left[\int_0^\infty dt\, e^{-ipt+i\omega t} \frac{2^{-\alpha} e^{-i\pi\alpha/2 - i\pi/4}}{\Lambda^{\frac{1}{2}+\alpha}(v-1)^{1/2} t^{\alpha+\frac{1}{2}}} \right.$$
$$\left. - \int_{-\infty}^0 dt\, e^{-ipt+i\omega t} \frac{2^{-\alpha} e^{-i\pi\alpha/2 - i\pi/4}}{\Lambda^{\frac{1}{2}+\alpha}(v-1)^{1/2} |t|^{\alpha+\frac{1}{2}}} \right]$$

when $p, \omega \ll \Lambda$, the main contribution comes from $t \gg 1/\Lambda$, so

$$G_2(p, \omega) = \frac{1}{2}(2\Lambda)^{-\alpha} \frac{1}{\sqrt{\Lambda(v-1)}}$$
$$\times \mathrm{sgn}(\omega - p)(e^{-i\pi\alpha} - \frac{i}{2}) \frac{\Gamma(\frac{1}{2} - \alpha)}{|\omega - p|^{\frac{1}{2}-\alpha}}.$$

In particular,

$$\mathrm{Im}\,G_2(p, \omega) = -\frac{1}{2}(2\Lambda)^{-\alpha} \frac{1}{\sqrt{\Lambda(v-1)}} \mathrm{sgn}(\omega - p)$$
$$\left(\sin \pi\alpha + \frac{1}{2} \right) \frac{\Gamma(\frac{1}{2} - \alpha)}{|\omega - p|^{\frac{1}{2}-\alpha}}.$$

For small U, the main contribution from the first term in $G(x, t)$ can be obtained by approximating v to be 1 and omitting $1/\Lambda$ terms. Define $x^+ =$

$(x + t)/2$, $x^- = (x - t)/2$, then

$$G_1(p, \omega) = \frac{1}{2\pi} \int\limits_{-\infty}^{\infty} dx \int\limits_{-\infty}^{\infty} dt\, e^{-ip(x^+ + x^-) + i\omega(x^+ - x^-)} 2^{-\alpha}$$

$$\frac{1}{2x^-}[\Lambda^2(x^- + i/2\Lambda)(x^+ - i/2\Lambda)]^{-\alpha}$$

$$= \frac{1}{2\pi} 2^{-\alpha} \Lambda^{-2\alpha} \int\limits_{-\infty}^{\infty} dx^- \frac{1}{x^-} e^{-i(p+\omega)x^-} (x^-)^{-\alpha}$$

$$\int\limits_{-\infty}^{\infty} dx^+ e^{-i(p-\omega)x^+} (x^+)^{-\alpha}$$

$$= \frac{1}{2\pi} (2^{-\alpha}) \Lambda^{-2\alpha} \frac{\Gamma(-\alpha)\Gamma(1-\alpha)}{|\omega + p|^{-\alpha}|\omega - p|^{1-\alpha}}$$

$$\times \begin{cases} i(1 - e^{-2i\pi\alpha}) & \omega - p > 0,\, \omega + p > 0 \\ -2\sin\pi\alpha & \omega - p > 0,\, \omega + p < 0 \\ 2\sin\pi\alpha & \omega - p < 0,\, \omega + p > 0 \\ -i(1 - e^{-2i\pi\alpha}) & \omega - p < 0,\, \omega + p < 0. \end{cases}$$

That is,

$$\text{Im}\, G_1(p, \omega)$$
$$= -\frac{1}{\pi} \theta(\omega^2 - p^2)\text{sgn}(\omega - p)(2^{-\alpha})\Lambda^{-2\alpha} \frac{\Gamma^2(1-\alpha)}{\alpha}$$
$$\times \sin^2 \pi\alpha \frac{1}{|\omega + p|^{-\alpha}|\omega - p|^{1-\alpha}}$$
$$\text{Re}\, G_1(p, \omega)$$
$$= -\frac{2^{-\alpha}}{2\pi} \Lambda^{-2\alpha} \Gamma(-\alpha)\Gamma(1-\alpha) \frac{\text{sgn}(\omega - p)}{|\omega + p|^{-\alpha}|\omega - p|^{1-\alpha}}$$
$$\times (\theta(|\omega| - |p|)\sin 2\pi\alpha + \theta(|p| - |\omega|)2\sin\pi\alpha).$$

For $p, \omega \ll \Lambda$, $G_2(p, \omega) \ll G_1(p, \omega)$, so in what follows, we only consider the contribution from $G_1(p, \omega)$.

For small but finite α, Im G does not contain a part of the form $Z\delta(\omega - \epsilon)$. This means the quasiparticle picture fails.

For general interaction, since three branch cuts exist in $G(x, t)$, the Fourier transform is rather involved. If we throw away all the $i\delta$'s first, we can show

$$G(\omega, p) \propto \frac{1}{(\omega - v_c p)^{1/2-\alpha}(\omega - v_\sigma p)^{1/2-\alpha}}$$

$$\times F\left(\alpha, \frac{1}{2} - \alpha; \frac{1}{2} + \alpha; \mu \frac{\omega - v_c p}{\omega - v_\sigma p}\right)$$

where $\mu = (v_\sigma + v_c)/2v_c$ and $F(\alpha, \beta; \gamma; z)$ is the hypergeometric function. The function has branch points at the $z = 0, 1, \infty$, corresponds to $\omega = v_c p$, $\omega = -v_c p$ and $\omega = v_\sigma p$, respectively. The singularities at the three points are

$$\frac{1}{(\omega - v_c p)^{\frac{1}{2}-\alpha}}, \quad \frac{1}{(\omega - v_\sigma p)^{\frac{1}{2}-2\alpha}}, \quad \frac{1}{(\omega + v_c p)^{-\alpha}}.$$

Since $F(\alpha, \beta; \gamma, x)$ is convergent for $0 < x < 1$, and apparently real, the Fourier transform is indeed real along the $\omega = 0$ axis:

$$G(\omega, p) \sim \frac{1}{(\omega - v_c p)^{1-\alpha}(\omega + v_c p)^{-\alpha}} \times F\left(\frac{1}{2}, 1 - \alpha; 1 + \alpha; \mu' \frac{v_c p + \omega}{v_c p - \omega}\right)$$

($\mu' = \frac{v_c - v_\sigma}{v_c + v_\sigma}$) and has nonzero imaginary part in other ω, p regions. We plotted the density of states given by the Green's function in fig 5.1.

PROPERTIES OF THE 1D HUBBARD CHAIN: CONDUCTIVITY

We have to consider the stability of the ground state against weak hopping between adjacent Hubbard chains. It is often believed that a cross-over happens at infinitesimal hopping, but this is not the case at all. Anderson has shown, in the case that the chain is not a Fermi liquid, a finite interchain hopping is required. Here the most important response is that the hopping of one electron creates a disturbance which propagates along the chain and causes a hole of equal momentum to hop to the same adjacent chain. The effect of the interchain hopping is to stabilize the state which would be obtained as a ground state of the single chain problem, until higher-order terms in the interchain hopping can become relevant. We may say that the one-dimensional repulsive Hubbard model has the property of "confinement." For details the reader is referred to Anderson's work (which is given in part II) and other chapters in this book.

One of the most important properties of the Luttinger liquid is that electrons are not eigenexcitations anymore, they are composite. There are three processes that we have to consider when we come to conductivity: the scattering of spinons, the scattering of holons, and the decay (recombination) of electrons into (out of) holons and spinons.

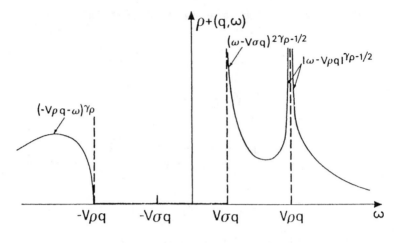

Figure 5.1

The holons have a very short relaxation time since they are the charge carriers and heavily scattered. But the spinon is not scattered by spin-independent potentials.

In one way of thinking, one can think of spinon excitations as the limit of renormalized electronic quasi particles as $Z \to 0$. This implies, since the spinon velocity

$$\frac{\partial \Sigma / \partial \omega}{\partial \Sigma / \partial k} = v_s$$

remains finite, that the spinons are neutral and incompressible. If they are neutral, they do not respond to simple potential scatterings and hence are not, to lowest order, scattered at all.

Equally one may argue from incompressibility: if the spinon gas is incompressible, its Fermi surface is not shifted by an external potential ($\partial \Sigma / \partial k = \infty$) and thus the potential causes no phase shift of the wave functions. Consequently the scattering cross-section vanishes.

Let's imagine we have a local weak potential acting at site j. Any random weak potential may be written as a sum of such terms, and we only consider those potentials that scatter up and down spins in the same way:

$$V_j = V_j(n_{\uparrow j} + n_{\downarrow j})$$

$$H_0 = \sum_k v_F(k - k_F)a_{1k}^\dagger a_{1k} + \frac{1}{2L} \sum_k V_k \rho_k \rho_{-k}$$

where

$$\rho_k = \rho_{1k} + \rho_{2k} = \sum_p a_{1p}^{\dagger} a_{1,p+k} + \sum_p a_{2p}^{\dagger} a_{2,p+k}$$

represents the Tomonaga-Luttinger model with no scatterings added, and a_{1k} and a_{2k} are annihilation operators for electrons on different sides of the Fermi surface. In the scattering term, we consider both impurity scattering and electron phonon scattering; the scattering term can also be separated into two parts: the small momentum part of the scattering or forward scattering part H_s', and a large momentum or backward scattering ($2k_F$) part H_L'.

$$H_s' = \sum_k g_s \varphi_{-k} (\rho_{1k} + \rho_{2k}) + \sum_{k,j} \lambda_s \exp(-ikR_j)(\rho_{1k} + \rho_{2k})$$

and

$$H_L' = \sum_k g_L \sigma_k \varphi_{-k} + \sum_{k,j} \lambda_L \exp(-ikR_j)\sigma_k.$$

Here g_s, g_L are electron-phonon coupling constants, and λ represents the effects of scattering from an impurity at site R_j. The density operator σ_k is defined as $\sum_p a_{1,k+p}^{\dagger} a_{2p}$ and describes excitations across the Fermi sea.

We rewrite the Hamiltonian in terms of boson operators

$$H_0 = \sum_k v_F(|k| - k_F) b_k^{\dagger} b_k + \sum_k \frac{V_k |k|}{2\pi} (b_k + b_{-k}^{\dagger})(b_k^{\dagger} + b_{-k})$$

$$\rho_1(k) = b_k \left(\frac{kL}{\pi}\right)^{1/2}$$

$$\rho_1(-k) = b_k^{\dagger} \left(\frac{kL}{\pi}\right)^{1/2}$$

$$\rho_2(k) = b_{-k}^{\dagger} \left(\frac{kL}{\pi}\right)^{1/2}$$

$$\rho_2(-k) = b_{-k} \left(\frac{kL}{\pi}\right)^{1/2}$$

$$[b_k, b_{k'}^{\dagger}] = \delta_{kk'}$$

and diagonalize it by changing boson operators in the following way:

$$b_k + b_{-k}^{\dagger} = \left(\frac{\omega_k}{E_k}\right)^{1/2} (\alpha_k + \alpha_{-k}^{\dagger})$$

$$b_k^{\dagger} - b_{-k} = \left(\frac{E_k}{\omega_k}\right)^{1/2} (\alpha_k^{\dagger} - \alpha_{-k})$$

where

$$\omega_k = v_F(|k| - k_F)$$

$$E_k = |k|(v_F^2 + \frac{2}{\pi} V_k v_F)^{1/2}.$$

Now

$$H_0 = \sum_k E_k \left(\alpha_k^\dagger \alpha_k + \frac{1}{2} \right)$$

$$H_s' = \sum_k [g_s \varphi_{-k} + \sum_j \lambda_s \exp(-ikR_j)](b_k + b_{-k}^\dagger)$$

$$= \sum_k [g_s \varphi_{-k} + \sum_j \lambda_s \exp(-ikR_j)](\alpha_k + \alpha_{-k}^\dagger)$$

we can see the forward scattering only amounts to a redefinition of the boson operators:

$$\alpha_k^\dagger \rightarrow \alpha_k^\dagger + \frac{1}{2} \left[g_s \varphi_{-k} + \sum_j \lambda_s \exp(-ikR_j) \right].$$

The forward scattering on the local potential does not cause resistance so it is only relevant whether backward scattering of an electron can take place, and that requires backward scattering of a spinon.

That this cannot take place may be made evident by using the large-U version of the Lieb-Wu equations described by Ogata and Shiba. In this version, which is correct to order $1/U$ for wave functions and energies, the wave function is explicitly factorized into spin and charge parts:

$$\Psi(x_1, \ldots, x_N) = \det|| \exp i\bar{k}_i \bar{x}_l || \times \Phi_H(y_1, \ldots, y_M).$$

This is an equivalent version of the Ogata-Shiba wave function expressed in hole coordinates $\bar{x}_i$'s. Now we can see why it is that the above potential cannot scatter spinons backwards: it has no effect whatever on the coordinates y, since it either gives zero or returns us to $\Phi(y_1, \ldots, y_M)$, so in this limit it cannot even modify the spinon wave function. This feature appears to hold even if *c $\Phi(y_1, \ldots, y_M)$ is modified by the neighbor corrections of the $t - J$ model.

If the spinon is not scattered, it may combine with a holon locally to form an electron. This means that the one-dimensional repulsive U Hubbard model does not respond to ordinary spin-independent scattering, and in particular weak randomness does not lead to any resistivity at absolute zero. As we can see above, the same can be said about the electron phonon scattering. This explains why phonons give no resistivity. This is a consequence of spin charge separation and will hold for any system with that property. This means the same might be true for the 2D repulsive U Hubbard model. This problem will be further discussed in the Appendix from a more rigorous point of view.

If impurity scattering and electron-phonon scattering do not give rise to resistivity, where does resistivity come from? The source of resistance is subtle; it is not a response to scattering, but rather scattering merely destroys the coherence of the accelerated preexisting excitations.

To calculate conductivity as a function of temperature, one needs only to consider the coherent propagation of the electrons. Rather simply, this is just given by the single electron Green's function $G_1(x, t)$ which describes the amplitude that holons and spinons arrive at x, t together. Any incoherent piece which is not in G_1 is scattered away by the strong disorder potential and will not contribute to the conductivity. For this purpose, the apparent breath of G_1, due to the electron's decay into holon and spinon, represents a true electronic mean free time satisfying $\hbar/\tau \simeq E_k^h \simeq$ the greater of T, ω. Or, more rigorously,

$$
\sigma = \frac{1}{\omega}\langle j, j\rangle
$$

$$
= \frac{1}{\omega}\sum_{\omega',k} G(k, \omega' + \omega)G(k, \omega')
$$

$$
= \frac{k_F^2}{\omega}\sum_k \int dv\, dv' \frac{\rho(k, v)\rho(k, v')}{[v - (\omega' + \omega)](v' - \omega')}
$$

$$
= \frac{k_F^2}{\omega}\int dv \sum_k \int dv' \frac{\rho(k, v)\rho(k, v')(\tanh \beta(v - \omega) - \tanh \beta v')}{[v - v' - \omega')}.
$$

We find $\sigma \propto 1/T^{1-4\alpha} \sim 1/T$ after inserting

$$
\rho(k, \omega) \simeq \frac{\omega}{(E_k^h)^{2-2\alpha}} \qquad \text{for} \quad \omega < E_k^h
$$

$$
\simeq 1/\omega^{1-2\alpha} \qquad \omega > E_k^h
$$

where $0 < \alpha < 1/16 \ll 1$.

PROPERTIES OF THE HUBBARD CHAIN: NMR RELAXATION

There is strong evidence of the existence of incommensurate fluctuating magnetic order in the normal state of high-T_c materials. The correlation functions convince us that the same kind of behavior is present in the one-dimensional Hubbard model. What we have learned in one dimension about the spin-spin correlation is that because of spin charge separation, the spin-spin correlation has two parts; one is "half" the free Fermi gas correlation, with a power law $1/r$, and only depends on spin-wave velocity. In the Heisenberg model, where the charge motion is frozen, this is the only term. The other part is the disruption by the hole motion, which has a power of $1/r$ at small U and $1/\sqrt{r}$ at large U. If we still believe in the charge-spin separation in 2D, we should still get some anomalous power-law for the spin-spin correlation.

In the standard arguments for the nuclear magnetic rates,[16] we are led to

$$\frac{1}{T_1} \propto g_N^2 A_{hf}^2 T N_0^2 \sum_q \frac{\text{Im}\{\chi^{-+}(q, \omega_0)\}}{\omega_0}$$

where g_N is the gyromagnetic ratio of the nuclear spin, A_{hf} the hyperfine coupling constant, and $\chi^{-+}(q, \omega_0)$ is the spin susceptibility at nuclear frequency ω_0.

$$\chi^{-+}(q, \omega_0) \sim i \int_0^\infty dt e^{i\omega_0 t} \langle \delta S_q^-(t) \delta S_{-q}^+(0) \rangle.$$

The temperature dependence due to the $2k_F$ spin-spin correlation is then given by

$$\frac{1}{T_1} \propto T^{e^{2\varphi}}.$$

For the copper atoms, deviation from the normal Korringa behavior is expected. The oxygen NMR does not see the $2k_F$ correlation due to the crystal structure and results in a rather different behavior.

*d

For large U, we would expect for the copper NMR

$$\frac{1}{T_1} \propto \sqrt{T}.$$

The fit of the NMR data of Barrett et al. to a power-law is remarkably good (fig 5.2).

Recently progress on the transport problem has been obtained in a different direction. Stafford et al. studied the optical conductivity of a Hubbard ring of finite circumference, from which they calculated the doping and system size dependence of the optical spectral weight. The reader is referred to his thesis[17] for the details. We believe that Stafford's calculation correctly gives the weight of the $1/\omega$ term for reasons given in the Appendix.

Now, let us briefly summarize what we have learned from the one-dimensional Hubbard model. In the 1D Hubbard model, $Z = 0$ but the Fermi surface in the sense of the Luttinger volume still remains at k_F. There are two kinds

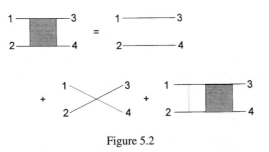

Figure 5.2

of excitations and two kinds of nonelectron Fermi surfaces; holons with $2k_F$
Fermi surface and spinons with k_F. The main feature of the electron Green's
function is the explicit charge-spin separation. As we discussed, the enhanced
$2k_F$ spin-spin correlation will have direct consequence in explaining the NMR
data. And most important, finite phase shift leads to a $Z = 0$ non-Fermi liquid
state, and correlation functions are directly decided by the phase shifts.

Appendix to Chapter 5

Physics of Conduction Processes in the 1D Hubbard Model

*e

The reader may have noticed that we have made almost no reference to a number of well-known studies of "resistivity" of the one-dimensional Hubbard model by Luther and Peschel, Schulz and Giamarchi, and others, as well as Fisher and Kane's work on the spinless interacting electron gas. We have also reinterpreted the results of Stafford and Phillips.

This is not because of any mathematical errors in their derivations—except that Luther and Peschel "for convenience" set $v_s = v_c$, hence reducing their system to a spinless Fermion one in principle—but to a problem with the actual physics of conduction. In order to use conventional perturbative techniques, specifically the time-honored "Kubo Formula," they assumed periodic boundary conditions and coupling of only the charge degree of freedom to an external E-field, presumably caused by a time-dependent flux through a ring.

This procedure is physically incorrect as a way of defining the conductivity in a real system. A real system is one in which electrons enter from an external electrode at one end, and are removed at the other. The chemical potential is higher at one end than the other, and the current is the response to the gradient of chemical potential for electrons, *not* to the electric field. The true conductance is correctly given by the *transmission coefficient* of the sample for *electrons* at the Fermi surface, i.e., by the Landauer, not the Kubo, formula

$$\sigma = \frac{e^2}{\hbar} \sum_{i,j} |t_{i,j}|^2. \tag{1}$$

i, j are "channel" variables. (Az'bel, Abrahams, and Lee, and many others have discussed the relationship of the two formulas, which agree in most instances—but not in general.) t_{ij} is simply a linear function of the interacting, scattered Green's function G_1, so that the derivation of the conductivity given in the main body of the chapter follows immediately, the formula being

$$\sigma = e^2 \int d^D \times dt |\nabla G_1(x, t)|^2 , \tag{2}$$

G_1 being the causal one-particle Green's function

$$\frac{1}{Z} \sum_n e^{-En/kT} \langle n|\psi(x, t)\psi^+(0, 0)|n\rangle . \tag{3}$$

It is of course necessary to impurity average the result (2), which in the conventional theory gives the well-known vertex corrections, but as is well-known in mesoscopic theory the conductivity of a particular sample can fluctuate from the mean appreciably.

A very simple and general theorem ensures the almost exact equivalence of (2) and Kubo's formula. This is the well-known fact that the *asymptotic* behavior of a two-particle Green's function such as enters into the Kubo formula is normally given by the product of two one-particle Green's functions.

Why then do the two expressions differ? The reason is that by enforcing periodic boundary conditions, the Kubo formula has prevented the electronic system from losing its coherence internally: the electron cannot decay into holon and spinon and hence cannot generate entropy, because the initial electronic state automatically regenerates: neither holon nor spinon can "get lost" in this configuration. Another way of saying it is that the real coupling is to electrons, which decay, not to the nondecaying charged bosons. What actually happens physically is that if the holon or spinon is scattered before regenerating the initial state, we can use the causal Green's function with outgoing boundary conditions in (2). But if, in fact, neither excitation is sufficiently rapidly scattered, we have a "drag" phenomena like phonon drag, and we must consider the process in detail. In that case we should look for the most effective momentum transfer process so long as it is slower than the electron decay process, and this will control the resistivity. This implies an inequality:

$$\sigma(\omega, T) \geq \max\left(\frac{1}{\omega^{1-\alpha}}, \frac{1}{T^{1-\alpha}}\right) \times \frac{ne^2}{\hbar m^*}.$$

(The coefficient is that determined by Millis and Stafford.)

The above is all for the Hubbard model—i.e., for repulsive interactions. It is not clear whether a "negative U" Hubbard model has any meaning physically; electrons do not have attractive interactions except via slow dynamic systems such as phonons which complicate the dynamics. Nonetheless, the spin gap characteristic of the negative U Hubbard model can arise through such couplings. As far as we can see such a gap will normally lead to *insulating* behavior if there is no pair condensation. The reason is again the same: true conduction is via *electrons*, not via charged bosons, and a gap for the spin implies a gap for the electron. (Sliding SDW or CDW conductivity is a separate matter.)

Thus we find ourselves in a very interesting paradox. There is no reason to disbelieve Fisher and Kane's RNG calculations for the spinless Fermion case, which shows that negative U is conducting, positive U insulating, and localized. For the spin case with $v_s \neq v_c$—which in the Hubbard model means except for $U = 0$—the opposite is the case: positive U implies $\sigma = \infty$ at $T = 0$ if T-symmetry is maintained while negative U is probably insulating. This is much closer to the experimental facts on unsaturated polymer chains.[18]

A very simple argument from scattering theory seems qualitatively to control most of these results without lengthy exercises in renormalization group methods such as have been de rigeur in the past. Let us at first simply consider the existence of a finite, short-range scattering potential $V(x_0)$, leaving aside the question of screening (or antiscreening, which is more usual) to be dealt with later.

We simply calculate $G_1(x \to \infty, t \to \infty)$ in perturbation theory

$$G_1(V) = G_1 + G_1 V G_1 + G_1 V G_1 V G_1 + \cdots$$

where

$$G_1 V G_1 = \int G_1(0, 0; x_0, t') V(x_0) G_1(\chi_0, t'; x, t) \, dt', \text{ etc.}$$

If G_1 is the free-particle Green's function $\frac{1}{x - vt}$, it is clear that the t' integral simply leads to a term of the form

$$\text{const} \times V_0 \times G(0, 0; x_j t)$$

which has the capability of interfering with G itself in the asymptotic region. This interference between the forward scattered ("shadow") wave and the direct wave is fundamental to all multiple scattering phenomena.

If the interacting Green's function is nonetheless of Fermi liquid type, and has a pole indicating asymptotically free behavior, again scattering is possible: a finite resistance can exist (except for multiple-scattering effects which may lead to localization).

But if the singularity in the Green's function is not a simple pole, conventional resistivity is never possible, because the scattered wave and the incident wave can never interfere asymptotically, one or the other being overwhelmingly larger. For the spinless case, the singularity is

$$G_1 \sim \frac{1}{(x, vt)} 1 + \alpha$$

with

$$\alpha > 0 \qquad U > 0$$

$$\alpha < 0 \qquad U < 0$$

and simply by dimensional analysis

$$G_1 V G_1 \sim \frac{1}{(x - vt)^{1+2\alpha}}.$$

This is more singular than G_1 in the asymptotic region if $U, \alpha > 0$, and hence overwhelms it: none of the incident waves makes it through, and the resistivity diverges upward. If $U < 0, \alpha < 0$ the opposite is the case: the scattered wave cannot interfere destructively with the incident wave and the conductivity is infinite.

This simple scattering argument seems to have the essence of the effect. The renormalization of the scattering amplitude may act in concert but does not control the phenomena.

Now when we go over to the spin case, we immediately change the nature of the singularity of G_1. It is a cut rather than a pole for all $U > 0$, and the effective α is $-1/2$: the scattered wave can never interfere with the direct one.

We have made, with Alexei Tsvelik, a more complete analysis, but it turns out again that the renormalization effect on V is not very important and the above considerations control. The conclusion then is:

(a) The spinless Fermion case is irrelevant because the particles which are moving are electrons having both charge and spin.

(b) The essence of the problem is spin charge separation. This is the fundamental incoherent (entropy-producing) step: it is essential in understanding transport in a Luttinger liquid. It occurs independently of scattering by external potentials.

(c) Scattering by external potentials is essential to avoid "drag" effects and is part of the problem, though the conductivity in the usual case is independent of it.

(d) It is essential to recognize that it is electrons only which carry current: all properties are functions of $G_1^+ G_1$, not of $G_c \times G_s$.

We are indebted to M. Ogata and A. Tsvelik for extensive discussion of these problems.

NOTES

*a This fact was recently rediscovered by Nozieres and Brazovsky in a preprint. It is the basis of the "holon non-drag" transport theory.

*b This "singlet pairing susceptibility" is not what is relevant to the interlayer tunneling mechanism, which is local in momentum space, not in real space. It also contains the effect of the repulsive U, which in 1D is unrenormalizable except by dynamic screening, which cannot be included in such a calculation. The appropriate pairing susceptibility for interlayer tunneling is correctly derived using $G_1^\ell G_1^h$, as in chap. 7.

*c This statement is true only in the straightforward perturbation theory sense. The conventional Boltzmann-type of theory simply gives no resistance, because, as we mentioned in a previous note, the spinon can carry current. However, two renormalizations intervene: the localization of charge carriers in 1d, which will then necessarily localize a spin; and the antiscreening of $2k_F$ fluctuations. Thus, while the standard papers purporting to solve the problem of 1D transport are incorrect, the whole problem remains to be solved. It is somewhat simpler in 2d, where $2k_F$ fluctuations are not as important.

*d This question is discussed elsewhere. (Chap. 6)

*e See note *c above. This Appendix, while not completely up to date and incorrect in some details, is left in because it makes some useful physical points.

REFERENCES

1. E. H. Lieb and F. Y. Wu, Phys. Rev. Lett., 20, 1447 (1968).
2. C. N. Yang, Phys. Rev. Lett., 19, 1312 (1967).
3. H. Shiba, Phys. Rev. B., 6, 930 (1972).
4. A. A. Ovchinnikov, Sov. Phys. JETP, 30, 1160 (1970).
5. F. Woynarovich, J. Phys. C, 15, 85 (1982).
6. M. Ogata and H. Shiba, Phys. Rev. B 41, 2326 (1990).
7. P. W. Anderson, Phys. Rev. 164, 352 (1967).
8. C. F. Coll. III, Phys. Rev. B, 9, 2150 (1974).
9. A. Luther and I. Peschel, Phys. Rev. B, 12, 3908 (1975).
10. V. J. Emery, in *Highly Conducting 1D Solids*, J. Derreese et al. eds., Plenum Press, NY (1978).
11. A. Luther and V. J. Emery, Phys. Rev. Lett., 33, 589 (1974).
12. J. Solyom, Adv. Phys., 28, 201, (1979).
13. F.D.M. Haldane, J. Phys. C 14, 2585 (1981).
14. F.D.M. Haldane, Phys. Lett. 81A, 153 (1981).
15. I. E. Dzyaloshinskii, Sov. Phys. JETP, 38, 1202 (1974).
16. B. S. Shastry, Phys. Rev. Lett, **63**, 1288 (1989).
17. Charles Stafford, Princeton Thesis, 1992.
18. Phillip Phillips and H. L. Wu, Science **256**, 1805 (1991).

6

"Normal" State of the 2D Hubbard Model: "Tomographic" Luttinger Liquid and $T_c = 0$ Superconductor

PART I. FAILURE OF FERMI LIQUID THEORY FOR 2D HUBBARD MODEL

In chapter 5 we discussed the physics of the one-dimensional Hubbard model at some length. Although this model was "solved exactly" twenty-two years ago, the physics has been very slow to be understood, and in fact in chapter 5 we found several new facts unknown prior to our work. Some of the crucial results are:

(1) "Confinement"—the fact that the coherent response to weak transverse hopping between two chains is blocked.

(2) ODLRO in the spin degrees of freedom, derived from a new hidden "order parameter" in the Heisenberg model.   *a

(3) "Perfect conductivity"—the fact that the charged excitations are weakly scattered by T-invariant potentials. $\rho \propto T$, in the presence of weak random time-reverse invariant scattering.   *b

(4) Antiferromagnetic-like Spin Fluctuations—the $2k_F$ response is enhanced over that of a Fermi liquid and leads to non-Fermi liquid real and imaginary parts of susceptibilities.

(5) Instrumental in understanding these and other effects is the one-particle Green's function.

We will here argue that all of these properties carry over into the two-dimensional Hubbard model. The detailed spectroscopy of 2d may be different, but these basic properties are transferable. The central reason for this fact is that, in 2d as in one, forward scattering of opposite-spin electrons is so strong as to make the Landau Fermi liquid state unstable, and the fixed point of the doped Hubbard model is a "Luttinger Liquid" in the sense of Haldane.

Let us first give a demonstration that the ground state is not a Fermi liquid, and then attempt to incorporate the same physics into a full, if schematic, solution including excitation spectra and wave functions.

The Landau theory is based on two very strong consequences of the exclusion principle, and on a simple perturbative renormalization scheme which is

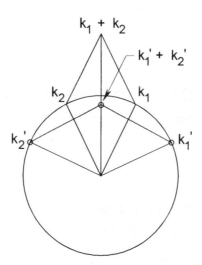

Figure 6.1. Total momentum cannot be conserved except for forward scattering.

hardly complex enough to describe as a "group" but is useful to think of in that way.

The two assumptions on which it relies are:

(1) that because of the restriction on energies, perturbative calculations of $\sum(k, \omega)$ will automatically give $Im \sum \propto \omega^2$ near $\omega = 0$, coming from the simplest skeleton electron-hole bubble diagram.

(2) Except for $k, -k$ scattering (the Cooper pair problem, which must be treated separately) scattering for k's sufficiently near the Fermi surface is forward scattering (or exchange, but retaining the same momenta). This is enforced in two dimensions by simple momentum conservation (see fig. 6.1). This means that the relevant interactions are of the simple form $f_{kk'} \, n_k \, n_{k'}$ or $f_{kk'} \, \sigma_{k'} \, \sigma_k$, and the momentum distribution characterizes the states. One may define quasiparticle occupation numbers 0 or 1 for renormalized quasiparticles, and these experience what Sutherland has called "non-diffractive scattering." We will expand on this later, but briefly, as Gallivotti and Benfatti have shown, the appropriate scaling in F.L.T. is to make the Landau energy expression for quasiparticles dimensionless by dividing by a cutoff energy $\Lambda \sim T$ or ω. Then the Fermi energy is relevant, the Landau parameters marginal, and scattering irrelevant. Thus all true inelastic processes vanish in the fixed point limit $\Lambda \to 0$. We will show that this fixed point is not stable in two dimensions, as we already know it is not in one. Nonetheless we will see that we can still hope that the fixed point is "non-diffractive" or Bethe-ansatz-like, as it necessarily is in 1d.

The simplest approach to this demonstration focuses on a type of diagram which is normally considered to be trivial in perturbation theory because it seems at first—and is in the many cases where FLT is valid—to be simply a correction to the chemical potential, or at most to the compressibility. This is the Hartree diagram giving the simple interaction corrections $f_{kk'}\, n_{k\sigma}\, n_{k'-\sigma}$ to the single-particle energies of occupied states. If the interaction is short-range or hard-core it has long been known that one must use for f not the simple Hartree $\langle V_{kk'}\,(q=0)\rangle$, but the multiple-scattered, resummed "scattering matrix" $T_{kk'}$, embodying multiple scattering, by the interaction U, of states k and k' into all possible other states $k+q$, $k'-q$, eventually returning to the exact momenta k, k' and energies E_k, $E_{k'}$. This multiple scattering corrects the mutual wave-functions of the two particles when they come close together, allowing for mutual recoil, and is the formal way one removes hard cores and replaces them by pseudopotentials. It is appropriate that this is where the projective aspect of the one-band Hubbard model should enter the problem. As PWA emphasized in a Phys. Rev. Letter, this resummation must eliminate from the physical, low-energy region a part of Hilbert space, namely, the "upper Hubbard band" composed of the antibound states in all particle-particle channels. In a recent article I summarized the difficulty with perturbation theory succinctly.

It is known but not often fully recognized that this resummation is not just convenient but necessary, because its physical nature is to include the effects of recoil upon scattering, without which every scatterer in a Fermi sea encounters the "orthogonality catastrophe" involving producing divergent numbers of pair excitations near the Fermi surface. Thus many-body perturbation theory is incomplete without a prescription for renormalizing the vertex for recoil; the effect of which appears already, in fact primarily, in the simple Hartree terms. (Note that the problems primarily occur for *antiparallel* spins, being cancelled to a great extent by exchange for parallel spins.)

One may run into difficulties in many scattering problems if one does not specify precisely the treatment of boundary conditions, and of energy conservation and the energy shell; and this is surely one of these. We visualize k and k' as *occupied* states in the quasiparticle sea, and hence as having precisely specified energies ω_k, $\omega_{k'}$, and satisfying boundary conditions of our choice, either real or outgoing. At this level of F.L.T. and of the renormalization theory this is correct.

Under real boundary conditions the energies of our states—or pair of states—will shift because of scattering, which means that they will not actually retain the original asymptotic k-values, quite. Alternatively, under outgoing boundary conditions the velocities will be modified in such a way also that the density of states in momentum space will be somewhat different, leading, in effect, to exactly the same result. This equivalence dates back to Friedel; a good discussion is in PWA, Varenna, 1966. It is useful to realize that our particles must be allowed to scatter not only against each other but against the boundary of the sample, in order to calculate wave-functions and energies correctly in the asymptotic regions. For instance, using real boundary conditions the states are real linear combinations of $e^{\pm ikr}$ so that forward and backward scattering are not separable; the same is true if we use outgoing b.c. because the scattering states mix ingoing and outgoing waves; this is characteristic of energy-shift terms. In Fermi liquid theory these effects are not visible because there is an overall renormalization of the mean potential which takes care of returning the momenta to exactly the original values at the Fermi surface. The Hartree self-energy is subtracted against a uniform mean potential which is added in to the bare Hamiltonian, in direct analogy to Dyson's renormalization scheme. But we shall see that this does not work if the scattering leads to a finite forward scattering phase shift for particles near the Fermi surface; the problem is modification of the "sea" of one spin by the addition of a single particle of the other spin. It is not possible to cancel this effect against a mean potential shift if the forward scattering is strongly dependent on the relative spin and momentum of the particles.

To make clear what happens, let us remind the reader of what happens in ordinary potential scattering. A local scatterer causes a phase shift of those partial waves which impinge on it, and the Friedel identity $\delta n = \sum_\ell (2\ell + 1) \times \frac{\delta_\ell (E_F)}{\pi}$ tells us that, correspondingly, particles move toward or away from this scatterer: the local density changes, or equivalently the energies and k-values of scattered states shift in such a way as to move particles through the Fermi surface. A local change in the scattering potential, as for instance a core electron transition, experiences the well-known overlap "catastrophe" which causes the x-ray edge singularities: the scattered many-particle state has a singular overlap with the original one:

$$(O_0 \mid O_V) = \text{const} \times \exp\left[-1/2 \sum_\ell (2\ell + 1) \left(\frac{\delta_\ell}{\pi}\right)^2 \ln |N|\right]. \qquad (I-1)$$

This is connected with the Hilbert space shift caused by the potentials adding or removing states from the occupied region of the band and localizing them near the potential. Correspondingly, states move through the outer boundary of the region.

Of course, if the potential is changed in a translationally invariant fashion, no net particles move into or out of any region and the chemical potential simply shifts to accommodate the potential shift, as remarked above. It is normally assumed that this is the correct way of treating the Hartree potentials of the different states acting on each other, and this assumption is embodied in Luttinger's theorem that the volume of the Fermi surface in k-space is invariant. As a matter of fact, Luttinger's theorem in some form seems to retain a remarkable force even in one-dimensional models in which Fermi liquid theory has clearly broken down completely: if one had to state this modified form, it seems to be that the low-energy Fermionic excitations continue to occupy a surface in momentum space which encloses the Luttinger volume, and that other low-energy excitations exist at multiples of Fermi momenta. If one guesses at the source of such a theorem, it probably is an S-matrix theorem equivalent to Friedel's identity, applied to free particles impinging on an interaction region as in the Langer-Ambegaokar theory.

We are liberated from the prejudice that the Hartree terms cannot cause net phase shifts when we contemplate the wide spectrum of one-dimensional models, in *all* of which Fermi liquid theory fails, and for the same reason: In one dimension, no two interacting particles can occupy the same asymptotic k-values having zero relative momentum, be they bosons or Fermions of opposite spin, because all interactions are equivalent to a change in boundary conditions and shift k-values by an amount comparable to their separation; hence finite Fermi surface phase shifts occur. The phase shifts are the basic parameters of the theory in the Luttinger liquid technique of Haldane.

Abandoning our prejudice that the particle k' can only lead to a modified effective potential for k, we ask whether it cannot also modify its wave function, in such a way as to exclude amplitude from their overlap region: in fact, we know this is precisely what we must do to do the hard core resummation. Fortunately, it can be shown that for weak coupling, high dimensions, or unbounded bands, this leads to no problem. In these cases the pseudopotential resulting from resumming the scattering is equivalent to a finite, or at least not too divergent, scattering length a, in the relative $l = 0$ channel at very low relative momenta $Q \rightarrow 0$. The finite scattering length a in 3 dimensions is the essence of the Lee-Yang pseudopotential theory for dilute Bose and Fermi gases, as Galitskii showed; and Bloom showed that a diverges only as $\frac{L}{\ln L}$ for low densities of free Fermions in two dimensions, where L is the sample size. These lead to phase-shifts in the $l = 0$ channel of $\eta \approx a/L$ in 3d and $\eta = 1/\ln L$ in 2d, vanishing with the sample size. The phase shift remains roughly constant for all Q in the $\ell = 0$ channel, and vanishes in higher angular momenta, so the total shift in Hilbert space is of order $1/L$ or $1/\ln L$ states per particle, negligible compared to the total Hilbert space of unity per particle. (But, we see, only barely so in 2d!)

What we now want to show is that in the two-dimensional Hubbard model, specifically, using inessential simplifications specific to low densities of particles, the scattering length diverges for particles of the same momentum k, leading to a momentum shift which is (as in one dimension) a finite fraction of the momentum eigenvalue spacing.

The calculation is very simple, but it must be interpreted quite carefully. We consider the scattering of two electrons near the Fermi surface. We choose real, not outgoing, boundary conditions. Under these boundary conditions the energy shift is the scattering phase shift we require. We choose to work at low energies at which the energies are simply

$$E_k = k^2/2m \; ;$$

band structure will modify higher energies, but that will just affect cutoffs.

We consider nearly forward scattering, with the relative momentum $Q \simeq 0$. We write the Schrödinger equation for a scattering state wave function

$$\varphi = \sum_Q a_Q \, \varphi_Q \qquad (I-2)$$

with

$$\varphi_Q = \varphi_\uparrow (K + Q) \cdot \varphi_\downarrow (K - Q) \qquad (I-2')$$

and

$$E_Q = \epsilon_{K+Q} + \epsilon_{k-Q} = 2\epsilon_K + \frac{Q^2}{m}$$

$$Q = \frac{\pi n_x}{L}, \quad \frac{\pi n_y}{L}$$

as,

$$(E - E_Q) a_Q = \frac{U}{L^2} \sum_{Q'}{}' a_{Q'} \, (1 - f_{k+Q'}) \, (1 - f_{k-Q'}) \qquad (I-3)$$

or,

$$\frac{L^2}{U} = \sum_Q{}' \frac{1}{E - E_Q} (1 - f_{k+Q'}) \, (1 - f_{k-Q'}). \qquad (I-3')$$

The treatment of the occupancy factors is obvious but tricky. Specifically in the case of the Hartree terms, there is no hole-particle symmetry, and we want specifically to calculate the interaction of two particles in occupied states, so that the pole in $\sum \frac{1}{E-E_Q}$ applying to the specific pair of states under consideration must be treated separately. We will discuss the meaning of this procedure shortly. These states may, however, *not* recoil into states already occupied by other electrons, which are eliminated by the exclusion factors. Thus the $\sum'$ involves omitting exclusion factors for one particular state Q, initially $Q = 0$, for which we are calculating the energy shift, since we are asking what would happen, if one particle were removed, to the others' wave functions. Of course, the particles can recoil into unoccupied states. The entire exercise has the effect of creating a Jastrow-like hole around the position of any up spin particle; the question is whether that hole fundamentally disturbs the wave function or not.

We solve the Schrödinger equation for the lowest eigenvalue, first neglecting the exclusion principle terms. In this case, it is quite independent of K and the lowest eigenvalue is obtained by placing E_0 close to $E_{Q=0}$ and replacing the integral by a principal part sum.

$$\frac{L^2}{U} = \frac{1}{E_0 - E_{Q=0}} - \frac{L^2}{2\pi^2} \int_{Q_1 \pi / L}^{\frac{\pi}{a}} \frac{2\pi Q dQ}{Q^2} \qquad (I-4)$$

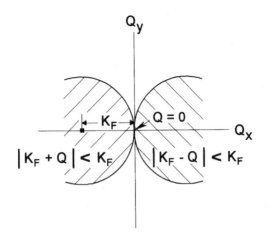

Figure 6.2. Region of recoil momentum Q excluded by Pauli principle.

$$E_0 - E_{Q=0} = \frac{\pi}{L^2 \ln \frac{L}{a}}$$

$$E_{Q_1} - E_{Q=0} = \frac{\pi^2}{L^2}$$

so

$$\frac{\delta}{\pi} = \frac{E_0 - E_{Q=0}}{E_{Q_1} - E_{Q=0}} = \frac{1}{\ln \frac{L}{a}}$$

which is negligible, as shown by Bloom, and causes no trouble for Fermi Liquid Theory.

The situation is very different when we take into account the limitation on recoil caused by the exclusion principle. For K inside k_F, the integral must always start at a finite Q, but the important point is the Fermi surface itself, since that is where we add and subtract particles. At the Fermi surface, the two excluded circles osculate and at low energy (see fig. 6.2) the fraction of the circle at Q which remains is $\frac{Q^2}{\pi k_F^2}$ so the relevant principal part integral may be approximated by

$$P - \frac{1}{2\pi^2} \left[\int_{Q_1}^{k_F} \frac{2\pi Q dQ}{Q^2} \frac{Q^2}{\pi k_F^2} + \int_{k_F}^{\frac{\pi}{a}} \frac{2\pi Q dQ}{Q^2} \right] \simeq \frac{1}{\pi} \ln(k_F a)^{-1} \quad ; \quad (I-5)$$

the lower limit Q_1, is replaced, effectively, by k_F! We find therefore that the energy shift is

$$E_0 - E_{Q=0} = \frac{\pi}{L^2} \frac{1}{\ln |k_F a| + \frac{1}{N(0)U}}$$

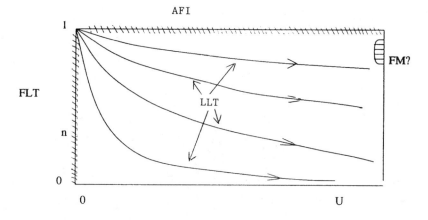

Figure 6.3. Phase diagram for the 2D Hubbard Model. Lines connect asymptotically equivalent Luttinger liquids: all are equivalent to strong coupling models, except possibly for a small region of ferromagnetism in the upper right-hand corner. The $n = 1$ line is all antiferromagnetic insulators; Fermi liquids occur only as n or $U \longrightarrow 0$.

and

$$\frac{\delta}{\pi} \simeq \frac{1}{\pi |\ln k_F a| + \frac{1}{N(0)U}} \qquad\qquad (I-6)$$

which is finite unless $U \to 0$. The simplifications appropriate to low densities are obviously inessential and the conclusion is a fortiori valid at higher densities, where the phase shift will become of order a finite fraction $\times \pi$. It will vanish for $U = 0$ and $k_F \to 0$, in contrast to the one-dimensional case, where low density is a strong coupling limit; the Hubbard Model flow diagram will be FLT on two sides, not one side, of the canonical rectangle (see fig. 6.3).

The above expressions are related to those of sec. 5 of AGD, where convergence of the pseudopotential to a finite scattering length a is assumed as a basis for *starting* perturbation theory. What we have shown is that in the 2d Hubbard model the scattering length is not finite, since as $Q \to 0$

$$\delta = Qa$$

which must converge to zero for a finite scattering length. That is, the difficulty does not lie within the usual framework of many-body perturbation theory but prior to it, in the derivation of the elementary vertex.

The procedure here is apparently unfamiliar to many-body theorists and must be explained quite carefully. So far, what we have done is not to calculate the vertex—or, equivalently, $f_{kk'}$—but to calculate the scattering phase shift which must be used to satisfy the boundary condition at the *origin* of a scattering event between particles at $k_\uparrow$ and $k'_\downarrow$. This phase shift δ is not the phase of a "vertex function" for quasiparticle scattering, which appears only at the next level of the theory: It need not be related to a Kramers-Kronig transform of some response function of the system as a whole, because we have not yet included interference among the scatterers and are merely trying to get the wave function right in the scattering region. The phase shift we calculated above could be derived from the standard expression

$$e^{i\delta} \sin \delta = T = \frac{U}{1 + U\Gamma_0} \qquad (I-7)$$

by writing

$$\Gamma_0 = i\pi n(0) + P \int \frac{dn(Q)}{E - E(Q)} \qquad (I-8)$$

but this Γ_0 is not related to any two-particle Green's function of the system, and the expressions are meaningless off the energy shell at general E.

What we must do to arrive at the physics is to treat the behavior near the actual two-particle pole of the Green's function properly, with care to satisfy the boundary conditions *both* at the origin and at ∞. As I have pointed out, energies may be assumed real: we are not concerned with decay processes—which in any case are $\sim \infty^2$ near the Fermi surface—but with wave-functions and energy shifts. If we had to express the process diagramatically, we would describe it this way. In deriving (I-6), we have calculated an irreducible vertex for the scattering process $k, k' \to k, k'$, i.e., a vertex in which the particles never return to precisely the same momentum state. This vertex can be expressed in terms of a boundary condition at the origin of relative coordinates. There are also processes involving $-k$ holes scattering against k' electrons, which must be included because the boundary at ∞ scatters $k \to -k$. We then use these vertices (plus boundary scattering) in a Bethe-Salpeter equation which necessarily includes anomalous diagrams (with holes and particles in state k, since we are after the Hartree terms in which k and k' are closed loops). These diagrams cannot necessarily be included properly simply by calculating at finite temperature T, as Kohn and Luttinger did, because we are not confining our attention to k and k' close to the Fermi surface, and the boundary scattering doesn't appear in ordinary perturbation theory in any case; but as Kohn and Luttinger showed, in diagram terms they involve "pinching" singularities between hole and particle terms.

Fortunately, we don't have to do the calculation using diagrams, since Schrödinger's equation does exist and we can simply solve for a wave function with given boundary conditions rather than adding up infinite sequences of diagrams. It is important, however, to make it clear what wave equation we wish to solve. Since we are including boundary scattering effects, we are not free to pick k and k', or $K = \frac{k+k'}{2}$ and $Q = k - k'$, at will. We can select a k-value, but then allowable values of k' are set by boundary conditions. That is, adjusting the overall potential to adjust k to the free particle value, the k''s are determined.

This process is very simple in execution, if complex conceptually. In the relative coordinates, the wave function is

$$\cos(Q(r - r') + \delta)$$

which vanishes at the boundary R if

$$Q \to Q_0 + \delta Q$$
$$\delta Q = \frac{\delta}{R}$$

and this δQ must all be added to k':

$$k' \to k'_0 + \delta Q = k'_0 + \frac{\delta}{R}. \qquad (I-9)$$

Actually, having resolved the scattering problem into partial waves, it is only the isotropic partial wave which is shifted. It is legitimate (using wave-packet arguments) to average over partial waves, of which there are a number $2\pi Q \Delta Q \times \pi R^2 = 2RQ$ at the same Q-value.

Thus the average energy shift is

$$\Delta E = \frac{1}{m} \left(k' \cdot \frac{\delta}{\pi R} \cdot \frac{\vec{Q}}{2RQ^2} \right)$$

$$= \frac{1}{2mA} k' \cdot \frac{(k - k')}{(k - k')^2} \delta$$

by adding an irrelevant constant we have

$$f_{kk'} = \frac{\delta}{\pi} \frac{k^2 - k'^2}{(k - k')^2}. \qquad (I-10)$$

This expression can be interpreted in a very simple way. What it tells us is that k' space is responding incompressibly to the fact that $Q = k - k'$ cannot be zero: no two opposite-spin particles may occupy exactly the same momentum state. The projective effect of short-range interactions is acting as an effective *fractional exclusion principle* between opposite spin particles.

*c

The breakdown of Fermi liquid theory in the presence of a finite $Q \to 0$ phase shift, or infinite scattering length, is not "marginal": the theory must take on very different properties. However, still within the low density, weak coupling limit, where we can assume the theory not too far from FLT, we can demonstrate simple divergences.

The effect of our partial "exclusion principle" is to open a hole in the Hilbert space of momentum states; and since Hilbert space is incompressible, other states must move out of the way to accommodate the missing one. The phase shift in any channel is a monotonically increasing function if there are no bound states (as there obviously aren't) so the shift in k space with volume δ/π is just passed up to the Fermi surface: the missing fraction of a particle reappears at the Fermi energy so that the Hartree self-energy is just proportional to $|\epsilon_F - \epsilon_k|$. This may be understood as a self-energy

$$Re\Sigma \propto |\omega| \qquad (I-11)$$

on the energy shell $\omega = \epsilon_k$. We cannot deduce $\Sigma(k, \omega)$ this way, but by Kramers-Kronig there must be an

$$Im\Sigma \propto |\omega| \ln \omega \qquad (I-12)$$

and $\frac{\partial \Sigma}{\partial \omega}$ must diverge logarithmically. It is clear that *both* $\partial\Sigma/\partial\omega$ and $\partial\Sigma/\partial k$ are anomalous, in contrast to "Marginal" FLT.

Another way to estimate the form of Σ is by direct calculation of the renormalization constant Z. As in our original Letter, we can calculate this directly from its physical definition as the amplitude of the quasiparticle pole in the Green's function in the limit as $\omega \to 0$, $k \to k_F$. In this limit the quasiparticle peak becomes sharp and Z is simply the square of a ground state to ground state overlap between a Fermi liquid ground state with N particles and the state with $N - 1$ particles with one bare particle added at the Fermi surface (hence with $\epsilon_k = \omega = 0$). Since one or the other state contains an odd number of particles they differ in momentum by k_F (and in spin by 1/2).

When we add an up-spin particle at the Fermi surface, it shifts the k-vectors in the $2k_F$ singlet particle-particle channel by $\delta Q = \frac{\delta}{L}$ so that, by the well-known "infrared catastrophe" theorem the overlap of the many-body wave function of all down spins is reduced by a singular factor

$$\exp\left[-\frac{1}{2}\left(\frac{\delta}{\pi}\right)^2 \ln N\right] = N^{-\frac{1}{2}(\frac{\delta}{\pi})^2}.$$

Thus this singular factor appears in

$$Z = (O_{N+1}, k_F O_N)| = \text{const} \times N^{-(\frac{\delta}{\pi})^2}. \qquad (I-13)$$

We can use well-known techniques (now related to the modern techniques of "conformal field theory") to convert an overlap as a function of N to Green's functions as a function of ω. That is, we can presume that the sensitivity to ω is given by inserting for the frequency variable simply the finite-size spacing of energy levels $\delta\omega \sim \delta E/\delta N \cdot \delta N \sim E_F/N$. Thus we get $Z \sim \omega^{(\frac{\delta}{\pi})^2}$, and since

$$Z = \frac{1}{1 - \frac{\partial\Sigma}{\partial\omega}}, \qquad (I-14)$$

$$\frac{\partial\Sigma}{\partial\omega} \sim -\omega^{-(\delta/\pi)^2}$$

$$\sum \sim \omega^{1-(\frac{\delta}{\pi})^2} \sim \omega\left(1 - \left(\frac{\delta}{\pi}\right)^2 \ln\omega\dots\right) \qquad (I-15)$$

which shows us that the perturbation theoretic approach is the first term of a perturbation series in $(\frac{\delta}{\pi})^2 \sim \frac{1}{(\ln n)^2}$, but that the true behavior is probably power-law in nature.

Now let us try to work out the consequences of this singular behavior. It is important for physical orientation to give a real space representation of what happens to the wave functions. We do this in two ways.

First, let us recall that the T-matrix resummation is in effect taking into account the solution of Schrödinger's equation for pairs of opposite spin electrons when they approach closely to each other. In the region $k_F r \ll 1$ this solution must be a superposition of the "indicial solutions" of

$$\left[-\frac{\hbar^2}{2\mu}\nabla^2 + V(r_1 - r_2)\right]\psi(r_1 - r_2) = 0. \qquad (I-16)$$

The solutions to this equation, when $V(r)$ is such as to make ψ vanish at $r_1 - r_2 = b$, are: (set $r = r_1 - r_2$)

$$1D: \ \psi = r - b$$
$$2D: \ \psi = \ln r/b \qquad (I-17)$$
$$3D: \ \psi = 1 - b/r.$$

To fit this onto an asymptotically free wave function of relative momentum Q, we must match this short-range behavior to the ψ'/ψ asymptotic behavior at $Qr \sim 1$, which is $\cos(Qr + \eta)$, which has $\psi'/\psi \sim Q\tan\eta$. Clearly, this gives us a phase shift η depending, as mentioned above, on the relative momentum Q, if other particles do not intervene.

When two particles come close together there is thus a Jastrow-like "hole" which we can think of as superposed on their asymptotically free wave functions.

In 3d, this "hole" is and remains only of order "b" in size; if, as appropriate to low densities, $b \lesssim r_s$, the interparticle distance, a low-density expansion can be based simply on compensating the average shift in k caused by the scattering with an overall shift in chemical potential.

For 2 and 1D, however, particles with small relative Q's will have "holes" extending out to a range $\sim 1/Q$. This cannot be meaningful if $Qr_s < 1$ because it implies the "hole" is using Hilbert space which is already occupied by other electrons of the same spin. Thus we must terminate the "hole" at roughly r_s, which implies, since the boundary condition gives a finite slope ψ'/ψ at this radius, that δ remains *finite* as $Q \to 0$: the scattering length δ/Q diverges and effectively opposite-spin particles cannot asymptotically occupy $Q = 0$.

A second remark is to consider what the pseudopotential necessary to describe scattering *in the presence of the exclusion principle restriction* would be—that is, if we resum the T-matrix diagrams *with exclusion principle included*, and use this as a vertex for doing perturbation theory ignoring these exclusions (i.e., for doing a Bloom-Galitskii low-density theory)—we arrive at a very interesting result.

The only potential which leads to a phase-shift finite as $Q \to 0$ is that which leads to a radial wave-equation whose solution is $J_p(Qx)$ where $p = 2\delta/\pi$, which is noninteger. In fact, the potential is just

$$V_{pseudo} = p^2/r^2. \tag{$I-18$}$$

The radial wave function is equivalent to that which would result from an angular momentum of p. This suggests that there is a core of truth in the "anyon" picture in the following sense: that the *opposite-spin particles* only can be described by a *relative* anyon statistics with $\pi/2 \ p = \delta$, to which is appended a compensating p flux to remove the angular factor. This does not resemble very much the kind of chirality assumed in anyon theories.

The basic renormalization process which gives us Fermi liquid theory is the successive elimination of virtual transitions into states in shells a distance

$$|\epsilon_k - \epsilon_F| = \wedge \to \wedge - d\wedge$$

away from the Fermi surface. For instance, the simplest self-energy bubble diagram for such a transition gives a contribution of order

$$d\sum = \frac{d\wedge}{\wedge} \times \wedge^2$$

with one energy denominator $\sim \wedge$ and two phase space integrals for free momenta $\sim \wedge$ each. This is highly convergent as $\wedge \to 0$, and also $\frac{\partial \Sigma}{\partial \wedge}$ is convergent, for conventional systems and finite interaction vertices. Thus the only relevant energy at the fixed point is the renormalized kinetic energy, and the only even marginal interactions are the Hartree, direct forward scattering terms, which we have just been discussing. This is, in a nutshell, the nature of Fermi Liquid Theory:

$$\text{Fixed point} \propto 1: \sum_k \epsilon_k \, n_k = \text{ kinetic energy}$$

$$\text{Marginal} \propto \omega: \sum f_{kk'} \, n_k \, n_{k'} = \begin{array}{l}\text{Hartree and Fock} \\ \text{forward scattering}\end{array}$$

$$\text{Residual interactions irrelevant}: < \omega \text{ as } \omega \to 0 : \left(\text{e.g., } \frac{\hbar}{\tau}, \text{ etc.} \right).$$

From this point on, there are several ways to go in two or higher dimensions. In one dimension, one finds that the Hartree terms and their long-wavelength extensions ($\rho_Q \, \rho_{-Q}$ couplings) are *always* relevant (at least for realistic models, i.e., all except the Tomonaga model which has only a one-sided Fermi surface). The residual interactions seem, on the basis of exact solutions as well as of arguments put forward by Haldane, to remain at most marginally relevant. Thus the fixed points can be identified by rediagonalization of the kinetic energy together with the coupling of long-wavelength density fluctuations, a rediagonalization which is carried out by the bosonization techniques originated by Mattis, Lieb, Luther et al., and formalized by Haldane in the "Luttinger Liquid" Theory. This rediagonalization involves a Bogoliubov transition mixing boson creation and destruction operators, and this means that while the fixed point Hamiltonian remains harmonic in bosonic variables, the exponents for physical correlation functions differ from one value of the coupling constant to another, implying a continuum of fixed points.

One may summarize the "Luttinger Liquid" point of view by referring to the same RNG which leads to the Landau liquid, effectively a "poor man's" RNG, but recognizing that now the fixed point Hamiltonian $\mathcal{H}$ is both kinetic energy and Hartree interactions, and must be rediagonalized, leading, among other results, to separation of charge and spin in $\mathcal{H}^*$. The divergent forward-backward scattering terms ensure that even at the fixed point level, and even where all scattering is purely forward, k-values outside the Fermi surface must be occupied, and the Fermi surface ceases to be discontinuous. Since this scattering is nondiffractive, at this level a Bethe ansatz is allowed, i.e., the distribution of k-values is not modified by scattering, and in fact this distribution has no singularity whatever at the Fermi surface. (This is the density of k-values in the Bethe ansatz real spacewave-function

$$\sum_{QP} (Q, P) \, e^{i \sum k_{Pi} \, x_{Qj}} ,$$

not the very different momentum distribution $\langle \sum_\sigma c_{k\sigma}^+ c_{k\sigma} \rangle$, which *is* singular at the Fermi surface, the singularity being much enhanced by the spin variables.)

Thus in several ways we can understand what the physics is in 1 dimension, and why it is reasonable to think of these different bosonized systems as "Luttinger liquids," by which we mean systems where the fixed point Hamiltonian still is equivalent to harmonic fluctuations about a Fermi gas state, with linearly dispersing bosonic density waves.

However, it is interesting to realize that contrary to conventional wisdom, it is not possible to get correct results for physical models (as opposed to the Luttinger model) by simple resummations. Nonperturbative methods such as Haldane's give essentially different results. (Metzner and di Castro have shown that perturbation theory can be made correct but only by use of certain Ward identities.)

To what extent is this concept generalizable to two dimensions? We really have no rigorous way, as yet, of doing so; on the other hand, we have three formal approaches, and we have the very strong *experimental* evidence summarized in Chapters 2, 3, and 4 that such a state describes the cuprate systems remarkably well. Finally, we have the argument of this chapter that the same kind of Hartree terms which cause trouble in one dimension are diverging in the two-dimensional case. The extra factor $1/Q$ makes them relevant rather than marginal in this case as well.

The two formal approaches which have a chance to describe the physics properly are the slave-particle gauge theory approach and the tomographic Luttinger liquid.

We want to emphasize strongly that since the gauge theory approach is exact in principle, and the tomographic approach has a sound physical and heuristic basis, they both are very likely to arrive at much the same results in the end, and must not be considered as opposing views. Nonetheless, they start out very differently.   *d

PART II. GAUGE-BASED THEORIES

At this point I would like merely to sketch the gauge theory and then go on to the tomographic alternative. (It was originally intended to include a chapter on it but this has been omitted: my apologies to X.-G. Wen, who prepared an excellent summary.)

We begin with a representation originated by Hubbard and applied here by Zou. (See also Kotliar and Ruckenstein, '88.) We write

$$c_{i\sigma}^+ = e_i s_{i\sigma}^+ + \sigma d_i^+ s_{i-\sigma} \qquad (II-1)$$

etc., where e_i and d_i are hard core bosons and $s_{i\sigma}$ a Fermion. e_i creates an empty site from $n_i = n_{i\uparrow} + n_{i\downarrow} = 1$, d_i a doubly occupied site. This highly overcomplete representation must be supplemented by a constraint,

$$e_i^+ e_i + d_i^+ d_i + \sum_{\sigma \pm 1} s_{i\sigma}^+ s_{i\sigma} = 1. \qquad (II-2)$$

In these terms the Hubbard Hamiltonian may be written

$$\mathcal{H} = U \sum_i d_i^+ d_i + \sum_{ij\sigma} t_{ij} \, (e_i \, s_{i\sigma}^+ + d_i^+ \, s_{i-\sigma}) \times (e_j^+ \, s_{j\sigma} + d_j \, s_{j-\sigma}^+). \qquad (II-3)$$

Zou pointed out that it might be possible to separate, perturbatively, the sector with finite values of $n_{di} = d_i^+ d_i$, by eliminating the terms in $d_i^+ e_i^+$ by a canonical transformation, which in effect reduces (c) to the "$t - J$" model derived by Rice et al. by a similar transformation

$$\mathcal{H} = \sum_{ij} \left(S_i \cdot S_j - \frac{1}{4} \right) n_i \, n_j$$
$$+ \sum_{ij} t_{ij} \, e_i \, e_j^+ \, s_{i\sigma}^+ \, s_{j\sigma}$$
$$(+ t'' \text{ exchange-hopping terms}) + \text{H.c.} \qquad (II-4)$$
$$+ U \sum_i d_i^+ d_i$$
$$+ \text{ other terms proportional to } d^+ d$$
$$= \mathcal{H}_{proj} + \mathcal{H}'.$$

This canonical transformation is convergent so long as there is a ("Mott-Hubbard") gap between the transformed states with finite d and the ("lower Hubbard") band in which the $t - J$ model operates. The earlier discussion of this chapter demonstrates that such convergence exists for all one or two-dimensional one-band Hubbard models, and for large U in other cases. Thus we write, for this separated subspace,

$$\mathcal{H} \to P \mathcal{H} P$$
$$= \text{const} + J \sum_{ij} S_i \cdot S_j + \sum_{ij} t_{ij} \, e_i^+ \, e_j \, s_{j\sigma}^+ \, s_{i\sigma} \qquad (II-5)$$
$$+ \text{ (possible exchange-hopping term)}.$$

With the constraint now reading

$$e_i^+ e_i + \sum_{\sigma = \pm 1} s_{i\sigma}^+ s_{i\sigma} = 1. \qquad (II-6)$$

This projection transformation is physically equivalent to the violent effects of singular forward scattering which we have been discussing: the transformation from the original model is *projective*, i.e., nonunitary, the Hilbert space has been massively restricted, and we expect that Fermi liquid theory or other simple perturbative methods are unusable in general: there can be no unitarily equivalent free Fermi gas. This theorem is widely ignored in the literature, so we may be forgiven here by emphasizing that it *is* a theorem. The existence of an exact identity between the Hubbard Hamiltonian, the $t - J$ Hamiltonian (5), and a local gauge theory means that the Hubbard model is simply outside the region of convergence of perturbation theory, as a mathematically demonstrable fact.

The constraint (II-6) can be enforced, since it is quadratic in field variables, by introducing a common gauge $e^{i\varphi_i(t)}$ for $s_{i\sigma}^+$ and e_i^+, and integrating over φ_i—i.e., introducing a $U(1)$ gauge field connecting the spinon and holon variables, which then carry a common conserved charge (not, of course, related to real charge, which is carried only by the e_i variables).

We then have a Hamiltonian which, as Zou and Baskaran showed, can be written in terms of a gauge field $\vec{a}$ along the links which couples to both e and s. It has been common to follow Baskaran (and Affleck in a different version) in assuming that the "spinons" $s_{i\sigma}$ minimize the exchange term by acquiring a pair amplitude

$$b_{ij} = \langle s_{i\uparrow}^+ s_{j\downarrow}^+ - s_{i\downarrow}^+ s_{j\uparrow}^+ \rangle \qquad (II-7)$$

attached to each bond. This is the "RVB" amplitude; and it is easy to see that mean field theory would lead to a large value for b at any temperature below $\sim J$.

The phase of b, however, cannot be fixed by any interaction because of the local gauge symmetry we have just discussed; only sums of the phases around plaquets are meaningful. Thus $|b|$ is taken (below a crossover at $T = J$) finite and the dynamics is controlled by phases. An extraordinarily complex literature has grown up around the problem of the coupled RVB phase-gauge field-holon problem. Some of the themes are as follows.

(1) Starting with Laughlin's early work, it has been established that in the absence of holons (i.e., the $n = 1$ case) locally stable "flux phases" exist in which the sum of the phases around a plaquet add up to π (for the square lattice) or similar states for triangular lattices. These do not seem to be globally stable vis-à-vis Néel states but are of great theoretical interest.

(2) Lee and coworkers have shown that the flux through a plaquet is related to chirality of the spin state, and that chirality is unfavorable for the mean kinetic energy.

(3) A number of heuristic gauge theories of the metallic state have had some success in describing experimental data.

(4) Finally, following along these lines, and using the assumption that in phases with finite flux a spin gap opens up to give a "short-range RVB," Laughlin, Affleck, Kivelson et al., Wilczek et al., and Wiegmann have all variously suggested chiral symmetry breaking superconducting phases. With some diffidence, and a consciousness that some experimental data may support some such phase, we tend to reject these speculations, at least in the form which they are normally couched, for several reasons.

   (a) Experimentally, the existence of constant χ_{spin} in the best superconductors, and of Korringa relaxations for most nuclei, indicates the presence of a free spin Fermi surface in the normal state. ARPES data support the presence of a Fermi surface for electrons as well. Thus experiment tells us there is no gap in the normal state. We reject the possibility that the superconducting gap is the RVB gap primarily because the scale is all wrong: RVB energies are $\sim J$ or t, i.e., $>1000°$ K, while the superconducting T_c fluctuates wildly from material to material. The suggestion that only the superconducting state is strange is completely contrary to the dogmas of chapter 2.

   (b) Kinetic energy seems to us clearly to dominate the physics in the "good" superconducting regime which is characterized by $t\delta > J$. Kinetic energy will clearly be disfavored by flux = chirality as much as spin ordering is favored, since the terms are of similar form and opposite sign.

(3) Perhaps most seriously of all, I believe most gauge-flux phase theories up to now are theoretically oversimplified in a crucial way, in that they are based on an oversimplified picture of the symmetries, as well as leaning far too heavily on the ubiquitous spin gap and local RVB assumptions. These assumptions are natural for $SU(N)$ spins for large N, but $SU(2)$ is qualitatively different in that because of the enormous overcompleteness of the pair-bond representation, pair bonding can be quite strong locally without opening up a gap—i.e., it can be *both* local and long-range.

The symmetry problem is the following. At $n = 1$ there is an extra symmetry caused by an extra redundancy of the spinon representation. For $n = 1$, it is clear that the operator $s_{i\sigma}$ has exactly the same effect as $s^+_{i-\sigma}$, so that one could actually use any linear combination of the two:

$$s_{i\sigma} \rightarrow u s_{i\sigma} + v s^+_{i-\sigma}$$
$$|u|^2 + |v|^2 = 1.$$

It is easily verified that this leaves $n_{i\sigma} + n_{i-\sigma} = 1$. This is an $SU(2)$ gauge symmetry, not a $U(1)$ gauge symmetry. Upon doping, only one component of this gauge field is mixed with the holon. In the half-filled, $n = 1$, case, this is the symmetry responsible for halving the number of spinon excitations, so that the spinon branches terminate at $k_F = \pm\frac{\pi}{2}$.

Upon doping, the exact 1D solution still retains the trace of this redundancy, in the exact spin-charge separation, and the continued restriction of the spinons to the original k_F, with higher k's obviously redundant. This does not follow in any obvious way from the present gauge theories. Spin-charge separation means that the spin excitation spectrum still has the redundancy which followed from SU2, but the SU(2) symmetry is not included in the group which is normally used. This is a vital feature of the real system and we wonder if the present gauge theories are complete until they describe it properly. A possibility is that in the doped system this becomes not an exact theorem but a symmetry of the ground state in some sense, in the same sense that the spin problem becomes the "squeezed Heisenberg" problem when looked at from the point of view of low frequencies and long wavelengths, but locally we have considerable granularity. Thus the equivalent linearized spin Hamiltonian may have SU(2) symmetry at the Gaussian fluctuation level, but not overall. Such a situation would be hard for gauge theory to sort out. (But see recent work of Wilczek and Zou.)

PART III

(A) Introduction: The Physics of the Hubbard Model

Before beginning part III I'd like to give an overall view of the physics of the 2D Hubbard model. It is worthwhile to think in terms of the "$t - J$" model, which we have not introduced elsewhere because the additional perturbative projective transformation which leads to it is not really necessary and introduces new parameters, but this transformation is a good way to visualize what is going on. A canonical transformation eliminating double occupancy leads to the Hamiltonian

$$P_G t \sum_{i\tau\sigma} c_{i\sigma}^+ c_{i+\tau\sigma} \, P_G + J \sum_{i\tau} S_i \cdot S_{i+\tau} - \frac{n_i \, n_{i+\tau}}{4} \qquad (III-1)$$

(+ second neighbor hopping which is normally treated as simply adding to that already present).

The perturbation theory leading to (1) converges if there is a gap to doubly occupied states. This form of the effective Hamiltonian brings out a vital aspect of the Hubbard model problem: its projective nature. (1) shows that U may always be considered to renormalize to ∞ so long as this perturbation theory converges; the J term is then an effective attractive neighbor interaction to correct for eliminating double occupancy entirely. This means that the theory is *always projective*: Hilbert space changes with the addition of particles. This in turn means that conventional many-body perturbation theory is *never* automatically convergent, because it contains no provision for variable Hilbert spaces;

equivalently, there can, if U is large enough, always be (anti)bound states, which represent essential singularities of the scattering problem. It is a sound conjecture that antibound states are a sufficient condition for nonconvergence.

The projective condition $n_{i\uparrow} n_{i\downarrow} \equiv 0$ introduces N constraints. This is equivalent to a new local gauge symmetry. Introducing an auxiliary field to enforce the constraint seems not to have been a very useful exercise, since it does not linearize the problem in Fermions. More useful have been the slave Boson or slave Fermion methods, but these too introduce gauge fields which we now realize *cannot* be validly treated in mean field theory, although that was the fashion a few years ago in the heyday of mixed valence theory. The essence of these gauge fields introduced by the constraint is that, in the end, they cause long-range interactions which modify the low-frequency physics in essential ways which cannot be mimicked by mean fields, just as electrostatic fields essentially modify ordinary physics. Our program here is to deal with the constraints directly, but gauge field methods might give correct answers if followed out correctly.

A second feature that will be best understood by thinking about (III-1) is the role of chirality. In a true Hubbard model, $J < \frac{1}{2}t$ and J is to be considered a relatively small parameter: spins are considerably softer than charges. The "chiral" theories of part II normally make the opposite assumption and are unphysical, at least in the cuprates. If the spin configuration is "heavy," we should treat the rapid motions due to "t" as taking place in a slowly moving spin background, and causing large differences in energy for different spin structures, hence effectively constraining the spin structure.

What property of the spin structure most effectively obstructs the free motion of the particles? It is clearly not simple antiferromagnetism, because in one dimension the particles move equally freely, independently of the spin structure, because only straight-line hopping paths are allowed. It turns out that the key property of the spin structure is chirality, because it can be shown that it gives the phases associated with the paths of electron motion. The phase factor accompanying the motion of an electron around a given path can be computed in the following way: we imagine a coordinate system in spin space following the electron's orientation from site to site, with the electron's local spin along the z axis. To reverse the spin this coordinate system will have to rotate around its x-axis, for instance. The sum of such rotations along a given path determines, in the usual way, the chirality of a path, which tells one how the kinetic energy coming from different paths is reduced by destructive interference. The kinetic energy is of course optimized by either having all spins parallel—with no chirality at all, but a very large extra kinetic energy because all states are only singly occupied—or by having a spin structure as close as possible to that of the Fermi sea, where chirality must be minimized. Chirality is also minimized by having as many singlet pairs as possible, since the chirality operator for three spins in a triangle is $S_i \cdot S_j \times S_k$, which vanishes if any pair is in a singlet. On the other hand, superexchange favors chirality, because then hopping into opposite-spin states is maximized—i.e., superexchange lives on the kinetic energy lost by creating chiral states. Thus exchange and kinetic energy are opposite in effect, and chiral states will only occur when the effect of kinetic energy is minimized in some way, in complete contradiction to speculations that doping favors chiral states.

What *is* relatively favorable for both superexchange and kinetic energy is singlet pairing as in the Bethe solution and the Lieb-Wu solution which closely resembles it. The secret of this compatibility is the absence of closed, nonre-tracing paths of electron motion, hence the absence of the possibility of chirality blocking such paths.

The physics of the Luttinger liquid ground state is now clear. In the un-perturbed Fermi liquid, kinetic energy is optimized under the constraint of the exclusion principle. Excitations far below the Fermi surface are pushed to very high energy and can be renormalized away; excitations near the Fermi surface experience only forward or backward scattering because of momentum conser-vation. Thus all of the *relevant* paths are, again, effectively one-dimensional retracing paths, and there is no conflict between exchange and kinetic energy— in effect, in this situation spin and charge motions are decoupled as they are in one dimension and *as they are*—except for irrelevant residual interactions—in the Landau liquid as well. That is: in essence the only difference between Landau liquid and Luttinger liquid is that in the former spin and charge move at the *same* velocity, even though effectively decoupled.

(B) Fixed-Point Level Theory of the Luttinger Liquid

The form of the Hartree terms which we derived in part I verifies our expectation that the breakdown of Fermi liquid theory occurs because these terms, which are marginal in Landau theory, become relevant. This is signaled by the fact that $f_{kk'}$ diverges as $1/|k - k'|$ and therefore its effect does not vanish as the cutoff $\wedge$ goes to zero. $f_{kk'}$ is part of the fixed point Hamiltonian $\mathcal{H}^*$ and must be diagonalized along with the free-particle part $\mathcal{H}_0$. To this problem of divergent coupling we can see two plausible solutions (in addition to gauge theory): I. a modified Bethe ansatz; and II. rediagonalization by Bosonization, or Luttinger liquid theory. Best of all, one may do both.

We need to convert the simple Landau parameter into an interaction. This may be done by wave-packet techniques. We form a wave-packet of size R of waves around $k \uparrow$, and ask what the modification in the potential for a wave-packet of states around $k' \downarrow$ is. Neither k nor k' can be defined better than $\sim \frac{1}{R}$. We define, then, a $\rho_{k\uparrow}(r)$ and a $\rho_{k'\downarrow}(r')$ where r and r' are coarse-grained variables defined to within R, and we have

$$V = \frac{\delta}{\pi} \int dr' \, dr \, f(r \mid r') \, \rho_{k\uparrow}(r) \, \rho_{k'\downarrow}(r') \, \frac{\widehat{k - k'} \cdot \widehat{(k + k')}}{\sqrt{(k - k')^2 + \frac{1}{R^2}}} \qquad (III-2)$$

where $f(r - r')$ is a δ-function broadened to width R. Fourier transforming this, we see that it is equivalent to

$$V \equiv \frac{\delta}{\pi} \sum_{kk'Q} \rho_{k\uparrow}(Q) \, \rho_{k'\downarrow}(-Q) \frac{\widehat{(k - k')} \cdot \widehat{(k + k')}}{|k - k'|}. \qquad (III - 3)$$

Now we use

$$\rho_{k\sigma}(Q) = c^+_{k+Q\sigma} \, c_{k\sigma}$$

and obtain

$$V = -\frac{\delta}{\pi} \sum_{kk'Q} c^+_{k+Q\uparrow} c_{k'+Q\downarrow} \, c^+_{k'\downarrow} c_{k\uparrow} \frac{\widehat{(k - k')} \cdot \widehat{(k + k')}}{|k - k'|}$$

$$= \frac{\delta}{\pi} \sum_{Qk} \sigma_{kQ} \cdot \sigma_{k'Q} \frac{\hat{Q} \cdot \widehat{|k + k'|}}{Q}. \qquad (III - 4)$$

Thus the Landau-type interaction of opposite spins may be converted into a
*e   divergent spin density-spin density interaction for purposes of bosonization.

Actually, this kind of manipulation is not necessary to get a pretty clear idea of how to bosonize. I like to return to the source of the effective interaction, namely, the principle of Hilbert space incompressibility. Putting our up spin at momentum k has the effect of shifting all s-wave scattering states of $\downarrow$ spin by $k \to k + \frac{\delta}{L\pi}$, which leads to an effective "incompressible" motion of k-values out in the radial direction by

$$\delta k' = \frac{\delta}{2\pi L^2} \left(\frac{1}{|k' - k|} \right). \qquad (III - 5)$$

Let us define

$$\vec{Q} = \vec{k}' - \vec{k} \qquad (III - 6)$$

and consider a k-value quite close to the Fermi surface, for which $Q \ll k_F$. Let

$$cos \, \theta = \frac{Q \cdot k}{Qk} \qquad (III - 7)$$

and we note that the energy shift of k' is proportional to

$$\delta E_{k'} \propto \frac{\delta}{\pi} \frac{cos^2 \theta}{Q}. \qquad (III - 8)$$

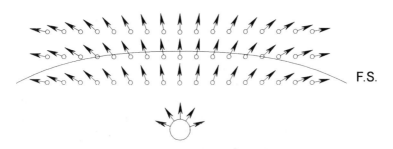

Figure 6.4. States are displaced incompressibly in k-space by the excluded volume around an added opposite spin particle.

If we could approximate the Fermi surface by a straight line the total energy shifts integrated over the Fermi surface amount to (see fig. 6.4)

$$\Delta E_{tot} \propto \frac{\delta}{Q\pi} \int dk'_\perp \cos^2 \theta$$

but $k'_\perp = Q \tan \theta$ so that

$$\Delta E_{tot} \propto \frac{\delta}{\pi} \qquad\qquad (III - 9)$$

independent of Q. Thus $\frac{\delta}{\pi}$ of a state is pushed out of the Fermi surface. This is valid in the limit $Q \ll k_F$, and in fact the curvature of the Fermi surface will introduce corrections of order $\sqrt{Q/k_\perp}$ only, as we now show.

The tomographic method is based on the idea of Luther, who has shown that a complete set of variables describing low-frequency fluctuations of the Fermi gas can be made up out of a set of "tomographic" density waves referring to each direction around the Fermi surface, called

$$\rho_\Omega^\sigma (Q). \qquad\qquad (III - 10)$$

These ρ_Ω's have the commutation relations, velocities, and dynamics of one-dimensional Tomonaga bosons, not 2-dimensional RPA density waves, because they are understood to be made up from Fermion operators only in the region of momentum space near the Fermi surface at direction Ω, hence have a unique velocity. We may think of the one-dimensional density waves $\rho_\Omega^\sigma(Q)$ as being made up from Fermion states which are one-dimensionally localized wave packets, transversely localized within a distance Δy which is restricted by the transverse mass and the uncertainty principle.

$$(\Delta y)^2 \sim \frac{1}{(\Delta Q_\perp)^2} \qquad\qquad (III - 11)$$

with

$$(\Delta Q_\perp)^2 \ll k_F Q \qquad (III - 12)$$

for a wave packet at $k_\parallel = k_F \pm Q$.

This means that the transverse kinetic energy, which measures the rate of diffusion of the wave packet in the y direction, is much less than the longitudinal energy if $Q \ll k_F$, so that the one-dimensional approximation is valid for excitations close to the Fermi surface and for long wavelengths Q^{-1}, in the sense that fluctuations remain one-dimensionally localized for a time long compared to their period of motion, and exactly so right at the Fermi surface.

From this point on, the procedure is just the same as the standard one-dimensional bosonization. The $\rho_\Omega^\sigma(Q)$ are recombined into bosonic variables representing spin and charge density waves which are the elementary excitations of the Fermi sea; at each point Ω on the Fermi surface these bosons move with two different Fermi velocities, v_s and v_c, one for charge and one for spin bosons. For these bosons the effective Hamiltonian is just harmonic with linear dispersion relations; but the physical operators such as electron fields $c_{k\sigma}^+$ are rather complex expressions because of the Bogoliubov transformation which connects the physical bosons with the bare ones.

To a lowest approximation and precisely at the Fermi surface, the Green's function is simply that of the corresponding one-dimensional system, since the operator $c_{k\sigma}^+$ for $k \simeq k_F$ is contained only in the $\rho_\sigma(\Omega)$ for $\vec{k}$ near Ω. Other, bosonic response functions such as density-density or spin-spin correlations are suitable angular averages over the different directions Ω, computed fairly straightforwardly as follows. We first consider, as Luther does, the real space-time correlation functions $C(r - r', t - t')$ resulting from exciting a spin or density fluctuation at point r and time t and measuring it at r' and t'. In propagating from r to r' we can assume that the waves travel as one-dimensional wave packets with velocities along the direction $\hat{\Omega}$ of $\vec{R} = \overrightarrow{(r - r')}$, but we must normalize by a factor $\frac{1}{2\pi R}$ because of the decreasing angular volume element with R. Thus we write

$$C(R, t) = \frac{1}{2\pi R} C_{1D}(\hat{\Omega}, R, t). \qquad (III - 13)$$

Then to get the correlation function as a function of Q and ω, we merely Fourier transform:

$$C(Q, \omega) = \int d^2 R \int dt \, e^{i(Q \cdot R - \omega t)} C(R, t) \qquad (III - 14)$$

$$= \int d\theta \, C_{1D}(Q \cos \theta, \omega).$$

This means that the singularities of the 1D models are mitigated in 2 and 3D by one-half and one power, respectively. Thus the $\frac{1}{\omega - Qv_s}$ singularity of the spin-spin correlation of the 1D Heisenberg (or half-filled Hubbard) model at $2k_F$ is mitigated in two dimensions and becomes $\frac{1}{\sqrt{\omega - Qv_s}}$, which is still a power-law singularity at all spanning vectors and corresponds to the observation that *all* half-filled Hubbard models are antiferromagnetic, irrespective of Fermi surface shape. The doped Hubbard model in 1D exhibits a $2k_F$ spin susceptibility singularity as $(\omega - Qv_s)^{-\eta}$ with $0 < \eta < 1/2$. $1/2$ is the value near half-filling for large U. This becomes a $\ln \omega$ singularity for small doping, which is of BCS type and will mean that *most* lightly doped systems will order magnetically. For heavier doping or weaker U, their singularity at $2k_F$ becomes a mere peak, and the system will have a tendency to antiferromagnetic fluctuations or ordering, with no actual symmetry breaking. The correspondence with the general run of experimental results is striking. It appears that in the best superconductors, however, the $2k_F$ fluctuations have become relatively harmless.

The same prescription does not work for the Green's function itself. The excitation to which the system is responding in this case is that of insertion of a Fermion of specific momentum, not a perturbing field acting on all the particles. Thus the initial excitation affects only the bosons associated with the particular direction $\hat{\Omega}$ along which the momentum of $c^+_{k_F+Q,\sigma}$ (say) is directed, and is localized in momentum, not position, space. However, for finite Q there are displacements of the opposite-spin Fermi surface out to $Q'_\perp$'s of order several Q, and as envisioned in some of our earlier work, this will cause holon and spinon excitations which are not quite colinear. This transverse coupling parallel to the Fermi surface is an effect which should be small but for which we have developed no useful formalism. The observed photoemission line shapes may be showing this effect in the low-frequency tail extending down to $\omega = 0$. A purely ad hoc fit to the observed curves may be obtained by weighting the integral (III-14) for the response function with an extra factor $\cos\theta$:

$$G(Q, \omega) = \int d\theta \cos\theta \; G_{1D}(Q\cos\theta, \omega). \qquad (III-15)$$

This fit is shown in the fig. 6.5 for $v_c \gg v_s$ [hence $G_{1D}(Q, \omega) \sim \omega^{-\frac{1}{2}+\alpha}(\omega - qv_s)^{\frac{-1}{2}}$] but we confess that it is purely a guess. It fits nicely within our guidelines: it reduces to G_{1D} as $Q \to 0$, and it is much more directed in Ω space than the simple angular average. But for justification—if that is needed—we shall have to await a theory at better than the fixed point level.

Many photoemission data show much more extensive structure at low energies over wide areas in k space. I believe this is also a manifestation of the holon-spinon structure of Im G. A paper is in preparation by S. Strong and myself. The essence of our argument is that noncircular geometry of the Fermi

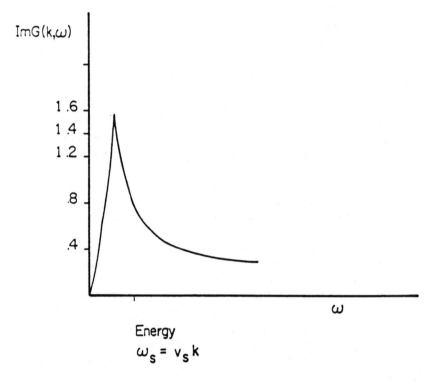

Figure 6.5. An estimate as described in the text of Im $G_1(k, \omega)$ after averaging over the angle between holon and spinon in 2D. The weighting factor is not calculated but chosen to fit some of the data.

surface, especially planar "nesting" sections, can lead to "ghost" Fermi surfaces
*f   and to structures of the type observed.

PART IV. TRANSPORT PROPERTIES OF THE 2D LUTTINGER LIQUID (WITH SOME REMARKS ON 1D); THE IMPORTANCE OF "HOLON DRAG" AND "HOLON PINNING"

A. Introduction

A bewildering variety of experimental anomalies of transport properties are characteristic of the states of materials which have been identified as "Luttinger Liquids." The most characteristic and most clearly anomalous properties are those of the $Cu O_2$ layers in high T_c superconductors, which show anomalies in resistivity, in infrared conductivity and Raman effect, in the anisotropy of resistivity, in the Hall effect, and in the thermopower.[1] Other less well characterized anomalies are seen in many quasi-one-dimensional materials, for instance, anomalous conductivity in certain polymers,[2] anomalous anisotropy of resistivity in TTF-TCNQ and Bechgaard salts, etc. Some 3d materials show anomalous infrared conductivity, and many heavy-electron materials show anomalous T-dependence of Hall effect and resistivity.[3]

We here focus on the one- and two-dimensional Hubbard model and specifically on the 2d case which is assumed to represent the CuO_2 layers. The properties of the layers are shown to follow from the separation of charge and spin at the Fermi surface along with the accompanying smearing of that Fermi surface which is caused by the anomalous dimension of the charge excitations. Similar calculations can be carried out for one-dimensional Luttinger Liquids, for which the Luttinger Liquid character is more familiar. However, in that case the calculations are more delicate and more restricted in the experimental regime to which they apply than the two-dimensional case because of the peculiar properties of one dimension, which cause even weak scatterers to renormalize to strong ones, and enhances localization effects. However, the theory given here provides a much better guide to one-dimensional physics than previous ones.

Even in two dimensions we must recognize that we can be sure of ourselves only in a restricted range of temperature and of strength of impurity scattering, which fortunately is the experimental area for most of the time. The elastic or quasi-elastic scattering must be strong enough to eliminate "drag" effects, which turns out to require

$$\frac{\hbar}{\tau_{e\ell}} > \frac{\hbar}{\tau_{\text{spinon-holon}}} \simeq \frac{(kT)^2}{t_{||}} \quad . \qquad (IV-1)$$

This sets an upper limit on T, which fortunately is not very tight in the cuprates, if not in 1d materials. (The conductivity is *higher* above this limit than we calculate). A lower limit on T is set by localization effects,[†] not very severe in 2d (though observed in $Bi\ 2\ 2\ 0\ 1$) but of order $T > \frac{\hbar}{\tau_{e\ell}}$ in 1d. Above this temperature we assume we can treat the scatterers perturbatively. Here $\tau_{e\ell}$ is the elastic or quasielastic scattering time which would result from random potentials in the absence of many-body effects.   *g

---

[†]A complication of localization effects in both 1 and 2d in the Hubbard model is that a localized state near the Fermi level is necessarily magnetic—it becomes an Anderson model—and introduces T-breaking scattering. In fact, it may be necessary, in order to resolve such questions, to reintroduce, at a mean field level, screening of long-range Coulomb forces leading to the Friedel sum rule, which would prevent simple scattering potentials from acquiring an odd number of electrons and becoming magnetic (but would ensure that wrong-valence impurities such as Ni and Zn substitutions for Cu were magnetic).

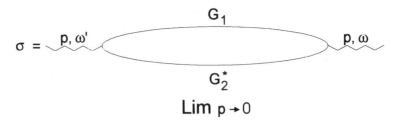

Figure 6.6.

B. Parallel Conductivity

The fundamental relaxation process appears most directly in calculating the conductivity along the chains or planes. This process can be thought of either as the decay of the electron into holons and spinons or as the scattering of holons on the gauge field of the spinons. (The latter has been the approach of Nagaosa and Lee,[4] Wiegmann et al.,[5] and others, who have attempted these calculations using gauge theory in a high-temperature approximation. At high T, where this may be valid, it gives similar results to our methods.)

The conductivity is of course correctly evaluated in terms of the diagram in fig. 6.6 for the d.c. limit of the current-current correlation function.

$$\sigma \propto \lim_{\substack{q \to 0 \\ \omega \to 0}} \langle j \; j \rangle (q, \omega) \qquad (IV-2)$$

where

$$j(q) = \sum_{k\sigma} k c^+_{k+q,\sigma} \, c_{k,\sigma}. \qquad (IV-3)$$

In the evaluation of this diagram we come across an apparent paradox. For reasonably low temperatures and pure samples, the momentum k must be very close to the Fermi surface, because the excitation energy spectrum either for the Landau or the Luttinger Liquid has zeroes only at k's near the Fermi surface. q is of course small. Thus when we draw the appropriate correlation function as a two-particle Green's function (see fig. 6.6)

$$G_2(k, k+q \; ; \; k' + q, k') = G_2(1, 2 \; ; \; 3, 4) \qquad (IV-4)$$

the vertices 1, 2, 3, and 4 have only one degree of freedom in relative momentum space and are therefore all very distant in real space (by the uncertainty principle). Writing σ as a current-current correlation function assumes that vertices 1 and 2 are relatively close, as are 3 and 4; this is the basic distinction among the various correlation functions. As pointed out by Yakovenko and Brazovskii,[6]

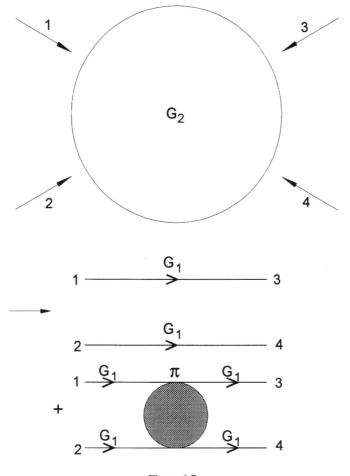

Figure 6.7

if the 4 vortices are at *general* points in space the two-particle Green's function is dominated by the product of two one-particle Green's functions

$$G_2(1, 2 \, ; \, 3, 4) \simeq G_1(1, 3) \, G_2(2, 4) \qquad (IV - 5)$$

(see fig. 6.7) because "vertex corrections" are in general less singular for small momentum transfer, i.e., less long-range in space.

Thus, apparently, we can calculate σ as an average of the simple product of two one-electron Green's functions, i.e., by using "Landauer formula" physics, emphasizing the propagation of single electrons, not "Kubo formula" physics which uses the current-current correlations. It is obvious that the Landauer formula leads, for the Luttinger liquid, to the fundamental decay process

$$\text{electron} \rightarrow \text{spinon} + \text{holon}$$

with decay rate $\approx \omega$ or $\frac{kT}{\hbar}$. This then determines the transport mean free time $\sim \omega, T$ *apparently independently* of extraneous scattering processes, and hence even in a perfectly periodic, pure sample.

In the pure sample it is possible in one dimension to solve exactly for the conductivity (Stafford and Millis)[7] and this is

$$\sigma = \frac{ne^2}{m^*} \delta(\omega) \qquad (IV-6)$$

where m^* is a density dependent constant calculated by them for the 1d Hubbard model (m^* goes to ∞ at half-filling). For the "Landauer" physics, on the other hand, one gets (IV-6) with

$$\delta(\omega) \rightarrow \alpha \omega^{-1+2\alpha}. \qquad (IV-7)$$

This discrepancy is very straightforwardly resolved as a "drag" effect: both answers are correct in the appropriate regime. In a perfectly periodic sample there are Luttinger-Ward identities which connect vertex corrections to the momentum derivative of the one-particle Green's function in just such a way as to restore translational invariance and the result (IV-6). (For this observation I am indebted to J. R. Schrieffer.) However, in a slightly impure sample, the vertex corrections fall off as

$$\exp -\frac{|x_1 - x_2|}{\ell_{elastic}}$$

which (we will show) makes them vanish where condition (1) is satisfied.

Let us try to understand these long-range vertex corrections. (Here we follow work of M. Ogata.)[8] It is easiest to visualize them for the phonon case, where in fact the same paradox arises: in a perfectly periodic system, the current cannot decay, yet the conventional calculation gives finite resistance. The reason is that the phonons propagate momentum until they are scattered by the electrons, returning their momentum to the current. The resulting correction is important if

$$\ell_{elastic}^{(phonon)} \gg \ell_{\substack{phonon-electron \\ scattering}} \qquad (IV-8)$$

which is the condition for "phonon drag" effects. Because the density of states for phonons is much greater than for electrons, $\ell_{elastic}$ can be much longer than the electron mean free path and still not satisfy (IV-8).

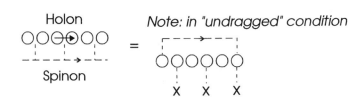

Figure 6.8. Decay of electron into holon and spinon.

In our case, the relevant criterion involves the mean free path of the elementary excitations into which the electron decays. The fundamental dissipative process is the decay (at rate $\sim \omega^{1-\alpha}$) of the electron into its constituent exact boson eigenexcitations, and these in turn can be thought of as scattering against each other in pairs, thus reconstituting the electron (see fig. 6.8). This process (which is simply spinon-holon scattering) is only taking place at a rate $\propto T^2$, because both spinon and holon are behaving as though they were Fermions. (This is to be distinguished from the holon-spinon scattering of Nagaosa and Lee, where the gauge field has not been eliminated. In our theory, the constraint is satisfied explicitly: the current is a current of physical electrons.) Another way to see it is that effectively all channels are wide open: all processes take place at rates essentially equal to phase space; but the holon-spinon combination has a larger phase space by a factor T^{-1}, hence the inverse process is lower by that factor.

This is a lot of argument to lead to a very simple result, namely, that by essentially dimensional analysis, the conductivity δ-function (IV-6) is replaced by using (IV-7)

$$\sigma = \frac{ne^2}{m^*}\,\alpha\omega^{-1+2\alpha} \qquad (IV-9)$$

so long as

*h

$$\frac{\hbar}{\tau_{el}} \gg \frac{T^2}{t_{||}}. \qquad (IV-10)$$

(Where here we of course think of ω as max (T, ω); it is the characteristic energy of the Fermi distribution of electrons.) But this *is* the key, central dissipative process which controls almost all of the transport properties of the Luttinger liquid. Most external perturbations act directly on the electrons, not on their charge or spin separately, so the fastest entropy-producing process after such a perturbation is the decay of the accelerated electron into charge and spin excitations. Again, it is this decay which disappears below T_c because of the reappearance of quasiparticles. (By exchanging charge with the holon condensate, spinons become charged quasiparticles.) This is the characteristic behavior of the high T_c materials, rapid decay by an *electronic* process above T_c, which disappears in the superconducting state with a very steep temperature dependence. And this is the process which controls conductivity along the planes (and when relevant, the chains) in the cuprate materials.

C. *C*-Axis Resistivity

Conductivity along the c-axis in the fully metallic cuprates occurs by interplanar tunneling processes. The simplicity of the situation is obscured by the fact that the hopping matrix elements "$t_\perp$" are very dependent on detailed structure. In $(LaSr)_2CuO_4$, $t_\perp$ is at least a single function of $k_\parallel$, since all CuO_2 planes are crystallographically identical. The best estimate for its value is $\frac{t_\perp}{t_\parallel} \sim .1$, from both interlayer exchange estimates and band calculations; numerical values are $t_\parallel \sim .5ev$, $t_\perp \sim .05ev$.

In the multilayer cuprates $t_\perp$ between close layers $[Cu(2) - Y - Cu(2)]$ is $t_\perp \gtrsim .1ev$, roughly twice the value for 2 1 4; between distant layers, it varies wildly. In Bi systems it must be very small, perhaps $t_\perp < 10^{-2}t_\perp$; where the intermediate layers are admixed near (or at) the Fermi level, as in Tl and $YBCO$ materials, it must vary widely but is always $< t_\parallel$. In all of these cases the c-axis resistivity is controlled by this smaller value.

The conductivity due to interlayer tunneling has two regimes with a crossover. The difference between the two regimes is whether the basic decay time is large or small compared to $t_\perp$:

$$kT \simeq \frac{\hbar}{\tau} \gg t_\perp \quad : \quad \text{incoherent regime} \qquad (IV-11)$$

$$kT = \frac{\hbar}{\tau} \ll t_\perp \quad : \quad \text{confined regime .} \qquad (IV-12)$$

The theory of tunneling conductivity in the incoherent regime has been given by N. Kumar and coworkers.[9] In this regime one calculates it using the Schrieffer formula appropriate to "Giaever" tunneling between two metal layers:

$$\sigma \propto \frac{e^2}{\hbar} \sum_{kk'\omega} |t_{\perp_{kk'}}|^2 \frac{f(1-f)}{kT} (\omega) \, \delta(\omega - \epsilon_k)\delta(\omega - \epsilon_{k'}) \qquad (IV-13)$$

and assuming, for the case of interlayer conductivity, that

$$t_{\perp kk'} = t_\perp \delta(k - k') \qquad (IV-14)$$

we get

$$\sigma \propto \frac{e^2}{\hbar} \mid t_\perp \mid^2 \sum_{k\omega} \mid Im \; G \; (k, \omega) \mid^2 \frac{f(1-f)}{kT}. \qquad (IV-15)$$

As Kumar shows, with G having an effective width of $kT \approx \omega$, this peaks up and gives a conductivity

$$\sigma \simeq \frac{e^2}{\hbar} \frac{t_\perp^2}{t_{\parallel}T} \left(\frac{T}{t_{\parallel}}\right)^{2\alpha}. \qquad (IV-16)$$

(The last factor is slowly varying, but almost certainly there; it corresponds to the slight nonlinearity in T of the parallel conductivity which is also observed.)

The reason why simple incoherent perturbation theory is correct is that in this regime the electron of momentum k which tunnels is so smeared out in frequency that the energy of tunneling is negligible in comparison, so that whether or not there is a Fermi surface displacement by $\delta\epsilon_k \simeq t_\perp$ is not relevant. Another description of the same concept is that the electron decays before it has a chance to be reflected coherently back to the first layer, so that its motion from layer to layer is purely diffusive. We emphasize that the T-dependence of the conductivity, while it mimics that of the parallel conductivity in the same region, does not imply any coherent, dynamical "Fermi velocity" in this direction: there is no persistence of motion through more than one layer. The regime is not "metallic." Note that

$$\sigma \ll \sigma_{Mott} = \frac{e^2}{\hbar} \qquad (IV-17)$$

by virtue of the condition $T > t_\perp$. The low-temperature boundary of this regime is at $T \sim t_\perp$ where

$$\sigma|_{crossover} \simeq \frac{e^2}{\hbar} \frac{t_\perp}{t_{\parallel}} \left(\frac{t_\perp}{t_{\parallel}}\right)^\alpha \qquad (IV-18)$$

which allows us in any given case to estimate $t_\perp$ from conductivity data.

Below the crossover the behavior of the conductivity may be estimated in the following way. As we have shown, at low temperatures there is no 3-dimensional structure to the Fermi surface and coherent electronic motion in the third dimension does not take place.[10] There is however, of course, virtual hopping and an energy response which results from it. If, in the course of a virtual excursion between planes, the electron is scattered inelastically, a conduction process can take place. Since the virtual processes spread over all

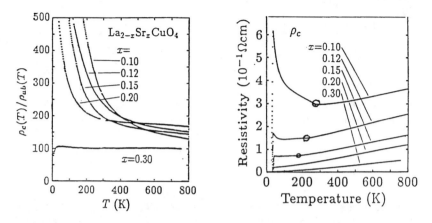

Figure 6.9. For $(LaSr)_2CuO_4$ the data are equivocal because the tetragonal-orthorhombic transition is probably also a 2d-3d one, and ρ_c curves have a pronounced break at that transition. Fig. 4a is probably the best for estimating $t_\perp$ as the temperature at which ρ_c/ρ_{ab} is around twice its asymptotic. For reasons given in the text, $t_\perp$ estimates from multilayers are not useful but the plots given in earlier papers of $\rho_c T$ vs T^2 are good evidence for the theory of the text.

energy scales, they are essentially operating continuously and do not depend on the magnitude of $t_\perp$ very strongly, and thus the conductivity is simply given by the rate of the basic inelastic process:

$$\sigma \simeq \frac{e^2}{\hbar} \left(\frac{T}{t_\parallel} \right)^{1-\alpha}. \qquad (IV-19)$$

*i   Another way of expressing this is that in this regime, because $t_\perp$ is relatively large compared with electron decay rates, the electron is equally likely to decay into a spinon and holon in the next layer as in its own. This is a *resistivity* process in-plane, but a *conductivity* process between planes.

We see that indeed the two formulas come together at the crossover conductivity (IV - 18). The crossover seems likely to occur at $T < t_\perp$ by factors of order unity. Fig. 6.9 is a plot of actual data. There seem to be no serious contradictions in the data except for one. BISCO in general has a very low, approximately activated conductivity ρ_c, even though $t_\perp$ is very small: Giaever tunneling according to (IV-16) seems to be absent. The probable explanation is that $\frac{t_\perp^2}{t_\parallel^2}$ is so small that conductivity is occurring primarily through defect regions with large local $t_\perp$; or else phonon-assisted tunneling is dominant. All
*j   other observations seem quite consistent with these ideas.

D. Hall Effect

I have given some preliminary qualitative discussion of the T-dependent Hall effect[11] but I have never written down the appropriate kinetic equations which are required to understand signs and magnitudes. Doing this is very enlightening and very much clarifies the situation.

First of all, there is one geometry in which the Hall effect result is extremely simple and straightforward. This is that in which current and magnetic field are parallel to the planes (but of course perpendicular to each other) and the Hall voltage is measured along the c-axis, i.e., between the stacked planes.

In this geometry, the voltage is measured between planes and one is reasonably sure that it measures only the forces on single tunneling electrons whose charge and spin are not separate excitations. Thus we may assume that an absence of current in this direction (which we will call the z-axis to avoid confusion) implies $\dot{p}_z = \dot{v}_z = 0$. Now

$$0 = \langle \dot{p}_z \rangle = n e \left(E_z - \frac{\langle v_a \rangle}{c} \times H_b \right) \qquad (IV-20)$$

so that, in turn, the Hall field tells us the mean velocity of the *electrons* which are carrying the current. If, as we expect, the Fermi surface is simple and electron-like in the planes,

$$R_\perp \approx -\frac{1}{ne} \qquad (IV-21)$$

where n measures the area of the *electron* Fermi surface; this is indeed what is observed. It would be fascinating to do the same experiment on "n-type" materials, where the band is more than half full: I predict that a positive c-axis Hall effect would happen.

*k

The Hall effect in the planes is much more interesting. It has long been a puzzle that in the best single crystal samples, $R_H \propto \frac{1}{T}$ over a wide range of temperatures. Even more striking is the more recently discovered generalization that $\theta_H \propto \frac{1}{T^2}$ quite accurately over a wide range for a large variety of pure samples, and that doping with pair-breaking impurities leads to $\theta_H \propto \frac{1}{a+bT^2}$, i.e., one may define a τ_{tr}, with

$$\theta_H = \omega_c \tau_{tr} \qquad (IV-22)$$

and

$$\frac{\hbar}{\tau_{tr}} = \frac{1}{\tau_{imp}} + \frac{1}{\tau_{int}} \qquad (IV-23)$$

and

$$\frac{1}{\tau_{int}} \simeq \frac{T^2}{t_\parallel}, \quad \frac{1}{\tau_{imp}} \simeq const. \qquad (IV-24)$$

The major key to these results is the distinctions between the motions caused by electric and magnetic fields. When a magnetic field is applied to any system, the equation of motion of the momentum is

$$\hbar \dot{k} = \frac{e}{c\hbar} \frac{\partial \epsilon_k}{\partial k} \times B = \frac{v}{c} \times B \qquad (IV-25)$$

which implies that $d\epsilon_k/dt = 0$.

This means that the states rotate along equal energy contours. This is true not only for free electrons but for the elementary boson excitations of the Luttinger liquid, or their semion equivalents, the holons and spinons. The entropy is also not modified, by Liouville's theorem, so there is of course no disturbance of thermal equilibrium caused by the magnetic field: $\delta E = \delta S \equiv 0$. All of this may be summarized by the observation that the singular interactions which cause the holon-spinon *commute* with the magnetic field. This may be seen in two ways: first, that their form is proportional to

$$\frac{\epsilon_k - \epsilon_{k;}}{(k-k')^2} \qquad (IV-26)$$

which vanishes for $\epsilon_k = \epsilon_{k'}$ which is the condition that two states be connected by the magnetic field; or, that the effective interaction is "statistical," equivalent to an exclusion principle, while the magnetic field only reshuffles states of the same energy and leaves the Fermi surface intact (to lowest order: we leave out of account diamagnetism and Landau level formation).

On the other hand, these singular interactions specifically do not commute with the E-field, which connects precisely the states with $\epsilon_k' \neq \epsilon_{k;}$ but $|k-k'|^2$ minimum, which are most strongly interacting. The equation of motion in an $\vec{E}$ field is

$$\hbar \dot{\vec{k}} = e\vec{E} \left(\frac{dE}{dt} = \frac{\delta E}{\delta \hbar k} \cdot e\vec{E} = e\vec{E} \cdot v \right). \qquad (IV-27)$$

The Fermi surface is displaced, and ordinary scattering will cause it to relax back to equilibrium as discussed in the previous section.

These contrasting behaviors mean that relaxation processes for momentum parallel and perpendicular to the Fermi surface (longitudinal and transverse) are different. The anomalous relaxation due to electron decay, discussed in the previous section, is ineffectual for the motion due to the B-field. We may divide the equation of motion for an electron into two pieces:

$$\frac{dp_\ell}{dt} + \frac{p_\ell}{\tau_\ell} = eE\cdot \qquad (IV-28a)$$

$$\frac{dp_{tr}}{dt} + \frac{p_{tr}}{\tau_{tr}} = \frac{evB}{c} \qquad (IV-28b)$$

where $\frac{1}{\tau_\ell} \propto T$ is the anomalous linear T-dependent relaxation rate, while $\frac{1}{\tau_{tr}}$ contains whatever other mechanisms can relax the momentum of individual spinons and holons. The two equations then give us *separate* expressions for conductivity and Hall angle: (IV-28a) leads to (setting $\frac{dp}{dT} = 0$)

$$\sigma = \frac{ne\tau_\ell}{m_\ell} \quad \text{with} \quad m_\ell = \frac{dp_\ell}{dv} \qquad (IV-29)$$

and (IV-28b) to, in the same way,

$$\theta_H = \omega_c \tau_{tr} = \frac{eH}{m_{tr}c} \tau_{tr} \qquad (IV-30)$$

where

$$m_{tr} = \frac{\delta p_{tr}}{\delta v}. \qquad (IV-31)$$

Both expressions are, of course, meant as suitable Fermi surface averages.

The relation between momentum and velocity is the second point where the physics differs from the conventional case. "m_ℓ" is relatively simple; it is valid to treat the anomalous resitivity process as we have done, as a decay of electron into holon and spinon. Detailed balance assures us that the average momentum and velocity of the spinon after decay is zero, since the same is necessarily true of the holon. Thus m_ℓ is just the average electron mass $\frac{\hbar k_F}{v_F}$. This is what enters into the z-axis Hall effect.

The transverse mass is more subtle. The momentum of a holon associated with an electron at $\vec{k}_F$ is $2\vec{k}_F$, that of the corresponding spinon $-\vec{k}_F$. Thus the two types of excitation are accelerated in opposite directions. Nonetheless, the Larkin identity, that charge and spin must move together, i.e.,

$$J_s = J_c = J$$
$$n_s v_s = n_c v_c \qquad (IV-32)$$

must hold: it is simply the statement that any net flux can only be a flux of actual electrons. But there is no corresponding rule for the momenta and in fact the momenta of charge and spin are not even colinear.

Let us first neglect impurity scattering (in which case, actually, as we already discussed, the anomalous scattering mechanism doesn't apply; but the Hall effect can still be defined). In this case the physics is that the electron is accelerated by E into a particular state of the continuum of spinon and holon quantum numbers. The individual spinons and holons Larmor precess in opposite directions until scattered (by a T^2 process) at which point they are equivalent to an electron whose precessed momentum is the average of the two precessions. But throughout the process the Larkin identity must hold, i.e.,

$$n_c v_c = n_s v_s. \qquad (IV - 33)$$

Thus the total momentum is

$$p = nv \left(\frac{p_c}{v_c} + \frac{p_s}{v_s} \right) = j m_L^* \qquad (IV - 34)$$

where m_L^*, the Larmor effective mass, is

$$m_L^* = k_F \left(\frac{2}{v_c} - \frac{1}{v_s} \right) \qquad (IV - 35)$$

which will in general be negative (hole-like) for reasonably strong coupling, in agreement with Shraiman and Shastry's[12] estimates and with experiment. Also, it goes, in the limit $U \to 0$ ($v_c = v_s$) to $m_L = m$, and in the limit $U \to \infty$ to $m_L = -\frac{k_F}{v_s}$, both entirely plausible results.

The actual physical situation, however, is rather different. We are in the "undragged holon" regime in general where the holon is almost immediately scattered inelastically while the spinons are unscattered by T-invariant potentials, having only a possible residual magnetic scattering resistance:

$$\frac{1}{\tau_{spinon}} = \frac{1}{\tau_{s-h}} + \frac{1}{\tau_{res}} = c_1 \frac{T^2}{J} + c_2. \qquad (IV - 36)$$

The charge current in this case is entirely carried by backflow and everything moves with the spinon velocity. Thus in this case the Larmor mass, m_L, which determines the cyclotron frequency and the Hall effect, is just $m_L = -\frac{k_F}{v_s}$ and

$$\theta_H = -\frac{eH}{c} \frac{v_s}{k_F} \left(c_1 \frac{T^2}{J} + \frac{1}{\tau_{res}} \right) \qquad (IV - 37)$$

where c_1 is of order unity. The magnitude of the Hall angle observed experimentally is rather small and this is partly accounted for by the rather large ratio which strong coupling may give ($\sim$ 3–4) between v_F and v_s. (The sign again denotes a hole-like [positive] Hall angle.) There may be vertex corrections which further reduce the Hall angle, but in the absence of a fully developed perturbation theory we cannot compute these (or even be sure they are there).

*1

E. Thermoelectric Power and Heat Conductivity

We have not worked out these phenomena. As pointed out by Stafford, they are useful in that, unlike the Hall effect, the effect exists and can, in principle, be computed in one dimension. Using the representation of the 1d Hubbard model by Woynarovich, he has been able to compute thermopower in the pure case as the sum of an electron-like spinon term and a hole-like holon one, with the net sign hole-like near half-filling.

Stafford's results bear a family resemblance to experiments but do not fit perfectly. Experimentally, one often sees a negative, electron-like slope but a positive sign. Clearly we are in the "undragged" regime of spinon flow, and the slope reflects the fact that holons do not move. But the overall constant background term, having the form of a positive *intercept* at $T = 0$ of Q vs T, must come also from the electron decay into holon and spinon on injection into the sample. This "intercept" is not as spectacular as the fountain effect but tells us a similar story of one component of the fluid being blocked from flowing, hence causing an anomalous backflow. This effect might be both spectacular and useful as a refrigeration mechanism if it were possible to isolate the individual plane and render it nonsuperconducting.

Thermopower along the c-axis should tell a similar story to conductivity. At high temperatures it should just reflect the slope of density of states and be electron-like. This may well persist to low temperatures, and be a telltale that the rise in resistivity as $T \to 0$ is not activated, which would entail a rise in thermopower.   *m

Thermal conductivity is simple: it reflects electrical conductivity via the Wiedemann-Franz law, which may have a factor of order unity in it. It should be constant in the ab plane, zero, effectively, along the c-axis, which allows us to calibrate the conductivity due to phonons. We have discussed the rise below T_c elsewhere. There will be no effect of T_c on the c-axis thermal conductivity.

F. Conclusions

We have given the outlines of a new type of transport theory which seems to explain well the behavior of the CuO_2 planes in high T_c material. The physical assumptions we have made do not necessarily apply to other Luttinger liquids as, e.g., the one-dimensional metals, which are often very pure and therefore have fewer anomalous properties. At all points we are frustrated by the unavailability of a reliable perturbation theory of the processes. We have a mixture of strong effects of interactions, leading to non-Fermi Liquid behavior, and large effects of impurities, which cut off the dragging of the holon excitations. Thus at every juncture we had to resort to special tricks and general conceptual understanding rather than the more easily understandable diagrammatic perturbation theory, which fails not just in detail but in principle, since also all vertex corrections are divergent. Nonetheless, we can produce a qualitative and sometimes quantitative theory of a wide variety of phenomena which leaves us with a measure of confidence in our results.

I acknowledge discussions with Y.-B. Xie, M.P.A. Fisher, C. Kane, D. J. Scalapino, H. Fukuyama, and especially N.-P. Ong. M. Ogata coauthored section IV. F.D.M. Haldane's helpful comments and penetrating insights were important throughout.

NOTES

*a   This result was not included in chapter 5 but is found in a reprint by Strong and Talstra in part II.

*b   Item (3) needs to be qualified. This is not true in 1D because scatterers are anti-screened, which essentially is to say that at low enough T they cause local states which become magnetic and the simple perturbation theory on which this statement is based fails. (3) holds in 2D (but see a later preprint about impurity scattering).

*c   The later paper of Anderson and Khveshchenko shows that (10) is not the best representation of the effect, which is better treated as a modification of the kinetics of the particles. The phase shift changes boundary conditions, hence modifies the Hilbert space, not the interactions. But many of the conclusions drawn are the same.

*d   I now believe that gauge theory is so profoundly modified by the Fermi surface that it may indeed not be possible to work it out from first principles. For instance, transverse gauge fields are meaningless and have no dynamics because a gauge field is simply a relabeling of the states within the Fermi surface. I also believe that the Anderson-Khveshchenko approach is, if not rigorous, at least capable of being worked out in an asymptotically rigorous way, so that I do not present the Luttinger liquid approach as tentative.

*e   See note, p. 11. I leave this section in because it gives yet another way of visualizing the NFL effect, and this seems the hardest argument for theorists to understand, so it should be stated many times.

*f   This paper was never completed. I believe the effect is real (and will be discussed later) but is not the primary reason for the "ghost" Fermi surfaces. These, where they are not experimental artifacts due to superstructure on the surfaces, are caused by the smearing of the Green's function by the *incoherent* hopping between planes. This gives a Lorentzian breadth to the G_1 peaks of order a bit smaller than $t_\perp^2 N(E)$, and makes the Fermi surface broad in k-space where $t_\perp$ is large.

*g   The low T crossover is expected to occur where $\hbar/\tau_{el} > kT$ or a little lower, because this is where the electron has no opportunity to decay into spin and charge before scattering and one must deal with interaction effects for randomly scattered, diffusing electrons. What happens below this crossover is treated in a preprint.

*h   This expression has been fitted to the infrared conductivity of Bontemps et al. in a reprint included in the book. It obviously gives $\omega\tau = \text{const} = \tan(2\pi\alpha)$ which fits an early observation of Schlesinger.

*i   This formula is incorrect, as shown by Strong and Clarke. The argument as given is incorrect: the virtual hops are much *shorter* than the time corresponding to $t_\perp$, so the process is "exchange narrowed": the probability of a real hop is $\sim t_\perp^2/t_{\parallel}$. The correct formulas are given in several of the reprints: the conductivity is proportional to $(\omega, T)^{2\alpha}$.

*j   BISCO's activated conductivity seems to be caused by the "spin gap." Observations are no longer in good agreement with $\rho \propto 1/T$; this seems to have been an artifact.

*k   The measurements were made on YBCO but it is not obvious that the samples were doped sufficiently highly to give an electron-like Fermi surface, contrary to my cocksure statement in the text. The experiment needs repeating on more samples and as a function of doping. The argument seems correct, as far as I can now see.

*l The sign of the Hall effect is still a bit puzzling. There is no current of holon excitations; the excitation which has a current is the spinon, but it clearly carries a backflow of the charge fluid Fermi sea which gives it a momentum $+k_F + v_s \delta k$, and velocity v_s, so it behaves like a simple electron. Thus in fact the Hall effect sign is not reversed in this case, even though the spinon excitations have (as eigenexcitations) momentum $-k_F$. But I do not consider the problem clearly solved. It is confused by the fact that the Fermi surface may close either way depending on doping. Coleman and Tsvelik have developed a two relaxation time theory which may (or may not) give adequate answers to some of these questions.

*m Again, I emphasize that, in samples doped optimally (around 20%) the Fermi surface is hole-like; I was under a misapprehension as to this sign when these pages were written.

REFERENCES

1. P. W. Anderson, Science, **256**,1526 (1992).

2. P. Phillips and H. L. Wu, Science, **252**, 1805 (1991).

3. Z. Fisk et al., Science, **239**, 33 (1988).

4. N. Nagaosa and P. A. Lee, Phys. Rev. Lett., **64**, 2450 (1990).

5. L. Ioffe and P. Wiegmann, Phys. Rev. Lett., **65**, 653 (1990); L. Ioffe and G. Kotliar, Phys. Rev. B **42**, 10348 (1990).

6. S. A. Brazovskii and V. M. Yakovenko, Sov. Phys. JETP **62**, 1340 (1985), and private communication.

7. H. J. Schulz, Phys. Rev. Lett., **64**, 2831 (1990), Int. J. Mod. Phys. B**5**, 57 (1991); N. Kawakami and S.-K. Yang, Phys. Rev. Lett., **65**, 3063 (1990), Phys. Rev B **44**, 7844 (1991); C. A. Stafford, A. J. Millis, and B. S. Shastry, Phys. Rev. B **43**, 13660 (1991); R. M. Fye et al., Phys. Rev. B**44**, 6909 (1991).

8. M. Ogata and P. W. Anderson, Submitted to Phys. Rev. Lett.

9. N. Kumar and A. M. Jayannavar, Phys. Rev. B **45**, 5001 (1992).

10. P. W. Anderson, Phys. Rev. Lett., **67**, 3844 (1991).

11. P. W. Anderson, Phys. Rev. Lett., **67**, 2092 (1991).

12. B. S. Shastry, B. I. Shraiman, and R.R.P. Singh, Phys. Rev. Lett., **70**, 2004 (1993).

7

Theory of Superconductivity in the High-T_c Materials

I. INTRODUCTION

From about September 1987 the mechanism for superconductivity in the cuprates has been clear. A paper of Hsu, Wheatley, and Anderson[1] proposing the mechanism, and papers of Hsu and Anderson[2] pointing out the strong heuristic evidence for interlayer tunneling in the T_c values, seemed to leave no alternative, especially as evidence piled up for the "confinement" explanation of the anisotropic transport properties. However, for most of the intervening period the theory of the normal state has been inadequate to the task of supporting any kind of detailed theory of the superconductivity transition. It is still true that the outlines of a theory which we now have are tentative and heuristic, with clear guidelines only on the basic mechanism; on the perturbative theory of the transition; and on the qualitative nature of the gap function. But we have reached the stage of sketching out a fairly complete model theory, which leaves us at least ready to lay out lines for further development and to make a stab at explanations of a considerable number of otherwise puzzling phenomena.

The key features of the superconducting mechanism appear to be the following:

(1) Absence of coherent c-axis electron motion in the cuprate layer compounds implies (via Kohn's identity) excess kinetic energy in this direction. This is due to the "confinement" property of incompressible quantum fluids such as the spinon gas; this could also be exhibited by chiral quantum fluids (and is, in the quantum Hall effect) and possibly various density wave systems.

(2) Josephson-type, two-electron transport is not blocked because the spinon fluid (whether short-range or gapless) is a pair condensate, so that singlet pairs tunnel freely. This motivates superconductivity as a $2 \to 3$ dimensional crossover. There is no necessity for dynamical screening of the BCS-Morel-Anderson-Eliashberg type[3] which restricts the T_c of all BCS pairing mechanisms as discussed by Anderson and Cohen[4] early on; therefore T_c can be considerably larger. However, dynamical screening can enhance an already high T_c.

(3) A simple perturbative self-consistency argument shows that T_c is proportional to $t_\perp^2/t_\parallel$ where $t_\perp$ is the frustrated interlayer tunneling matrix element. The same perturbative approach shows that the excitations regain quasiparticle character in the superconductor. The real reason for the high T_c turns out to be an interesting paradox having to do with the momentum-conserving nature of the interlayer mechanism. The different k-values are uncoupled from each other and hence no high kinetic energy states need be paired: we use only the pair susceptibility $\chi(k)$ for a given k, which is much greater than the average χ over all k's even though each $\chi(k)$ for the Luttinger liquid is slightly smaller than the corresponding BCS χ_{pair}.

(4) Finally, we recognize that the actual nature of the pairing wave function is determined not by the basic interlayer mechanism which raises T_c but by the "residual interactions," be they caused by phonons, spin-fluctuations, or other source. Thus the observation of any particular gap structure is no proof of mechanism. In some cases we predict strong anisotropy in magnitude of the gap independently of its symmetry.

Most of the features of this theory are specific to the cuprate superconductors. Many aspects of it, however, may turn out to be useful in theories of other superconductors, such as the organics, the BKB and "buckeyball" types, and even heavy electron superconductivity. These options are merely mentioned here and will bear further investigation.

II. CONFINEMENT AND ITS CONSEQUENCES

First it is important to bring forward an identity due to Kohn[5] for transport in a single tight-binding band (brought to my attention by Shastry), comparable to the well-known conductivity sum rule. Of course, the standard rule

$$\int \sigma(\omega)\, d\omega = \frac{1}{i}\frac{e^2}{\hbar}\sum_i (x_i, \dot{x}_i)$$
$$= \frac{4\pi\, ne^2}{\hbar\, mi}(x, p)$$
$$= \frac{4\pi\, n\, e^2}{m}$$

holds if one integrates over all interband transitions, but if one is measuring simply the Drude and interaction contributions from a single band it is important to have a one-band version. It is also, as Kohn pointed out, important to understand why a filled band gives no contribution even to $\sigma(\omega)$ at frequencies below interband edges.

Let us then consider the intraband version of the sum rule, where now x and $\dot{x}$ are intraband operators defined on a set of Wannier functions $W_n(r)$ (a number of theorems on Wannier functions are to be found in Blount's review[6]).

Let us introduce a set of field operators for the Wannier functions, $c_{n\sigma}^+$. In terms of these the position operator becomes

$$\vec{x} = \sum_n c_n^+ c_n \vec{r}_n \,,$$

and the velocity operator

$$\vec{v} = \dot{\vec{x}} = \frac{1}{\hbar} \sum_k \frac{\partial \epsilon_k}{\partial k} \cdot n_k$$

$$= \frac{1}{\hbar} \sum_{m,n} m \, c_{m+n}^+ \, c_n \, \mathcal{H}_m^{1-\text{electron}}$$

where the one-electron energy is

$$\epsilon_k = \sum_m \cos k \cdot r_m \, \mathcal{H}_m^{1-\text{electron}} \,.$$

That is, the one-electron hopping integrals $\mathcal{H}_m$ between Wannier functions $\vec{r}_m$ apart are the Fourier coefficients determining ϵ_k.

Again the equal-time commutator of x and v gives us the $t = 0$ correlation of current, but now the equation reads ("gap" is the interband gap)

$$\int_0^{Gap} \sigma(\omega)\, d\omega = \frac{4\pi e^2}{\hbar^2} \sum_{n,m} r_m^2 \langle c_n^* c_{n+m} \rangle \mathcal{H}_m.$$

The total one-electron energy is

$$\langle E_{1-\text{electron}} \rangle = \sum_{n,m} \mathcal{H}_m \langle c_{n+m}^* c_n \rangle$$

and if we note that $\mathcal{H}_0$ is the mean energy in the band, we realize that

$$\sum_{m \neq 0} \mathcal{H}_m \langle c_{n+m}^* c_n \rangle$$

must be negative and represent the *gain* in energy due to delocalization via band formation. If the tight-binding band has nearest-neighbor interactions only, then

$$\int \sigma(\omega)\, d\omega \propto \mathcal{H}_0 - \langle E_k \rangle \,;$$

in any case, vanishing conductivity implies no band formation and vice versa.

In the c-direction of the cuprate crystals the nearest neighbor condition $m = 1$ only is obviously satisfied (note that this theorem holds for each direction separately).

Thus we must conclude from the absence of coherent c-axis conductivity (which holds for practical purposes right into the infrared) that the hopping integrals in the c-direction are frustrated in the sense that the system has failed to gain the kinetic energy (of order $t_\perp^2 \rho(\epsilon)$) which would be available from them. The mechanism is of course confinement due to correlated spin liquid formation, as explained in chapters 5 and 6.

The hopping integrals between close planes in the multiplane YBCO, BISCO, and Tl compounds are non-negligible (band calculation estimates put them at greater than $\sim .1$ ev, as do estimates from exchange integrals in the insulating phases), and the most striking experimental proof of confinement is the absence of the expected infrared absorption from these close pairs. An estimate gives

$$\int \sigma(\omega)\, d\omega = \frac{4\pi e^2}{\hbar^2} \frac{2t_\perp^2}{t_\parallel} a^2$$

$$\sim .1 \cdot \frac{4\pi e^2}{m}$$

i.e., the total integral should be $\sim 10\%$ or more of that in the $a - b$ plane. The interband splittings caused by $t_\perp$ are ~ 500–1000 cm^{-1}, so one would expect an absorption band in that region comparable in intensity to the ab plane conductivity in the same region (which is $\sim \frac{ne^2}{m\omega \ln \omega/\omega_c}$ with $m \sim 2m_e$). Nothing of the sort is seen.

A weaker bound is set by the observed c-axis penetration depth λ^{-2}, weaker because in YBCO and other two-layer materials there are two sets of $t_\perp$, one for the long ("chain") intervals and the other between the close planes, and the former controls the anisotropy both of DC conductivity and of λ^{-2}.

III. PERTURBATION TREATMENT OF T_c; GREEN'S FUNCTIONS
OF THE SUPERCONDUCTING STATE

The principle of the T_c calculation is as follows. We know the Green's functions of the *interacting* Luttinger liquid in the normal state. In the spirit of the calculation of other responses of the Luttinger liquid, we recognize that we are very unlikely to get correct answers by first perturbing the free particle system with an anomalous self-energy, and then turning on the interaction U; the correct order is to allow the particles to propagate with the *interacting* Green's functions in response to the anomalous self-energy as a perturbation. We imagine, then, that we have two Luttinger liquid layers in close proximity with a tunneling matrix element $t_\perp$ between them: as in a confinement calculation

$$\mathcal{H}_{pert}^{(1)} = t_\perp \sum_{k\sigma} c_{k\sigma}^{+(1)} \, c_{k\sigma}^{(2)} \,. \tag{1}$$

(More correctly, in general $\mathcal{H}_{pert}^{(1)} = \sum_{k\sigma} t_\perp(k) \, c_{k\sigma}^{+} \, c_{k\sigma}^{(2)}$.) To begin with, we assume that an anomalous self-energy has been introduced by some external influence into layer (1); later we will calculate this self-consistently using the interaction (1). So we introduce

$$\sum_{anom} = \mathcal{H}_{pert}^{(2)} = \Delta_k \, c_{k\uparrow}^{+} \, c_{-k\downarrow}^{+} + \Delta_k^* \, c_{-k\downarrow} \, c_{k\uparrow} \tag{2}$$

where Δ_k is an assumed anomalous self-energy part in one particular momentum state k.

Later, when we come to attempt to write a self-consistent gap equation, we will realize that it is not totally consistent to introduce (2) without recognizing that, like any perturbing potential, it must be corrected for mean-field Landau parameter-like responses. In a repulsive U Hubbard model, there is a straightforward mean field response to a pairing potential which corrects for the repulsion of two electrons on the same site, of the form

$$\Delta \rightarrow \Delta - U \sum_{k'} \langle c_{-k'\downarrow} \, c_{k'\uparrow} \rangle. \tag{3}$$

We assume $\sum_{k'} \langle c_{-k'\downarrow} \, c_{k'\uparrow} \rangle \simeq 0$ as a constraint which will be imposed on our self-consistency problem, and for the time being assume this takes care of U. In general we will have various other interactions and the full interaction problem will be dealt with in the next section.

The remainder of the calculation is straightforward. Having taken the mean field interaction into account, in lowest order perturbation theory we can assume that the pair $k_\uparrow$, $-k_\downarrow$ is independent of all others and that the two particles don't otherwise affect each other's propagation except to order Δ^2. We must calculate the pair susceptibility

$$b_k = \langle c_{-k\downarrow} \, c_{k\uparrow} \rangle = \chi_{pair}(k) \Delta_k$$

and

$$\chi_{pair}(k) = \int d\omega \; G_1(k, \omega) \, G_1(-k, -\omega). \tag{4}$$

The diagrammatic representation is given in fig. 7.3. We are simply going to require self-consistency:

$$\Delta_k = t_\perp^2 F_k$$

where F_k is the propagator for converting an electron k into a hole in $-k$. To convert b_k to a propagator we have to divide by the volume element in space-time, $k_F \, v_F$, where the appropriate velocity is the geometric mean $v_F = \sqrt{v_c v_s}$. Thus

$$F_k = \frac{b_k}{E_F} \qquad \text{and} \qquad \Delta_k = \frac{t_\perp^2}{E_F} \chi_{pair}(k) \Delta_k \ .$$

Thus T_c is determined by

$$\frac{t_\perp^2}{t_\parallel} \chi_{pair}(k) = 1. \tag{5}$$

To illustrate how it goes, let us imagine that somehow a simple free electron system had become confined, so that we could use the Josephson tunneling mechanism to achieve a T_c. The BCS $\chi_{pair}(k)$ is

$$\chi(k) = \frac{1}{\epsilon_k} \tanh \frac{\epsilon_k}{2k_B T}$$

$$= \frac{1}{2k_B T} \qquad (\epsilon_k \to 0)$$

so (3) above would read

$$\frac{t_\perp^2}{t_\parallel} \cdot \frac{1}{2k_B T} = 1$$

$$k_B T = \frac{t_\perp^2}{2t_\parallel} \ .$$

For conventional BCS pairing, one has a very different mechanism so that the relevant χ_{pair} is the one which is local in real space, not momentum space:

$$\overline{\chi}_{pair} = N(0) \int \frac{d\epsilon}{\epsilon_k} \tanh \frac{1}{2} \beta \epsilon_k \tag{6}$$

which has the well-known $\ln T$ form. From this point of view, it is clear that the size of T_c is determined by this locality in k-space for high T_c's, not by a large pair susceptibility; we shall find that χ_{pair} is actually not as large as for free particles, surprisingly to me.

Now we will need to calculate $G_1(k, \omega)$, which is nontrivial even though we know the space-time version of G_1 quite well enough in the relevant asymptotic region $x \gg a, v_f t \gg a$. This Green's function, which we have used repeatedly in previous calculations, is the one-dimensional one[7]

$$G_1 \sim \frac{e^{ik_F x}}{(v_s t - x)^{1/2} (v_c t - x)^{1/2} (v_c^2 t^2 - x^2)^\alpha} \qquad +x \to -x \tag{7}$$

with α a small exponent $0 < \alpha < 1/16$. The signs are chosen so that G_1 is real for x positive and $t > \frac{x}{v_s}$ or $< -\frac{x}{v_c}$ which are not causally available regions.

In Fourier analyzing G_1 we must consider three interesting regions: essentially, k well above, at, or well below the Fermi surface. The relevant criterion is whether $v_F |k - k_F| \lesssim \omega_0$ where ω_0 is a lower frequency cutoff at $\sim T$ or a measuring frequency. For $v_F |k - k_F| < \omega_0$ we may ignore k in the exponent of the Fourier integral

$$G_1(k, \omega) = \int dx \, dt \, e^{i|kx - \omega t|} G_1(x, t)$$

and we have

$$G_1(0, \omega) = \int dt e^{i\omega t} \int dx \frac{1}{(v_s t - x)^{1/2} (v_c t - x)^{1/2} (v_c^2 t^2 - x^2)^\alpha}$$

$$= \int \frac{dt e^{i\omega t}}{t^{1+2\alpha}} \quad F(v_c/v_s)$$

where $F = \int \frac{dy}{-2(v_s - y)^{1/2} (v_c - y)^{1/2} (v_c^2 - y^2)^\alpha};$

$$G_1(0, \omega) = \frac{1}{\omega^{1-2\alpha}} F\left(\frac{v_c}{v_s}\right) \int \frac{du e^{iu \operatorname{sgn} \omega}}{u^{2\alpha}}. \tag{8}$$

(The "sgn ω" in the integral means that the appropriate branch of $u^{-2\alpha}$ is chosen to make the result pure real for the "wrong" side of the Fermi surface. This approximation replaces the $\frac{1}{2}(1 \pm |k - k_F|^{2\alpha})$ singularity at E_F by a step function, which it resembles for $\omega_0/E_F > (2)^{-1/2\alpha} \sim 10^{-4}$.)

$G_1(|k - k_F| v_F > \omega_0)$ will not be needed in this section but under the approximation $v_c \gg v_s$ it may be shown to be of the form, roughly,

$$\frac{1}{(k v_c - \omega)^{1/2} (k v_s - \omega)^{1/2 - 2\alpha}}$$

with a cusp at $\omega = k v_s$ and a finite imaginary part only on the "right" side of the Fermi surface. (Neglecting the small amplitude which appears above $\omega = -v_c(k - k_F)$ on the "wrong" side.)

We are now ready to calculate the pair susceptibility for use in finding T_c. Clearly the most unstable k-values are the small ones with $v_s |k - k_F| \ll T_c$, so that we use (8) as our $G_1(k, \omega)$. An important point which was missed in the Los Alamos paper where this derivation first appeared[8] is that it is not $Im\, G_1$ (which satisfies a sum rule) but $Re\, G_1$ which is relevant. $Im\, G_1$ contains a factor $\sin 2\pi\alpha$, appropriately because in order to satisfy the sum rule

$$\int d\omega \, Im\, G_1(k, \omega) \sim 1 \tag{9}$$

with

$$Im\ G_1 \sim \omega^{-1+2\alpha}.$$

$Re\ G_1$, on the other hand, is $\propto \cos 2\pi\alpha$ which is not small. Otherwise the calculation is the same as in that reference.

We may introduce a finite temperature cutoff in ω simply by multiplying by the factor $\tanh \frac{\omega}{2k_B T}$ and

$$\chi_{pair}(k) = \dot{\wedge}^{-4\alpha} \int d\omega \tanh \frac{\omega}{2k_B T} \frac{1}{\omega^{2-4\alpha}}$$

$$\simeq \left(\frac{T}{\wedge}\right)^{4\alpha} \times \frac{1}{2k_B T}$$

where $\wedge$ is an upper cutoff of order $t_{\|}$. The exponent 4α is at most $1/4$ so the cutoff factor is of order $1/2 - 1/3$ giving $k_B T_c \simeq 1/5\ t_\perp^2/t_{\|}$, which is the right order of magnitude.

A second very interesting result follows from this exercise in perturbation theory and from our assumption—eminently plausible if not proven—that to lowest order in Δ there are no vertex corrections to the pair response. This latter assumption is just the Landau theory type assertion that holons and spinons *are* the elementary excitation spectrum of the Luttinger liquid, noninteracting to a lowest approximation except for mean motion of the entire substrate fluid (i.e., backflows and mean fields). In the Varenna notes[9] we pointed out that in the presence of a condensate, spinons would acquire charge by emitting and absorbing quasiparticle pairs and become indistinguishable from conventional quasiparticles; so that T_c is not only a 2D-3D crossover but a restoration of quasiparticle character. This may be seen very simply by using Nambu spinor notation where the Green's function is a 2×2 tensor, the two components being $k\sigma$ and $-k, -\sigma$ holes. The conventional kinetic energy and self-energy are written

$$\left(\epsilon_k + \sum(k, \omega)\right) \underset{\sim}{\tau_3} \tag{10}$$

and the off-diagonal or anomalous self-energy is

$$\Delta_1 \underset{\sim}{\tau_1} + \Delta_2 \underset{\sim}{\tau_2}.$$

Conventional BCS theory, then, has a tensor Green's function

$$\underset{\sim}{G}^1 = \frac{1}{\omega - (\epsilon_k + \sum)\underset{\sim}{\tau_3} - \Delta\underset{\sim}{\tau_1}} \tag{11}$$

with poles at $\omega^2 = (\epsilon + \sum)^2 + \Delta^2$.

Another way to do the same manipulation is to consider the tensor $\underset{\sim}{G}^1$, which is then simply

$$\underset{\sim}{G}^1 = \begin{pmatrix} (G_1^{-1})_{electrons}^{+k} & \Delta \\ \Delta^* & (G_1^{-1})_{holes}^{-k} \end{pmatrix}^{-1}$$

which is the same as (6). We know $G_{electrons}^{-1}$ and G_{holes}^{-1} in our case, and so we get

$$\underset{\sim}{G}^1 \simeq \begin{pmatrix} \omega^{1-2\alpha} & \Delta \\ \Delta^* & \omega^{1-2\alpha} \end{pmatrix} \tag{12}$$

and G now has poles at the points

$$\omega^{2-4\alpha} = |\Delta|^2 \tag{13}$$

which very much resemble the quasiparticle poles of BCS. This remarkable effect is unique to the Luttinger liquid–interlayer tunneling mechanism and is one of the crucial observations which can have no other reasonable explanation: namely, the sudden appearance of a sharp quasiparticle peak in photoemission below T_c (see figs. 7.1a, b).[10]   *a

One very interesting fact which confuses some other possible observations of this phenomenon is that the quasiparticles which are thereby created (at all k where a Δ appears, although (7) only applies near k_F) are still not mobile in the c-direction, since the Δ self-energy term merely connects holes and electrons in the same plane. There is still no c-axis velocity. Thus electronic thermal conductivity, for instance, is still not possible in the c-direction, and no singularity in K_c will appear at T_c. Also, no new infrared absorption will appear at T_c. In fact, the data of Uchida on $(La - Sr)_2CuO_4$ show no quasiparticle conductivity, only an imaginary part, at $\omega = 0$.

c-axis tunneling will not have enhanced quasiparticle peaks, contrary to a suggestion in one of my previous papers. There may be additional density in BCS-like peaks in ab tunneling, but the ab tunneling process involves complexities which have yet to be worked out.

When we attempt to follow this formal derivation beyond lowest-order perturbation and include Δ in G to high order we find that the Green's function becomes unphysical with negative imaginary parts, etc. This is not surprising: spin-charge separation and the bosonization procedure that leads to it are incompatible with the superconducting gap. The diagonal part of G_1 must resume its simple pole character. How this transition or crossover takes place has not yet been solved; but it seems perfectly possible for spin-charge separation to persist for $v_F|k-k_F| > \Delta$ even though it is absent for $v_F|k-k_F| < \Delta$, because the commutation relations for bosons of momentum Q do not depend on the precise structure of the momentum distribution for $|k - k_F| < Q$.   *b

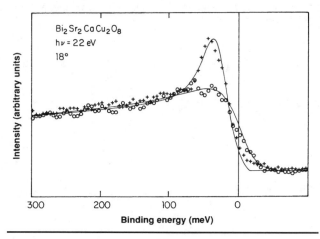

(a)

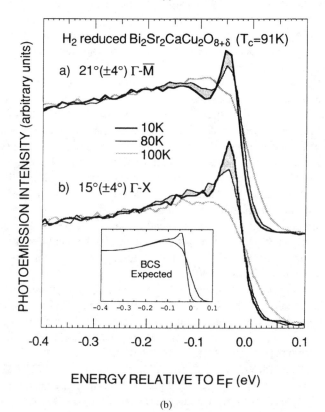

(b)

Figure 7.1. Illustrations of the appearance in ARPES of the quasiparticle peak below T_c, from Refs. 10 and 15.

IV. HEURISTIC GAP EQUATION FOR THE INTERLAYER
MECHANISM

It is possible to set up an equation which encompasses all of the physics of the interlayer mechanism. Most but not all of this equation can be estimated quantitatively. The resulting physics is remarkably complex and multifaceted. In one way this is unfortunate: in contrast to BCS, there appear to be few simple, characteristic behaviors in the *superconducting* state, in marked contrast to the near-universalities of the normal state. On the other hand, this leaves us with a richness of phenomena which is in fact necessary to confront the many unexpected observations about the superconductor.

We begin with the assumption that, to a good approximation, different momentum values are dynamically independent. This is equivalent to a mean field theory, and an identical assumption underlies standard BCS theory.

In BCS theory this step can be justified by diagram theory arguments: it amounts to assuming that the anomalous term is a self-energy part without serious final state interactions—i.e., no "vertex corrections." In the Luttinger liquid, in general, separation of diagrams into self-energy, vertex, and polarization parts is invalid: each correlation function, in general, has its own different singular asymptotic behavior. We justify our approach on two separate grounds.

(1) For the interlayer tunneling terms, since momentum is conserved ($\Delta k = 0$) the x-coordinates of the two tunneling events are widely separated ($\Delta x \to \infty$) so that all four vertices of the effective interaction are widely separated in space-time, and whatever vertex corrections may exist vanish with distance: essentially, 4-point correlations of widely separated points x_1, x_2, x_3, and x_4 are simply products of pairs of G_1's.*

(2) The conventional interaction terms such as U, phonons, etc., are averages over the whole Fermi surface with little contribution from the nearby regions in k-space where there are strong correlations. These interactions are local and don't depend on the asymptotics. That is, the Fermi velocities at the other points on the Fermi surface quickly dephase correlation effects, leaving the effective interactions local in space and, except for the well-known temporal structure due to phonons, in time. This is the identical "Migdal" argument used in BCS.

All of this is somewhat clearer in diagrams. The BCS gap equation is shown in fig. 7.2; the assumption is that there are no vertex corrections of the type shown in fig. 7.1b, because the two lines (A) and (B) would generally have very different momenta and hence little effect at low energy (big denominators or low phase space). Our gap equation is shown in fig. 7.3. There are two

*I am indebted to V. Yakovenko and Berezinskii for this observation.

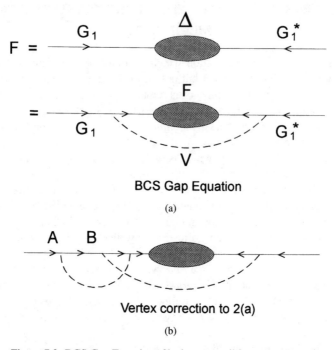

BCS Gap Equation

(a)

Vertex correction to 2(a)

(b)

Figure 7.2. BCS Gap Equation: 2b shows possible vertex correction.

contributions; in the first, the two "t"'s occur at very distant points, hence there are weak interactions between the G_1's or between G_1 and Δ. We do, however, accept that G_1 has an unknown (as yet) modification due to Δ which can only be estimated for large or small Δ. The second is free of vertex corrections for exactly the same reasons as in BCS.

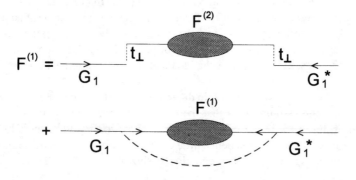

Figure 7.3. Gap Equation of this chapter.

Using this approximation we deal initially only with a pair of electrons of momentum $k \uparrow$, and $-k \downarrow$, which have nominally four states, $n_{k\uparrow} = \pm 1$, $n_{k\downarrow} = \pm 1$. In BCS the unperturbed Green's function is diagonalized by these occupation numbers, but in Luttinger liquid theory there is still the holon-spinon degree of freedom. The excitation energy in the "normal" system has a "cut" spectrum spread over a continuum. On the other hand, once paired by an anomalous self-energy Δ, as we showed in the last section there is a unique quasiparticle energy. Thus pairing entails a loss of entropy.

In rough confirmation, we found in our perturbation theory that the linear response to an anomalous self-energy Δ is somewhat smaller than that of a free electron state, by a factor $\simeq (\frac{T}{\Delta})^\alpha$. The simplest description we can give of the facts is that there is a general functional dependence of $\langle b_k \rangle$ on the k-dependent anomalous self-energy Δ_k

$$ b_k = F\left(\frac{\Delta_k}{T}, \frac{\epsilon_k}{T} \right). \tag{14} $$

which is very similar to the BCS expression

$$ b_k^{BCS} = 2 \tanh \frac{\beta E}{2} \cdot \frac{\Delta_k}{E_k} \tag{15} $$

(where $E_k^2 = \epsilon_k^2 + \Delta_k^2$), but has a somewhat smaller initial slope. Like the above expression, it must saturate to unity if $\Delta_k \gg \epsilon_k$, T. Except for the slope, which is the perturbative χ_{pair}, its precise shape is relatively unimportant: it will look something like fig. 7.4. As a "hunch" (and my hunches have not had perfect accuracy in this field) I suspect that $F(\Delta/T)$ will have less negative curvature, relative to BCS, giving a sharper, more nearly first-order intrinsic transition.

The inevitability of the nature of F follows from the fact that the singlet paired "full" and "empty" states for a given k are simple and unique, and if $\Delta \gg \epsilon$, T are well-separated from the manifold of quasiparticle states. We know the linear dependence of F on Δ, and its asymptotic value: how can it be more complicated? To this we add the somewhat stronger assumption of rough independence of the different k's which we already justified in terms of vertex corrections (or the absence thereof) and there is very little freedom of choice.

The entire process is essentially to confess that *except for confinement* the Luttinger and Fermi liquids are physically very similar, which is the source of much of the confusion in the field of high T_c.

Now that we have equation (14) for b_k given Δ_k, we must close the self-consistency loop by calculating Δ_k from the b_k's and the interactions. There are three different kinds of energy dependent upon b_k. First, there is the interlayer pair tunneling energy we discussed in the previous section. This is momentum-

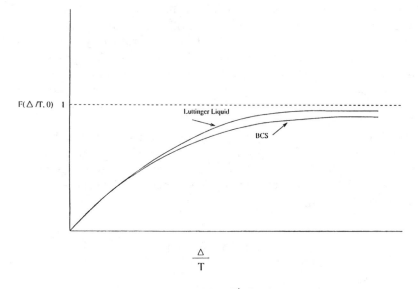

Figure 7.4. $F(\frac{\Delta}{T}, 0)$.

conserving, and as discussed there it takes the form

$$E_J = -\frac{t_{\perp ij}^2}{t_{\|}} \sum_{\substack{kk' \\ ij}} \left(b_k^{(i)}\right)^* b_{k'}^{(j)} \, \delta(k - k') \tag{16}$$

*c   where i and j are different planes. This δ-function is of course not infinitely sharp, because the momentum k is not infinitely sharp; there is a breadth $\delta k = \hbar/\ell$ with ℓ the inelastic mean free path, which is probably that of the spinon as defined by the Hall angle. In the absence of magnetic impurities this decreases as T^2 or faster as we go to low temperature. We understand the δ-functions in (16) as broadened by this and any other relevant broadening mechanism. By definition, we determine the anomalous self-energy from:

$$\Delta_k = -\frac{\partial E}{\partial b_k}. \tag{17}$$

There are still electron-electron interactions including those which are responsible for the Luttinger liquid ground state. Specifically, insofar as we produce a mean amplitude $b_k^+ = \langle c_{k\uparrow}^+ \, c_{-k\downarrow}^+ \rangle$ it will experience the short-range repulsion energy

$$U \sum_{\substack{sites \\ R}} n_\uparrow(R) \, n_\downarrow(R) = U \frac{1}{N} \sum_{k,q,k'} c_{k\uparrow}^+ \, c_{-k+q\downarrow}^+ \, c_{-k'+q\downarrow} \, c_{k'\uparrow} \tag{18}$$

and the relevant term in the mean fields is

$$U/N \sum_{kk'} \langle b_k^+ \rangle \langle b_{k'} \rangle \qquad (19)$$

which, as we saw in the previous section, can be evaded completely if we simply set

$$\sum_k \langle b_k^+ \rangle \equiv 0. \qquad (20)$$

However, this may not be the optimum T_c.

This will not, in fact, be the only interaction term. There will be attractive second-order electron-phonon terms, as well as an inevitable k-dependence of the repulsive interactions. We may include all of these interactions in a general form

$$U = \sum_{k,k'} U(k, k') b_k^+ \, b_{k'} \qquad (21)$$

and this interaction kernel may be factorized into a sequence of eigenvalues and eigenfunctions

$$U = \sum_\lambda U_\lambda \left| \sum_k \alpha^\lambda(k) b_k^+ \right|^2 \qquad (22)$$

where

$$\sum_{k'} U(k, k') \alpha^\lambda(k') = U_\lambda \alpha^\lambda(k).$$

The U_λ will in general be both positive and negative. At least the most important one, U_0, belonging to the most nearly uniform, particle-hole symmetric, eigenfunction $\alpha^0(k) \approx$ const, will be positive. But it is not true that all eigenvalues are repulsive whether particle-hole symmetric or not; in fact, usually negative ones exist. In fact, it is not unlikely that an isolated single layer (as in BISCO 2 2 0 1) might be superconducting at a low T_c.

V. A ROUGH PICTURE OF SOLUTIONS OF THE NEW
GAP EQUATION

It turns out that the self-consistency ("Gap") equation which follows from (14), (17), (16), and (22) can be put in a remarkably simple form closely resembling the BCS-Eliashberg equation but with one very large difference: the resulting T_c's are all enhanced by the "Josephson" interlayer term. The formal steps necessary to do this are quite simple. We rewrite equation (15):

$$b_k = F\left(\frac{\Delta_k}{T}, \frac{\epsilon_k}{T}\right)$$

as

$$\Delta_k = F_k^{-1}(T, b_k) \tag{23}$$

where F_k^{-1} is a T, k dependent "inverse susceptibility" or "stiffness" function. (It is $\chi_{pair}^{-1}(k)\Delta_k$ in the linear approximation.) F and F^{-1} represent the response of state k to a pair field in the absence of interaction.

We now write Δ_k using (16) and (21) as

$$\Delta_k^{(i)} = E_j^{ij} \sum_{k'j} b_{k'}^{(j)} \delta\,(k - k') - \sum_{k'} U_{kk'}\, b_{k'}^{(i)} \tag{24}$$

where i, j refer to layers and E_j^{ij} is the Josephson coupling $\propto t_\perp^2/t_{11}$ between i and j. Now we average over the layers and write our equation strictly in terms of 2-dimensional $k's$, assuming Δ_k and b_k do not depend on i, j. (A simple diagonalization is necessary if Δ is dependent on i as in 3-layer compounds.) We introduce a general coefficient T_J for the sum (or eigenvalue) of the Josephson coupling, and get

$$\Delta_k = T_J\, b_k - \sum_{k'} U_{kk'}\, b_{k'}. \tag{24'}$$

Then equation (23) becomes

$$T_J\, b_k - \sum_{k'} U_{kk'}\, b_{k'} = F_k^{-1}(T, b_k)$$
$$F_k^{-1}(T, b_k) - T_J b_k = -\sum_{k'} U_{kk'}\, b_{k'}. \tag{25}$$

We now redefine an effective "stiffness" or "inverse susceptibility" function as

$$F_k^{*-1}(b_k) = F_k^{-1} - T_J b_k \tag{26}$$

and using (26) equation (25) becomes an equation of the same form as the BCS-Eliashberg equation:

$$F_k^{*-1}(b_k) = -\sum_{k'} U_{kk'}\, b_{k'} = \Delta_k^* \tag{27}$$

(defining Δ^*) or,

$$b_k = F^* \left(\frac{\Delta_k^*}{T}, \frac{\epsilon_k}{T} \right). \tag{28}$$

Here Δ^* can be obtained from the b_k by 27, so 27–28 is a self-consistent equation for $b_k(T)$. To see what this means, we write it out in the linear approximation which determines T_c:

$$F_k^{-1} \simeq [\chi_{pair}^0(k)]^{-1} \cdot b_k$$

$$(F_k^*)^{-1} \simeq \{[\chi_{pair}^0(k)]^{-1} - T_J\} b_k$$

$$F_k^* \simeq \frac{\chi_{pr}^0(k)}{1 - \chi_{pr}^0(k)T_J} \Delta_k^*.$$

And, finally,

$$b_k = -\frac{\chi_{pr}^0(k)}{1 - \chi^0 T_J} \sum_{k'} U_{kk'} b_{k'}. \tag{28'}$$

The vital difference between (28') and the usual BCS equation is that χ_{pr}^0 diverges only at $T = \epsilon_k = 0$ (where it behaves like const$/T$ or const$/\epsilon_k$), whereas $\chi^* = \frac{\chi^0}{1-\chi^0 T_J}$ diverges at $\chi_0 T_J = 1$, that is, at $T \simeq T_J$. Thus *all* transition temperatures must be greater than T_J if there *are any* negative (attractive) eigenvalues U of the interactions.

We see, however, that although T_J controls the magnitude of T_c in most cases, it has little influence on the form of the gap Δ_k in 2d momentum space. If the most attractive interaction is phonons, this will be BCS-like; if it is spin fluctuations, it has been proposed that $\Delta_k \sim k_x^2 - k_y^2$ and that will be what is observed; if the best eigenvalue is odd in ω or in $|k - k_F|$, that will be the observed form of gap. The interlayer mechanism has no prejudice on this matter, and any one or all may occur—as for instance, we may speculate that the "55° K" YBCO could be $d_{x^2-y^2}$ and the "95°" one phonon (BCS)-like or vice versa.

Solving the gap equation in this form is not at all simple, and we encounter some surprises. For orientation only (it may or may not be very realistic) we work out a case where we use the simple BCS model for $U_{kk'}$.

$$U_{kk'} = -V \quad (\epsilon_k, \epsilon_{k'} < \omega_D). \tag{29}$$

Then Δ^* of (27) is

$$\Delta^* = V \sum_{k'} b_{k'} = \lambda \int_0^{\omega_0} b(\epsilon_k) \, d\epsilon_k \tag{30}$$

(defining $\lambda = N(0)V$) and

$$\Delta_k = T_J b_k + \Delta^*. \tag{31}$$

For simplicity we approximate F by the two straight lines

$$b_k = \frac{\Delta_k}{E_k} \qquad \beta E_k > 1$$

$$b_k = \frac{\Delta_k}{T} \qquad \beta E_k < 1.$$

This will mean that our coupling constants T_J and λ must be adjusted downward by a factor of 2 or so, perhaps. The result for T_c is the equation

$$\frac{1}{\lambda} = \frac{T_c}{T_c - T_J} + \ln \frac{\omega_D - T_J}{T_c - T_J}. \tag{32}$$

If λ is small, the first term dominates and

$$T_c - T_J \simeq \frac{\lambda}{1 - \lambda} T_J \qquad x \to 0 \tag{33}$$

which vanishes only linearly with λ as $\lambda \to 0$: $T_c - T_J$ is much bigger than $(T_c)_{\text{BCS}}$: the effects of phonons are *enhanced*. If T_J is small, we recover a BCS-like result:

$$T_c \sim \omega_D \, e^{-(1/\lambda - 1)} \qquad T_J \ll T_c. \tag{34}$$

This is modified if we include the repulsive interaction. As for BCS, this repulsive interaction may be partially removed by ladder summation:

$$U \to \mu^* \simeq \frac{U}{1 + U \ln \frac{\epsilon_F}{\omega_D - T_J}} \tag{35}$$

and the effective attractive interaction is

$$\lambda \to \lambda - \mu^* \tag{36}$$

which is likely to be of the same order as in normal materials, $\sim.25$–$.5$. This leads then to a very enhanced $T_c > T_J$ by factors of 50–100%.

*p

Note that if λ is fairly large and T_J is not $\ll \omega_D$, the phonon enhancement *is not very sensitive to ω_D*: the isotope shift *even of the phonon enhancement*

$T_c - T_J$ is *small*. Thus the isotope effect interpolates from normal to nearly zero as T_c rises, as is observed.

One final point—in this case, applicable to all of the three ways of avoiding or reducing the effect of U. (That is, conventional dynamic "screening"; d—or other—angle-dependent gap; and odd-gap as originally proposed by me and followed up by Mila and Abrahams and by Balatsky et al.) This is that any interlayer tunneling effect is expected to be highly anisotropic around the Fermi surface. This is because $t_\perp$ is not independent of k, in fact strongly dependent on it, as can be seen from even the excessively literal-minded band calculations. The reason is a very simple fact of tight binding theory: the primary coupling between close CuO_2 layers is via indirect hopping to the Y (or other unique) ion separating these layers.

The actual band structure is for practical purposes a 2-parameter function

$$t(\cos k_x a + \cos k_y a) + t'[\cos k_x \, a \, \cos k_y \, a].$$

The dominant parameter is t, and because the orbital is a $d_{x^2-y^2} - O_{p\sigma}$ hybrid, with $O_{p\sigma}$ *odd* about the midpoint of two nearest neighbors, Cu's $t > 0$. Thus the band maximum for the antibonding band occurs not at Γ but at $k_x = k_y = \pi$, and one has a rough "topological" circle about this point. (Chain or other bands are irrelevant and do not seriously hybridize.)

Now the Y or (in BISCO or $TlCO$) Ba or Ca sits at the center of a square of Cu's. Thus the hopping integral will nearly vanish if k_x or $k_y = \pi$, because then the band amplitude on Y will nearly vanish (except for weak p-orbital contributions: s and d will vanish). Thus we expect the energy T_J to be small at these points, and thus $\Delta(k)$ will also be small there. This anisotropy has been observed by Shen et al. in exactly this form. It unfortunately resembles a $d_{x^2-y^2}$ gap but does not change sign, if motivated by phonons. Frankly, for 95° K material the almost miraculous impurity independence of T_c was hard to reconcile with a sign-changing gap, and one's prejudices are in favor of a simple phonon theory with T_J enhancement.

*d

VI. SURVEY OF THE EXPERIMENTAL PICTURE AND CORRELATION WITH THEORY

A great deal of the experimental description of the superconducting state can be expected to follow a highly anisotropic but conventional Landau-Ginsburg phenomenology, with values of coherence lengths and penetration depths which do not depend very much on the underlying theory (except for the one instance that c-axis conductivity does not predict the corresponding penetration depths). We take this phenomenology as given and focus on more microscopically relevant measurements.

It is hardly possible to find any microscopic phenomenon which behaves in a "conventional" way in the high-T_c materials. That this is so is partly concealed by the natural tendency to minimize the difference between the results and one's conventional prejudices. Sometimes this has amounted to the use of data selection and the "pencil finesse"—the summary and abstract which describe data in terms which don't represent the actual measurements. I shall try to call attention to some of the more common misconceptions about the data where it seems worthwhile.

I will list the various kinds of data in some kind of order of the importance I place on them. The emphasis is completely on the superconducting state; we covered the normal state in chapters 3 and 6. I give first a list of the kinds of data which I will discuss.

1. Photoemission
    Angle-averaged; angle-resolved
2. Heuristics of T_c
3. Infrared
    c-axis data; the "2-fluid" model
4. Specific Heat (incl. magnetic field effect)
5. NMR
    Evidence for zero gap; 2-fluid; Hebel-Slichter absence
6. Tunneling
    Asymmetrics; zero-bias anomaly: c-axis differences
7. Normal Fluid Conductivity
    Below T_c: Thermal and microwave
8. Penetration Depth
9. Scattering Effects
10. Isotope Effect: Some Speculations
11. Miscellaneous

1. Photoemission

This is the key experiment in high-T_c materials. Unlike low-T_c superconductors, the effects of superconductivity are striking both in angle-averaged and angle-resolved experiments. Photoemission measures directly the density of states $\rho(\omega) = Im G_1(\omega)$, either averaged over k or, for ARPES, $\rho(k, \omega) = Im G_1(k, \omega)$.

Angle-averaged: Here the best—but not only—experiments are those of Petroff.[11] It is quite evident from fig. 7.5 that a spectacular shift of intensity takes place at T_c in BISCO. What is striking is *(1)* that this shift cannot be modeled at all by BCS; *(2)* that for angle-averaged photoemission conventional theories of BCS type unequivocally satisfy an overall sum rule, because of

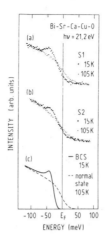

Figure 7.5. Illustration of contrast in sum rule satisfaction in all-angle photoemission between BCS and high T_c. Lower figure is a calculated BCS, upper is actual measurement. Note shift of weight *out* of gap region in BCS is bigger than peak; vice versa in experimental data from Ref. 14.

the identity

$$n_{k_F+q} + n_{k_F-q} = 1$$

which holds above and below T_c.

A simple calculation shows that, in fact, in a limited region of width $n\Delta$ where Δ is the energy gap, an amplitude of order $\frac{2}{n}$ should be *missing*, shifted to higher energy: the loss in the gap is *undercompensated* [as we note in the BCS fit of fig. 7.5]. What occurs experimentally is clearly an *overcompensation* of the gap (by a factor of about 3 in these experiments) as opposed to an expected *undercompensation* by a factor 2. Thus the discrepancy between BCS and experiment is a factor 6. The interlayer theory result is that a fraction of order $\frac{1}{2}$ of the amplitude is shifted from the long-tailed density distribution

$$\rho = \alpha\,\omega^{-1+\alpha}$$

to a peak at Δ, over a range of energies $\sim \Delta$. This is true only for ϵ_k within Δ of the Fermi energy. With Δ of order $\frac{1}{5}$ to $\frac{1}{10}$ of the cutoff, the factor 6 is easily accounted for.

This angle-averaged result is one of the firmest arguments in favor of the interlayer theory; yet the experimentalists have claimed agreement with BCS theory in their descriptions of the data, I think primarily as a result of misunderstanding the nature of the disagreement.

Another point can be made here. In the BCS theory an old theorem due to Chester demonstrates approximate equality of the decrease in phonon frequencies ω_{ph} and hence of the ion's zero point energy with the condensation energy $-N(O)\Delta^2$, as may be expected on very general grounds, though it appears to be a peculiarity of BCS that the electronic energy changes cancel. The photoemission data show clearly that, unlike the BCS case, the electronic kinetic energy *decreases* below T_c, suggesting that this *electronic energy lowering* provides the source of the condensation energy.

Angle-resolved photoemission: in several measurements both in YBCO and in BISCO samples, this provides a detailed picture of these large shifts in the density of states below T_c. The original published data of Arko et al. (fig. 7.4) are very striking in the sharpness of the superconducting peak, a sharpness incompatible numerically with the explanation given by the experimenters in terms of "clumping" of amplitude at the gap and of poor momentum resolution. As we remarked, the amplitude decreases to half as $k \rightarrow k_F$, and BCS gap theory cannot be smeared out to resemble fig. 7.1. Again invoking the sum rule, one must assume the peak amplitude to have come from the whole "tail" of the normal state distribution.

Measurements of Mitzi, Shen et al.[12] confirm this picture in more detail (fig. 7.1a), and bring out a new feature: the dip in amplitude below $\sim .09 \, ev$ leaving a sharp edge at .09 in addition to the peak at Δ with $\Delta \sim .03 \, ev$. $.09 \, ev$ is suspiciously close to the expected magnitude of $t_\perp$, but as yet we have not been able to solve for a Green's function for a pair of layers this far below T_c. It is also possible that a 3Δ effect could appear in the nonlinear gap equation as suggested by some authors, but I feel this is equally likely to be a feature of the interlayer effect. The shift in density of states is equivalent to a very anomalously rapid decrease in the "quasiparticle" scattering rate at T_c, which is seen in many other experimental situations as emphasized by Littlewood et al., and as we shall discuss later. As discussed in section V, the photoemission evidence for anisotropy is in excellent agreement with the interlayer mechanism. Some of the experts on photoemission (especially N. V. Smith and Y. Petroff) have argued that the interpretation by Olson et al., Varma et al., and by myself is erroneous and that a large broadening due to the uncertainty in velocity in the c-axis direction is present. This can be resolved experimentally in two ways: simply by observing that σ_c is related to $\langle v_c^2 \rangle$ and that σ_c is very small, or by using the width of the superconducting peak to calibrate the energy and angle resolution. Both are consistent, as we argue elsewhere, with the bands dispersing *only* in the ab plane. If that is the case, the resolution depends only on the angular aperture. Allen (and also Sawatsky) have ably argued that there is no appreciable excess broadening.

2. Heuristics of T_c

Heuristics of T_c values were a primary hint leading to the superconductivity mechanism. The normal state properties of cuprate layers are remarkably in-

TABLE 7.1

Type of pair	V_{ij}	T_C's predicted by V_{ij}'s
Cu "close layers"	80°	2201 2212 2232
Through chains	15°	Bi 10° 85° 112°
Through Tl_2O_4	30°	Tl 60° 110° 125° "∞-layer": 160–170°
Through Bi_2O_4	5°	YBCO 95°
La_2CuO_4 structure	20°	La_2CuO_4 40°

dependent of the intermediate "reservoir" layers but this is clearly not true of superconductivity and particularly of T_c values. The Bi materials illustrate this rule best of all. BISCO 2 2 0 1 ($Bi_2Ca_2CuO_6$) can be prepared with T_c from 0 to 10°K, depending on the amount of residual "magnetic" scattering in very well characterized crystals; "2 2 1 2" ($Bi_2Ca_2SrCu_2O_8$) is the best character-ized and easily prepared crystal of all, and is normally 85° K, and "2 2 2 3" with three layers is 110°. Tl compounds show a similar rapid rise with number of close layers, starting however at about 60° for one-layer materials. Apparently the "Tl_2O_3" layers have quite low-lying bands hybridized in.

Hsu,[2] particularly, made a reasonable fit of all the (maximum) T_c's for given nominal compositions. He defined a coupling constant "V_{ij}" $\propto t_\perp^2/t_{||}$ for each pair of layers i, j and solved the simple set of equations

$$\Delta_i = \frac{V_{i,i+1}}{T_c} \Delta_{i+1} + \frac{V_{i,i-1}}{T_c} \Delta_{i-1} \tag{31}$$

for the eigenvalue T_c. One can easily find a consistent set of V's which fit all T_c's to within 10% or less, if the composition is optimized. A table of a few of these is given as Table 7.1 and the T_c's deduced are listed.

The dominant interaction in all the highest T_c cases is that between "close" CuO_2 layers, i.e., those with no "apical" oxygens and only a divalent ion in the intervening layer, which can be counted on to provide $\sim 80°$ for a pair of layers, otherwise isolated (and in mean field theory, fluctuations are ignored).

It is clear that we have too many parameters to pretend to a rigorous fit but the striking correlation with the $Bi - Tl$ numbers and the qualitative overall fit to many other data which are given by assuming that close pairs dominate is quite convincing to me. The numerical agreement with estimates of $t_{||}$ and $t_\perp$ is also very good. $t_\perp$ between the close planes is about .1–.15 ev (from effective interlayer exchange integrals or estimates of c-axis curvature in band calculation). $t_{||}$ is about .5 ev, giving $\frac{t_\perp^2}{t_{||}} \gtrsim .02$ ev and the power-law reduction accounts for the slightly low T_c's relative to the similar parameter $J_{interlayer}$.

This table need not be very seriously modified when we take phonon enhancement into account. It is possible that $Bi\ 2\ 2\ 0\ 1$ is almost entirely a phonon transition temperature so the $5°\ V_{ij}$ is probably really zero. Otherwise, since where T_J is reasonably large $\Delta T_c \propto T_J$, we simply assume phonon enhancement to be a constant factor. It improves the numerical agreement, in fact makes
*e   it fit very well.

3. Infrared

The infrared spectrum contains a number of useful indications but, unlike its role in the normal state, is not totally conclusive. The best data by far are on YBCO single crystals, with both reflection and transmission available. A little reflection data are available on BISCO.

On pure $Y Ba_2 Cu_3 O_7$ the gap in σ_{ab} appears strikingly at $\sim 500\ cm^{-1}$, and the total missing conductivity matches very well with measurements of penetration depth for in-plane currents. Schlesinger's work on untwinned crystals[13] seems to separate well into a plane contribution perpendicular to the chains, with a clean $500\ cm^{-1}$ gap, and a residual finite conductivity feature which may be due to a chain band which does not seem to be gapped to the same extent.

As we move toward $O_{6.7}$ with $60°\ K\ T_c$'s, the $500°$ gap-like feature remains, to some extent, even above T_c. Below T_c there seems to be a decrease in
*f σ_{ab}, primarily below the smaller gap, which accounts for λ^2 according to sum rules.

Strikingly, in the O_7 crystals the decrease in σ_{ab} does not take the form of a gap opening but of a gap deepening, a phenomenon which has been called an effective "2-fluid model" since the same T-dependence appears to fit λ^2, infrared conductivity, NMR $1/T$, and Knight shift (see later). It is too early to see whether this behavior is a consequence of our new form of gap equation; this must await numerical solutions of the rather complex problem which emerges when we include anisotropy of $t_\perp$. These will certainly not resemble the simple BCS curve and will depend very much on the detailed physics of each particular substance. We ascribe this qualitatively to the locality in k-space possible with the interlayer mechanism (see fig. 7.2 in previous section) which implies that the gap may jump quickly to a rather large value for k's with a large $t_\perp\ (k)$ and near the Fermi surface, while remaining nearly zero for other values of k. But computations, as we emphasized in that section, are almost impossible to do reliably. The key point is that we predict the type of non-BCS behavior seen,
*h   in sign and approximate magnitude.

The c-axis data which exist suggest a very striking fact: that the c-axis *infrared* conductivity continues to be small in spite of the development of strong c-axis *super*conductivity. This reflects the fact that the pairing self-energy which connects the layers is entirely of anomalous type and does not transport single quasiparticles in the c direction. This is unique to the interlayer mechanism. D.C. measurements in large ab axis magnetic fields, which decouple the phases of different layers and thus turn off c-axis supercurrent, also show no c-axis quasiparticle conductivity.   *i

4. Specific Heat

The specific heat data is very substance-dependent. Good samples of YBCO show a pronounced, steep specific heat jump, while BISCO, for instance, has never shown a jump but instead has an apparently "third-order" character as though the gaps appear quite gradually. This sample—and structure—sensitivity is itself in sharp contrast to BCS superconductors, where homogeneous samples always show very sharp thermal transitions and an almost universal transition shape. Part of the contrast, of course, is caused by the short coherence lengths of high T_c, but the resulting fluctuations should steepen, not smear out, the specific heat and other thermal properties. In the interlayer theory, there is no reason to expect a universal behavior, since the nature of the transition will be strongly affected by anisotropy around the Fermi surface and by the residual interactions, even though T_c will be relatively insensitive to structure as will the superconductivity at lower temperatures.   *j

One effect which is fairly universal and highly visible in YBCO is the dependence of the specific heat peak on magnetic field.[14] The onset T_c does not shift with field, but the specific heat "leans over"—becomes broader, and the vertical jump becomes a linear slope inversely proportional to field.

Detailed calculations are not in order but this is, again, evidence of considerable variation in superconducting properties from one momentum to another.

5. NMR

NMR data are, again, not at all of BCS form. There is no Hebel-Slichter peak of T_1^{-1}, below T_c. A subtle point makes a contrast between NMR T_1's and Knight shifts vis-à-vis transport properties such as thermal or electrical conductivity: as far as all spin properties are concerned, the spinons behave like ordinary quasiparticles at all momenta except near $2k_F$, where they have an enhanced susceptibility in the normal state which enhances the Cu nuclear relaxation. Thus quasiparticle formation, as far as spinons are concerned, merely reattaches the charge to the spin, which was responding much as though it were a conventional Fermion in the first place. Thus the reappearance of the quasiparticle peak causes no anomalous spin responses. The spin properties are controlled by conventional spin-spin correlation functions, not the anomalous responses which accompany any macroscopic current flow.

The "2-fluid model" occurs for the same reason that it does for infrared response: the momentum values seem to be going superconducting serially rather than all together, with different T_c's, so coherence factors have little effect.

A very important series of experiments by Takigawa[15] demonstrate that the gap function probably has zeros, or small values, in that he shows that susceptibilities are nonlinear in field even at low T. The nonphonon possibilities such as "d-wave" or odd gaps are somewhat favored by this observation and others suggesting actual gap zeros: the experimental situation is still not clear. $t_\perp$ anisotropy may be the simple explanation. A puzzling phenomenon is the correspondence of T_1^{-1}'s and Knight shift K, not K^2. I see no reason for this yet.

*k

6. Tunneling

The situation with regards to tunneling in high-T_c superconductors is as anomalous and frustrating as ever. This may not be surprising in that the state is so very nonclassical, actually especially the normal state, but even in the superconductor quasiparticles have no c-axis velocity among other anomalies. The observations can be summarized more or less as follows:

I. Evaporated junctions: Good evaporated normal metal to high-T_c junctions can be produced, unequivocally giving single-particle tunneling because they show the counterelectrode's appropriate superconducting tunneling spectra. Most such junctions show similar tunneling spectrum with the following features:[16–18]

   (1) In the normal state, a clear | $\vee$ | conductivity usually larger than the zero-bias term.
   (2) In the superconducting state, a sequence of "humps"—surely not BCS "peaks"—at voltages compatible with reasonable gaps, the highest being in reasonable agreement with infrared or other measurements.
   (3) The "| $\vee$ |" persists at high voltages.
   (4) There is a large background conductance which often—usually—shows a very sharp "zero-bias anomaly": a dip which can be extremely sharp and very temperature (and magnetic field) dependent, even at very low temperature.
   (5) The structure near the "BCS" humps is often asymmetric in positive vs negative voltage, an effect *never* observed in ordinary superconductors.

II. STM measurements often show clearer gaps and more definite peaks, but are very nonreproducible. The STM gaps are the large gaps of the IR measurements.

*l

III. Break junctions sometimes show very clear two-superconductor characteristics but irregularly and with the peaks almost too pronounced.

IV. Evaporated junctions deliberately made to avoid the c-axis show larger background current and no "| ∨ |" characteristic.

V. Under some circumstances clear Andreev effects are seen but seldom is the Andreev "gap" more than a smallish fraction of what the true gap must be.

This hodgepodge can be sorted out a bit by conjecturing that c-axis and a-axis tunneling may be very different, and that also, as suggested by Littlewood et al., along the c-axis the transverse momentum of the tunneling electrons in the barrier is too small to match the 2D Fermi surface and direct tunneling is inhibited. A few conjectures follow:

(1) The "| ∨ |" is a characteristic of c-axis tunneling and may well be, as originally conjectured, the effect of spin-charge separation: perhaps no true c-axis tunneling at zero bias and zero temperature (in the normal state) is possible.

(2) The background current and zero-bias anomaly come from tunneling in the a or b direction, either at crystal edges, etch pits, or other defects. Both may well be consequences of the vanishing of the gap at some or all points of the Fermi surface.

The strangest observation of all is the very sharp zero bias anomaly (seen in tunneling into normal metals: it *cannot* be a condensate effect). A direct derivation of the zero bias anomaly and of its puzzling sharpness and power-law character has not yet been even attempted. The possibility that this is a chain effect should not be overlooked—is it observed for Bi or 2 1 4 compounds?

An important second anomaly is the hole-particle asymmetry in tunneling curves. This is as yet unexplained.

In summary, the main point which should be made is that the theory of this chapter contains considerable richness to match the puzzling and complex phenomenology of high T_c tunneling. Nonetheless, we cannot say that we have yet explained the most peculiar observations such as background current and zero-bias anomalies. It is worth noting that all theories so far ignore the chains, which surely have open Fermi surfaces. It would be useful to see how many of these anomalies recur in BISCO and 2 1 4 samples, on which little tunneling work has been done.

7. Normal Fluid Conductivity

A very important set of experiments indicates that the normal fluid, i.e., quasi-particle conductivity, has a strong peak below T_c, at least in the planar direction; quite the opposite is true of the c-axis quasiparticle conductivity. Various slightly controversial and difficult microwave measurements of this phenomenon exist, which are sufficiently numerous and diverse that the existence of the peak is strongly supported; but the clean measurement is that of thermal conductivity by Ong and Wang,[19] Kapitulnik,[20] and Salomon[22] et al. The upturn (see figs. 7.6a, b) has been ascribed to a reduction in phonon scattering by electrons, but as we know there is no evidence for phonon scattering of electrons, at least of the low-frequency acoustic phonons which are responsible for thermal conductivity, so that it is questionable to invoke the inverse effect. The temperature dependence and magnitude of ab plane thermal conductivity is consistent with a Wiedemann-Franz relation to electrical conductivity. (In the c direction there is *no* T_c anomaly and the conductivity must be all phonons; Wiedemann-Franz gives a negligible electronic contribution.) There is, of course, in the Fano interference of optical phonons with the electronic Raman background, evidence for interactions between electrons and optical phonons. That such features are relatively weak in the optical conductivity, though present, shows that the transport processes are not much affected by electron-phonon scattering. In essence, the electron gas is immune to static impurity scattering as shown in chapter 6. "Frozen phonons" cause no scattering, so the scattering rate is proportional to ω_{ph}/J and is repressed by at least an order of magnitude in the normal state. The phonon thermal conductivity is in any case too small to account for the peak, if we believe Wiedemann-Franz.

The thermal conductivity is a particularly clean measure of the quasiparticle contribution since the superconducting pole contributes nothing. Both conductivities are given by the integral of the square of the appropriate interacting, scattered one-particle Green's function, which Green's function is the quantity which shows such a marked quasiparticle peak in the superconducting state in photoemission. Strong has estimated the effect of the rapid rise of this peak—equivalent to the rapid decrease of quasiparticle scattering rate referred to by Littlewood—and shown that this can be correlated with the peak in thermal or electrical conductivity.

These phenomena are one of the cleanest demonstrations of the charge-spin separation physics.

8. Penetration Depth

The main effect having to do with the penetration depth is simply the considerable discrepancy between the missing c-axis conductivity in the gap and the observed c-axis penetration depth. This we have mentioned repeatedly and needs no further discussion.

The T-dependence of the penetration depth has been often adduced as an argument against gap zeroes because the low-temperature penetration depth does not deviate from zero more rapidly than, at best, T^2 and probably more slowly than that. However, it is notable that even quite precise measurements have never been able to reach into the exponential regime which is quite evident in the BCS, lower temperature superconductors. Some authors are quite

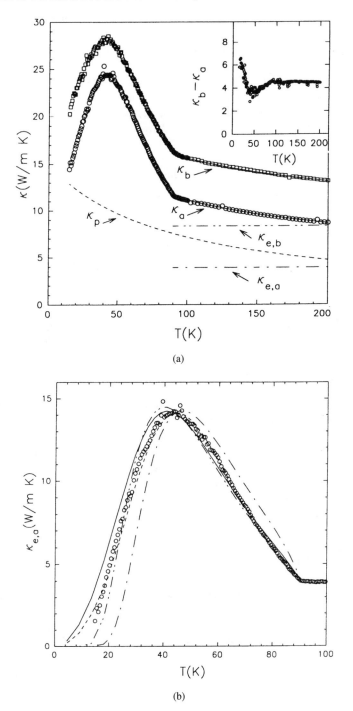

Figure 7.6. (a) Thermal conductivity peak below T_c, from Wang and Ong Ref. 22. (b) The same from H. B. Salomon et al., Phys. Rev. Lett., **69**, 1431 (1992).

persuaded that there are gap zeroes and a T^2 dependence of λ, but we shall
reserve judgment.

*m

The penetration depth, the background infrared conductivity, and the NMR χ and $(T_1 T)^{-1}$ all vary very congruently with temperature (the "effective 2-fluid model" of Schlesinger and others). This is understandable given the relative independence of different momentum values; there will be a tendency for a large fraction of the momenta to be either gapped to a fairly large extent, or not at all, and for the energy gap to be relatively uniform, both of which lead qualitatively to this result.

9. Scattering Effects

Ordinary impurity scattering, as we have seen, does not affect the normal state, because, in effect, it does not scatter the spinons which determine the Fermi surface (or, an alternative description, because charge-spin separation prevents elastic scattering of any kind). This carries over at least to T_c which is determined by a perturbation theory, and in fact the coherence length once established is so short that scattering has little effect in any case. "Magnetic" scattering which scatters spinons, however, reduces T_c proportionally to the scattering rate, even if the magnetic moment is Kondoized away. The decrease in T_c in the Ni and Zn doped YBCOs (see figs. 7.7a, b)[21] is directly and almost quantitatively proportional to the corresponding effect on the spinon relaxation rate, which can be measured with the Hall effect. The effect may be estimated simply by recognizing that the effect of magnetic scattering on the pair susceptibility at a given k will be virtually identical, except for small corrections, to that in BCS. It would be interesting to follow this effect out in more detail with photoemission and infrared measurements on doped single crystals. In one-layer-BISCO, Simon has studied resistivity in the region where scattering is very close to reducing T_c to zero, and it is clear that at $T = 0$ the magnetic scattering is increasing the resistivity somewhat more effectively. This may well be caused by disruption of the Luttinger Liquid correlations within $1/\tau$ of the Fermi surface leading to reappearance of ordinary scattering. The behavior in this region is quite complex and very worth studying.

One form of scattering which is quite effective is the grain boundary. In fact, quite small-angle grain boundaries are used as Josephson links in the high-T_c materials.[22] This is an excellent confirmation of the momentum-conserving nature of the pairing potential. A grain boundary in an ordinary metal with a BCS isotropic gap reflects and refracts the incoming electron from one state in one grain into a new state with the same gap and kinetic energy. Unitarity of the scattering and the finite transmission coefficient of the grain boundary mean that it is only a slight weakness in the pair stiffness, a slight local decrease of the effective ρ_s.

Quite a contrast in behavior of the grain boundaries is observed in high-T_c materials. The grain boundary disconnects the gap functions of the two grains

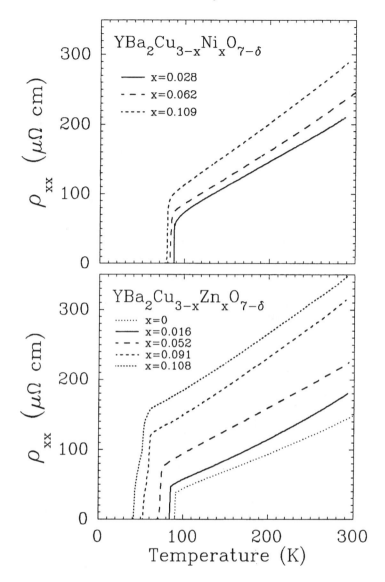

Figure 7.7a. Doping dependence of T_c for Ni and Zn doping in the planes, also showing enhancement of residual resistance without effect on linear resistance.

very effectively, with a very sharp dependence on the angle between the two grains. In fact, a reasonable description of the behavior would be that for sufficiently small angles the grain boundary causes no scattering at all, while at a larger angle it becomes a barrier. This is the "critical strength of scattering"

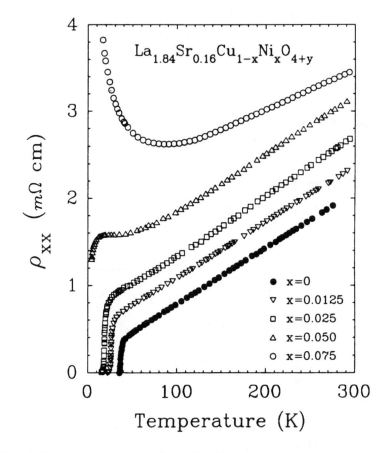

Figure 7.7b. Doping dependence of T_c for Ni and Zn doping in the planes, also showing enhancement of residual resistance without effect on linear resistance.

phenomenon which we deduce from the basic Luttinger liquid analysis: a weak perturbation has no scattering effect, a strong one becomes a barrier with very small transmission—i.e., *confines*. It is important to measure the normal state resistance of the grain boundaries to see whether this conjecture is correct.

10. Isotope Effect: Some Speculations

One of the strong predictions of the interlayer theory is that momentum states with $\epsilon_k > \Delta$ will be mostly unaffected. Thus when a BCS peak is seen—in tunneling or photoemission, for instance—it should have an upper as well as lower edge. This is in fact rather visible in photoemission and, occasionally, is suggested by tunneling data. Unlike the extensive tail and "phonon bumps"

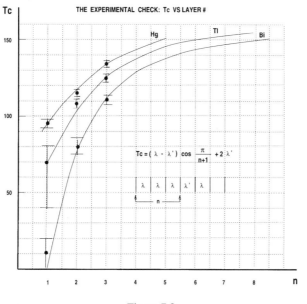

Figure 7.8

of the BCS-Eliashberg theory, higher energies should in general show little structure (but see the specific structure in photoemission possibly ascribable to the interlayer energy).   *n

The isotope effect in these materials may not be as definitive as in the usual ones because, as Fisher et al. have suggested, the conventional "Grüneisen" effect of zero-point motion may not be negligible. However, there is a clear trend for $\alpha = \frac{d \ln T_c}{d \ln m}$ to decrease with T_c; it is finite if small for 2 1 4, within experimental error of zero for $T_c > 80°$.

The modified gap equation of section V does have this behavior *even* when an appreciable fraction of T_c is contributed by phonons. When T_J is large, $(\Delta T_c)_{phonons} \simeq \frac{\lambda^*}{1-\lambda^*} T_J$, which is *independent* of $\hbar\omega_D$. On the other hand, for small T_J the BCS expression for T_c is more nearly right and α should be finite but not as big as 1/2. Strong is calculating this dependence.   *o

One should be on the lookout for other manifestations of the "third length (or energy) scale": the relaxation rate of spinons $\equiv$ the breadth of the k-space or frequency δ-function in the gap equation.

In summary, so far many data can be uniquely correlated by means of the interlayer theory, but many more remain to be explained. There are no completely incomprehensible effects but many complex and interesting ones. One may straightforwardly reject simple BCS or even strong-coupling Eliashberg on the basis of many data, but the present theory has no claims to be comprehensive as yet, only not to be excluded by the many otherwise puzzling anomalies.

NOTES

*a This statement is overly dogmatic; as Varma pointed out, this merely confirms that the scattering causing the T-dependent resistivity is electronic in nature and hence gapped by superconductivity. In chapter 4 we remark however on the modern measurements of $\hbar/\tau$ which do suggest spin-charge separation, and go beyond what is possible by merely gapping the spectrum.

*b Chakravarty and Lin, in a long preprint, have made some progress in this direction. I do not agree, however, with their model for G.

*c This expression is derived and discussed in a paper by Chakravarty and Anderson reproduced among the reprints. The main point is that the energy integral involved comes mostly from very high energies, hence an effective Hamiltonian without dynamic effects is appropriate. As remarked in the chapter 2 epilogue, there are also "superexchange" interactions of similar form, which have little effect on most results.

*d This discussion is incorrect in almost every possible respect. First, $t_\perp$, or at least the gap, is observed to be maximum at $M = o, \pi$ and $\overline{M} = \pi, 0$, not as stated in the text. Second, O. K. Andersen's band calculations as interpreted by Chakravarty et al. show that the coupling is not through the Y, which does not hybridize much with any of the relevant states; the $0p\sigma$ interacting only with f-levels on the Y. Instead, Cu 4s is the orbital which carries the coupling which is direct plane to plane and the 4s admixture cancels along the lines $\Gamma - Y$ and $\Gamma - X$ from 0,0 to π, π. (The two neighbors have opposite sign because of the $d_{x^2-y^2}$ sign change.) Third, the gap is certainly of d symmetry (which fits well with $t_\perp$'s variation) and "one's prejudices" were just wrong. The miraculous impurity independence is still just that. Is the holon gap s-wave, the spinon gap $d_{x^2-y^2}$? Is this possible?

*e The phonon enhancement of course does not exist for d symmetry. The Hg and Pb superconductors discovered recently fit in reasonably well with Bi and Tl, the bands from the charge reservoir layers coming successively closer to the Fermi level and presumably hybridizing into the planes increasingly strongly. The inter-bilayer coupling remains smaller than that within the bilayers. An intriguing question is whether there are actual pockets of Fermi surface in the charge reservoir layers, as appears to be the case for the chains. I have included a figure (7.8) showing T_c vs. layer number n for $n = 1, 2, 3$ ($n > 3$ has not been adequately doped in most cases.) This shows the broad agreement between Hsu's formula and T_c's for Bi, Tl, and Hg compounds.

*f This is the "spin gap" which we have discussed elsewhere. The spin gap-superconductivity competition is a complex subject about which I have only preliminary, quite vague ideas. The true gap may open primarily at nonzero ϵ_k and couple the spin gap phase only partially.

*g There is good evidence that the spin gap is not an SDW. The discussion here seems to have been prescient, according to our latest ideas.

*h The fact that the solutions have these characteristics is not widely understood. A rough description is given in chapter 2 (epilogue). Experimental observations confirming these anomalies are constantly appearing.

*i This work is discussed in detail in chapter 4.

*j As I suggested at the Les Diablerets Gordon Conference, Sept. '95, the difference between BISCO and $YBCO_7$ can well be the difference between spin gap and non–spin gap materials. The spin gap will shift a considerable part of the condensation energy to above T_c, and change the character of the transition, if our theory of it is correct.

*k It is, of course, now clear that gap zeros—even regions of zeros—actually occur.

*l Recent measurements by Fischer on BISCO seem very reliable. They show a single large peak with no tail at higher energies but asymmetry above the peak, and a small but definite conductivity extending all the way to zero voltage, fitting the photoemission gap shape rather than "d-wave."

*m There is no longer any need to reserve judgment; the early measurements were simply not as accurate as claimed, and it is quite clear there are gap zeroes. The extended regions of zeroes, which photoemission shows, should not affect the T-dependence of penetration depth, which depends only on the shape of the rising portion of $\Delta(k)$.

*n And probably not, but to the motion of amplitude from the Green's function tail into the quasiparticle peak.

*o The isotope effect is still a puzzle both experimentally and theoretically. Many doped or otherwise impure or anomalous samples show large isotope effects. On the other hand, the claim of a finite isotope effect for pure, optimally doped 2 1 4 has become subject to some controversy. The samples were very close to a well-known anomaly at $\delta = .125$, which is probably a commensurate CDW of some sort (in fact, the "stripe n phase"). Long experience with isotope effect measurements leads me to reserve judgment almost completely—this is the least reliable of all experimental probes.

*p One is tempted to suggest that the moderate T_c of NdCCO could be an example of the theory of this section.

REFERENCES

1. J. M. Wheatley, T. C. Hsu, and P. W. Anderson, Nature **333**, 121 (1988); Phys. Rev. B **37**, 5897 (1988).

2. T. C. Hsu and P. W. Anderson, Physica C **162–64**, 1445 (1989); also T. C. Hsu, *Thesis*, Princeton University, 1989.

3. W. L. McMillan and J. M. Rowell, "Superconductivity," R. D. Parks, ed. (Dekker, N.Y., 1969), p. 561.

4. M. L. Cohen and P. W. Anderson, AIP Conf. Proc. #4, "Superconductivity in d+f Band Metals" (1972), p. 17.

5. W. Kohn, Phys. Rev. **133**, A171 (1964).

6. E. I. Blount, "Solid State Physics, Adv. in Res. and Appl.," vol. 13, Seitz & Turnhall eds. (1962), pp. 306–70.

7. G. D. Mahan, "Many Body Physics" (Plenum, N.Y., 1981), p. 347.

8. P. W. Anderson and Y. Ren, Proceedings of the International Conference on the *Physics of Highly Correlated Electron Systems*, Los Alamos, December 1989,

"High Temperature Superconductivity," K. Bedell et al., eds. (Addison-Wesley, 1990), pp. 3–33.

9. P. W. Anderson, Proceedings of the Enrico Fermi International School of Physics, Varenna July 1987, "Frontiers and Borderlines in Many Particle Physics" (North-Holland, 1987).

10. C. G. Olson, R. Liu, D. W. Lynch, A. J. Arko, R. S. List, B. W. Veal, Y. C. Chang, P. Z. Jiang, and A. P. Poukilar, Science, **245**, 731 (1989).

11. Y. Petroff et al., Phys. Rev. Lett. **62**, 336 (1989).

12. D. S. Dessau, B. O. Wells, Z.-X. Shen, W. E. Spicer, A. J. Arko, R. S. List, D. P. Mitzi, and A. Kapitulnik, Phys. Rev. Lett. **66**, 2160 (1991).

13. Z. Schlesinger, L. D. Rutter, and R. Collins, et al., Physica C **185**, 53 (1991); Z. Schlesinger, R. T. Collins, et al., Phys. Rev. Lett. **65**,801 (1990).

14. S. E. Inderhees, M. B. Salamon, J. P. Rice, and D. M. Ginsberg, Phys. Rev. Letters **66**, 232 (1992).

15. M. Takigawa, J. L. Smith, and W. L. Hults, Physica C **185–89**, 1105 (1991).

16. M. Gurvitch et al., Phys. Rev. Lett. **63**, 1008 (1989).

17. L. H. Greene et al., Proceedings of "Elect. Structure & Mech. of High-T_c Superconductivity," Ashkenazi and Vezzoli, eds. (Plenum, 1991).

18. M. Lee, A. Kapitulnik, and M. R. Beasley, "Mech. of High-T_c Superconductivity," H. Kamimura and A. Oshiyama, eds. (Springer, Berlin, 1989), p. 220.

19. S. Hagen, Z. Z. Wang, and P. Ong, Phys. Rev. B **40**, 9389 (1989).

20. K. Mori et al., Physica C **162**, 572 (1989).

21. D. A. Bonn, P. Dosaijh et al., to be published, Phys. Rev. Lett; Bonn, Liang et al., Submitted to Nature (1992); M. C. Nuse et al., Phys. Rev. Lett. **55**, 3305 (1992).

22. T. Chaen et al., Phys. Rev. Lett. **67**, 1088 (1991).

PART II
Supplementary Material

A

"Luttinger-Liquid" Behavior of the Normal Metallic State of the 2D Hubbard Model

P. W. Anderson

(Received 13 December 1989)

ABSTRACT

Analysis of interacting fermion systems shows that there are two fundamentally different fixed points, Fermi-liquid theory and "Luttinger-liquid theory" (Haldane), a state in which charge and spin acquire distinct spectra and correlations have unusual exponents. The Luttinger liquids include most interacting one-dimensional systems, and some higher- (especially two) dimensional systems in which the band spectrum is bounded above: systems with Mott-Hubbard gaps and an upper Hubbard band. We give a theory which is useful in calculating normal-state, and some superconducting, properties of high-T_c superconductors.

PACS numbers: 71.25−s, 71.30+h

Haldane[1] has characterized the behavior of a large variety of one-dimensional quantum fluids by the term "Luttinger liquid," showing that they can all be solved by common techniques based on transforming to phase and phase-shift variables for the Fermi-surface excitations (a procedure often called "bosonization" even though some of the Luttinger liquids start out as Bose systems). These systems are characterized by fractionation of quantum numbers—e.g., in the Heisenberg spin chain the excitations are spin-$\frac{1}{2}$ fermionlike, while in the Hubbard model they are spin-$\frac{1}{2}$ chargeless spinons and $\pm e$ spinless holons with fermionlike properties—and, often, a Fermi surface with nonclassical exponents, and unusual exponents for correlation functions, but the correct "Luttinger" volume.

I will here restrict the term, for my purposes, to systems based on fermions– preferably ordinary electrons—and argue that the Luttinger liquid is a fixed point, or a manifold of fixed points, of the same renormalization group which, "usually," leads to the Landau-Fermi liquid as a unique fixed point. (The interaction parameters of Landau-Fermi-Liquid theory are well known[2] to be marginal operators around a single fixed point, the effectively free Fermi liquid.)

Some years before, Luther[3] showed that the bosonization techniques used to solve these one-dimensional models are equally applicable to d-dimensional Fermi gases, and he claimed that they describe certain facts slightly more accurately than Fermi-liquid theory—the existence of $2k_F$ singularities in correlation functions, for instance, for the free-particle systems. But Luther did not consider the possibility that the interacting d-dimensional problem could lead to new physics.

The first new point I want to make is that two of the reasons usually given for the unique nature of one-dimensional Fermi systems are untenable. The first is that in 1D one has only forward scattering, or backward scattering where the momentum of one particle is maintained, if not its spin. This is indeed the correct reason for viability of the Bethe *Ansatz*. But after renormalization the Landau theory has only forward or exchange scattering, and the renormalized particles indeed obey a Bethe *Ansatz* of the simplest form. This is the essence of Luther's argument, that the excitations can be bosonized in each direction around the Fermi surface.

Second, it is argued that particles cannot be interchanged in 1D without encountering phase-changing interactions, and hence statistics are meaningless in 1D: but none of Haldane's arguments seem to fail in the slightest if we introduce weak long-range hopping integrals in any of the examples, and such hopping integrals can allow a Berry process. No one argues, in fact, that real electrons living in 3D space in the presence of a chain of ions, which know perfectly well that they are fermions, will not obey the models and show the fractionization effects.

The unique effect in 1D is one which is also present in a class of higher-d models, specifically 2D repulsive Hubbard models, and in some strong-coupling higher-d cases. This is the presence of an *unrenormalizable Fermi-surface phase shift*. Such a phase shift signals that the addition of a particle changes the Hilbert space for the entire system of particles—it requires a net motion of field amplitude through the distant boundary of the system, or a net change of wavelengths. The effects of such phase shifts were explored thoroughly in connection with the "x-ray edge problem"[4] and are summarized in the "infrared catastrophe"theorem:[5]

$$\langle VAC(V) \mid VAC(0) \rangle \propto \exp[-\tfrac{1}{2}(\delta/\pi)^2 \ln N], \qquad (1)$$

where $|VAC(V)\rangle$ is the noninteracting Fermi sea in the presence of a potential V which causes a phase shift δ. The singularity is the result of the shifting of the entire spectrum of k values (in the presence of fixed boundary conditions) or of the displacement of wave-function nodes (for scattering boundary conditions), and is independent of the finite contribution which may ensue from local modifications of the wave functions.

In the conventional higher-d, free-electron-gas cases to which Landau-liquid theory applies, it is implicitly assumed—and indeed self-consistently so—that the phase shift caused by adding or removing a single particle can be made to vanish in favor of a renormalization of all the quasiparticle mean-field energies. It is assumed that there is an effective mean-field energy whose eigenstates are the precise k states of the appropriate free-particle system.[6] The formal result of this process is that the wave-function renormalization constant Z is finite: That is, the overlap integral

$$Z = \langle c_{k\sigma}^{\dagger} \Psi_G(N) \mid \Psi_{k\sigma}(N+1) \rangle > 0, \tag{2}$$

where Ψ_k is the exact wave function of the $(N+1)$-particle state with one quasiparticle added, and, in particular,

$$\langle c_{k_F\sigma}^{\dagger} \Psi_G(N) \mid \Psi_0(N+1) \rangle > 0 \tag{3}$$

[where in a Fermi system, the $(N+1)$-particle system necessarily has one particle added near the Fermi surface, and hence its ground state is quasidegenerate]. Equation (3) cannot be true if there is a phase shift due to the addition of $c_{k\sigma}^{\dagger}$.

In one dimension, for interacting particles, such a phase shift is unavoidable, since the effective range of interactions is necessarily (for real interactions) of the order of the wavelength ($\delta = k_F a$). Thus in all the realistic one-dimensional systems, $Z = 0$, the Fermi-liquid fixed point is excluded, and the phase shifts due to interactions must be taken into account as relevant variables—in fact, in many cases renormalization invariants. $Z \equiv 0$ implies that the Fermi-sea excitations—which may still exist—do not carry charge (but may carry spin and be spinons). I will summarize Haldane's analysis of the spectrum of the 1D Hubbard model shortly.

In 2D, the scattering length for free particles and repulsive interactions diverges only as $1/k \ln k$ as $k \to 0$, and for higher dimensions it is $\sim 1/n \sim k_F^{-1/d}$; in both cases no serious problems need ensue for shorter wavelengths. But there is one type of problem where finite phase shifts are inevitable, namely, systems with a single-particle spectrum bounded above and below in energy. In this case the introduction of an extra particle may cause a bound state to split off from the top of the spectrum (an "antibound state").[7] By Levinson's theorem,[8] the presence of a bound state, either above or below the band, is signaled by a difference π in phase shift in the appropriate channel between top and bottom of the band. This corresponds to the fact that one state must be removed from the band to make up the bound state, and to Friedel's identity[9]

$$\Delta n(k) = \sum_l (2l + 1)\delta_l(k) \tag{4}$$

for the change in number of states to be found below a wave vector k due to a phase shift $\delta(k)$. Continuity—or the fact that any bound state must be a superposition of all states in the band—tells us that some δ must remain finite at all energies in the band.

For any dimension, a repulsive interaction U sufficiently strong to split off an upper Hubbard band adds one state to that band for each added electron; the "upper Hubbard band" can be thought of as equivalent to the manifold of antibound states and where it is present we must have $Z \equiv 0$ in the occupied lower Hubbard band, since the Hilbert space changes when we add a carrier. In the 2D Hubbard model (with one band, the generalized Hubbard models recently introduced are an irrelevancy) any repulsive potential whatever will split off bound states above the band, because of the well-known fact that in two dimensions all potentials bind. Thus, although for very low occupancies or very weak interactions the relevant singularities come in with small coefficients which are nonanalytic in interaction ($\sim e^{-t/u}$) or density ($e^{-n^2 \ln n}$), $Z = 0$ in all cases. These terms will not be picked up in series expansions.

We can identify the relevant interactions by thinking of the upper Hubbard band as a kind of "ghost" condensate in a channel of $2k$ total momentum and zero total spin reflecting the fact that each particle of down spin prevents some state of the same momentum and up spin from being occupied, so that the "condensate" represents both states being occupied.

For each (conserved) total momentum $K = k + k'$, the final-state energy is a d-dimensional function of relative momentum Q which has some maximum value, above which the antibound state for momentum K appears for all U in two dimensions. Every scattering state in the K channel must be orthogonal to this state, which defines the eigenvalue of the S matrix which has finite phase shift. When we restrict ourselves to excitations near the Fermi surface by renormalizing away everything but a shell of states, there still must be a finite forward-scattering phase shift in the $K = 2k_F$ channel at each point in the Fermi surface. This finite phase shift means that up- and down-spin particles cannot occupy *exactly* the same k state. Thus, as in one dimension, one relevant interaction is forward scattering of up or down spins.

A second relevant interaction channel may be identified as the precisely backward scattering of up versus down spins, which comes from the $2k_F$ bound state of up-spin holes to down-spin electrons and vice versa; that is, the channel containing a $+k\uparrow$ electron and a $-k\downarrow$ hole. Viewed from the other hole-particle channel, this is a backward scattering.

It is worth discussing the resulting state in terms of a "renormalized Bethe *Ansatz*" picture. If we, following Benfatto and Gallivoti,[2] use a "poor man's renormalization-group" procedure to eliminate k states far from the Fermi surface, we will end up with a shell of low-energy excitations with momenta near the Fermi surface. Even if $Z = 0$ and even in the presence of our new interactions, for the thin shell of states near k_F every real scattering is nondiffractive because of momentum conservation in that the two k-vectors never change: Charge is always scattered forward. When $Z = 0$, however, the original k

states of the Fermi liquid are not adequate to contain all the particles, and the Bethe *Ansatz* wave function contains a continuous spectrum of k's through the Fermi surface, exactly as in one dimension, where the large-U Hubbard model solution may be written

$$\sum_Q (-1)^Q \det \|e^{k_i x_{Q_j}}\| \times \text{(spin function)},$$

and the spectrum of k_i's extends continuously to $|Q|$, $Q > k_F$. We presume that the same form is valid for renormalized particles in 2D, near the Fermi surface.

Under bozonization the backward-scattering interaction turns into a term proportional to $\nabla\theta_\uparrow \nabla\theta_\downarrow$ in the Hamiltonian for the phase-shift variables θ_σ, defined by $\nabla\theta_\sigma = 2\pi\rho_\sigma$. This term in the effective boson Hamiltonian must be transformed away by a Bogoliubov transformation. But as in one dimension, when transformed back into fermion variables the new dynamical variables, even though their equations of motion have linear dispersion relations, do not correspond to simple fermion or boson excitations, and have Green's functions with nonclassical exponents. They can be thought of as two spinless fermions (which is, after all, what we started with) two semions, or whatever; but all physical response functions correspond to fermion or boson combinations. Charge and spin separate, the low-energy spin excitations being like fermions at the original k_F, the charge excitations centering around the spanning vectors $2k_F$.

In a previous paper,[10] we attempted to derive charge and spin separation by solving the double-occupancy problem with a slave-boson technique and a constraint; the basis of such a theory is undoubtedly correct but the mean-field treatment of the gauge variable, which results from the constraint, was not.

The actual correlation functions and Green's functions in the two-dimensional case have not yet been calculated: Fortunately, many experimental data can be calculated by using the photoemission data[11] to describe a semi-empirical fit, and by using the one-dimensional Hubbard model as an appropriate guide to understanding.[12] The actual calculation of physical properties will be described separately.

At present, all experimental observations seem compatible with this point of view, and many puzzling ones receive almost unique explanations.

I acknowledge vital discussions with B. S. Shastry, Y. Ren, J. Yedidia, A. Georges, D. H. Lee, F.D.M. Haldane, and S. Girvin; also the hospitality of the Aspen Center for Physics and the IBM Yorktown Heights Laboratory. This work was supported by the NSF, Grant No. DMR-8518163, and AFOSR Grant No. 87-0392.

REFERENCES

1. F.D.M. Haldane, J. Phys. C **14**, 2585 (1981).
2. G. Benfatto and G. Galivotti, "Perturbation Theory of the Fermi Surface ..." (to be published); P. W. Anderson, in Proceedings of the Cargese Advanced Research

Workshop on Common Trends in Statistical Physics and Field Theory, Cargése, France, 1988 [Physica (Amsterdam) Suppl. (to be published)].

3. A. M. Luther, Phys. Rev. B **19**, 320 (1979).

4. G. D. Mahan, Phys. Rev. **163**, 612 (1967); P. Noziéres and C. de Dominicis, Phys. Rev. **178**, 1097 (1969).

5. P. W. Anderson, Phys. Rev. **164**, 352 (1967).

6. This is the essence of the procedure of A. A. Abrikosov, L. P. Gorkov, and I. Dzialo-shinskii [*Methods of Quantum Field Theory in Statistical Mechanics* (Prentice Hall, Englewood Cliffs, 1963), Sec. 20], which in the conventional case renormalizes the forward-scattering phase shift to $\delta \sim [\ln(\Delta p)]^{-1}$ by use of the Cooper phenomenon. This renormalization procedure cannot work if there are antibound states present, I presume, because the assumed "nonsingular" parts of the vertex are not harmless, but infinite.

7. T. Hsu and G. Baskaran (unpublished).

8. N. Levinson, Kgl. Dan. Vidensk. Selsk, Mat.-Fys. Medd. **25**, No. 9 (1949).

9. J. Friedel, Adv. Phys. **3**, 446 (1954).

10. Z. Zou and P. W. Anderson, Phys. Rev. B **37**, 627 (1988).

11. D. W. Lynch, in Proceedings of the International Conference on Materials and Mechanisms of Superconductivity and High-Temperature Superconductors, Stanford, California, July 1989 [Physica (Amsterdam) (to be published)]; G. Olson et al., Phys. Rev. B (to be published).

12. P. W. Anderson, "'Experimental' Demonstration of the Mechanism for High Superconducting Transition Temperatures" (to be published); in Proceedings of the Materials Research Society Fall Meeting, Symposium M, Boston, November 1989 (to be published).

B

Singular Forward Scattering in the 2D Hubbard Model and a Renormalized Bethe Ansatz Ground State

P. W. Anderson

(Received 23 July 1990)

ABSTRACT

In a recent Letter we argued that the existence of an upper Hubbard band necessarily would lead to Luttinger liquid ($Z = 0$) properties for a strongly interacting electron gas, as opposed to Fermi liquid. In this paper we identify the singular scattering diagrams and make a hypothesis about the form of the ground state of the 2D Hubbard model.

PACS numbers: 75.10.Lp, 74.65.+n, 74.70.Vg

In a recent Letter[1] we argued that the existence of an upper Hubbard band necessarily would lead to Luttinger liquid ($Z = 0$) properties for a strongly interacting electron gas, as opposed to Fermi liquid. In this paper we identify the singular scattering diagrams and make a hypothesis about the form of the ground state of the 2D Hubbard model.

In order to demonstrate the singularity, we assume that the state is a Fermi liquid and show that forward scattering modifies the state in a singular way. The most extensively studied 2D problem is the low-density limit, studied by Galitskii[2] and Bloom,[3] and known by expansion in n to go to a Fermi-liquid limit as $n \to 0$. We do not quarrel with this but identify new singular terms proportional to $(\ln n)^{-1}$ which are controlling for all finite n. The relevant terms must be treated outside conventional many-body perturbation theory because they enter not in the standard perturbation series as conventionally used but in the initial determination of the scattering vertex or "pseudopotential" to be used in that theory, i.e., in replacing the bare scattering potential V by a "scattering matrix" T which describes the local response of the wave function of one particle to the potential of another. Unfortunately, T has very complex low-energy singularities on the "energy shell" which depend crucially on boundary conditions, and once boundary conditions are included forward and backward scattering are no longer clearly distinguishable. The correct treatment of the singularity can always be managed by directly calculating the energy shift in the particle-particle channel, which we now do.

Schrödinger's equation for an eigenstate with energy E of the two-particle scattering problem in a channel of momentum $2k$ for a pair of opposite-spin particles reduces to[4]

$$L^D/U = \sum_Q (E - \epsilon_{k+Q} - \epsilon_{k-Q})^{-1}. \tag{1}$$

Let $\epsilon_{k+Q} + \epsilon_{k-Q} = E_Q$, where $2Q$ is the relative momentum of the particles in a given intermediate state. There will be an eigenvalue E of (1) above every value of E_Q and below the next one; the lowest one, $E(Q = 0)$, will be that for forward scattering, and we may write $[E(Q = 0) - E_0]/(E_1 - E_0) = \delta/\pi$, where δ is the phase shift in the isotropic channel. (We use the low-density limit to justify the simplifying assumption that the problem is approximately Galilean for small Q. The higher energies E_Q will not satisfy this, but they only enter in upper limits of integrals. The arguments go through, but less simply, for any k.) The sum in (1) diverges at low Q in 1D, and as a result in 1D, $\delta = \pi$ independently of U and of the upper limit. This means that at low density opposite spins effectively obey an exclusion principle in 1D as is well known from the Hubbard model literature. In 3D the sum converges at the low end, and is controlled by the large values of Q, so that in general even for large U

$$1/(E - E_0) \sim L^3,$$

and since $E_1 - E_0$ is $\sim L^2$, $\delta \to 0$ as $1/L$; this is conventional scattering length theory and leads to the Fermi liquid. However, in this case the $l = 0$ channel must be treated more carefully, which we have not yet done. In 2D, the sum diverges at both ends logarithmically, and the eigenvalue equation reduces to

$$\frac{1}{E - E_0} - L^2 P \int_{\pi/L}^{\pi/a} \frac{d^2 Q}{E_Q - E} = \frac{L^2}{U},$$

$$\frac{1}{E - E_0} \simeq L^2 \ln \frac{1}{a^2 Q_{\min}^2},$$

so that $\delta \sim 1/\ln L$ which, as Bloom showed, represents a divergent scattering length but is sufficiently small to allow a Fermi-liquid theory. Note the relevance of the upper cutoff π/a; the problems do not occur for free particles, with $a \to 0$.

Something quite new happens if we go to finite (if small) density and consider particle-particle scattering at the Fermi momentum k_F ($k = 2k_F$). Now we must exclude occupied states from the sum in (1) and it reads

$$\frac{L^2}{U} = \sideset{}{'}\sum_Q \frac{1}{E - E_Q},\tag{2}$$

where $\sum'$ implies that neither $k + Q$ nor $k - Q$ is inside the Fermi surface. Figure 1 shows the effect of this exclusion: It removes all but a fraction $Q^2/\pi k_F^2$ of the phase space at a given small $|Q|$. This eliminates the "recoil" effect which we might expect would prevent singularities. Now the principal part integral converges nicely, behaving like $(1/k_F^2)\int Q^2 Q\, dQ/(E_Q - E_0)$ for small Q. For small k_F the singular term becomes $\ln(1/k_F^2 a^2)$. Therefore the phase shift becomes finite, of order $\delta \sim \pi/2 \ln(k_F a)$. As k_F increases, we have in general a finite phase shift δ somewhere between 0 and π. A finite phase shift for forward scattering has very severe consequences for Fermi-liquid theory. In particular, it means that two particles (or quasiparticles) of opposite spin may *not* occupy the same plane-wave state, as is assumed in Landau theory, in which it is assumed the fixed point is the free Fermi gas. It also implies that the renormalization constant

$$Z \sim e^{-(\delta/\pi)^2 \ln L^2} \to 0.\tag{3}$$

Recognizing that for finite time or frequency, $(L^2)_{\text{eff}} \sim v_F^2/\omega^2$, this gives us a form for the singularity,

$$Z(\omega) \sim -\left(\frac{\partial \Sigma}{\partial \omega}\right)^{-1} \sim \omega^{(\delta/\pi)^2};\tag{4}$$

for small δ, this closely resembles the $\ln \omega$ behavior proposed on experimental grounds.[5]

We should reiterate that Landau's Fermi-liquid theory is based on the consideration of relative energy scales, equivalent to a renormalization-group theory. The relevant, fixed-point Hamiltonian $\mathcal{H}^*$ is the free Fermi gas; the Landau interactions $f_{kk'} n_k n_{k'}$, which embody only the renormalized Hartree forward and backward scattering terms, are marginal, i.e., of order ω or T, and all of many-body perturbation theory is irrelevant at the fixed point, i.e., of order ω^2 or T^2, which makes it hard to pick up the difficulties in that theory. Haldane's "Luttinger-liquid" theory,[6] which is the one-dimensional response to diverging Hartree terms, rediagonalizes the first two terms in the hope that scattering will still be irrelevant, since one has done as little damage to Fermi-liquid theory as possible. We now try to carry out the same program in 2D, but of course, unlike the above, we must now move into speculative territory.

To do this we follow Fermi-liquid theory in doing first a conventional renormalization eliminating high-energy virtual scatterings, and assume that the remaining "quasiparticle" states involved in low-energy excitations all have

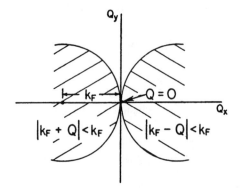

Figure B.1. Region of recoil momentum Q excluded by the Pauli principle.

momenta near a Fermi surface. This elimination will only have made the effective singular forward scattering bigger.

At this point we are talking about renormalized, "quasiparticle" states, which have singular density-density interactions but all virtual scatterings into high-momentum states are renormalized away. Whether Fermi liquid or not, simple geometry shows that all real scatterings of pairs of particles near a Fermi surface are "nondiffractive" in Sutherland's sense: The momenta are conserved, only spin can be exchanged. Thus for the effective particles, a Bethe ansatz may be assumed:

$$\Psi = \sum_{P,Q} \exp\left(\sum k_{Qi} x_{Pj}\right) [Q, P],$$

with the set of k's invariant. The $[Q, P]$ determine the spin state; the k's determine how charge moves. But the effect of singular forward scattering is to tell us that the set of k's occupy a volume in k space greater than that of the Luttinger Fermi surface; this is precisely what happens in the one-dimensional case. In this event the distribution of k's has *no singularity* near the Fermi surface and all low-energy charge motion takes place by collective "sliding" motions of the k distribution. (As in 1D, the k's are not real particle momenta and do not determine the Fermi surface which depends on the spin motions alone.) Following Luther,[7] we can rewrite the free-particle Hamiltonian in terms of charge- and spin-density waves moving in a given direction Ω with Fermi velocity v_F. The spin-density waves encounter no singular scattering (triplet pairs do not forward scatter) but the charge-density waves must be rediagonalized by the standard Bogoliubov transformation to take into account the singular scattering, thus leading to different charge and spin velocities and to charge-spin separation. The collective charge modes may be thought of as "holons" but are not particlelike. In particular, they are not affected by elastic scattering, by an argument similar to the "dirty superconductor" theorem.[8]

Residual resistance can be caused only by spin scattering. We suggest that the state may be considered to be a $T_c = 0$ superconductor, in the absence of spin scattering. More detailed discussions of specific properties will be published elsewhere.[9]

I acknowledge stimulating discussion with S. Sorella, E. Lieb, and constant input from Y. Ren. This work was supported in part by the National Science Foundation, Grant No. DMR-8518163, and by the U.S. Department of the Air Force, Grant No. AFOSR 87-0392.

REFERENCES

1. P. W. Anderson, Phys. Rev. Lett. **64**, 1839 (1990).
2. V. M. Galitskii, Zh. Eksp. Teor. Fiz. **34**, 151 (1958) [Sov. Phys. JETP **7**, 104 (1958)].
3. P. Bloom, Phys. Rev. B **112**, 125 (1975).
4. The argument here is in a form suggested by A. Parola, S. Sorella, M. Parinello, and E. Tosatti (to be published).
5. P. W. Anderson, in *Frontiers and Borderlines in Many-Particle Physics*, International School of Physics, "Enrico Fermi," Course CIV, edited by R. A. Broglia and J. R. Schrieffer (North-Holland, Amsterdam, 1987).
6. F.D.M. Haldane, J. Phys. C **14**, 2585 (1981).
7. A. M. Luther, Phys. Rev. B **19**, 320 (1979).
8. P. W. Anderson, J. Phys. Chem. Solids **11**, 26 (1960).
9. Y. Ren and P. W. Anderson (to be published).

C

Fractional Statistics without Anyons:
The Luttinger Liquid*

P. W. Anderson

As I tried to explain in my Kanazawa lecture and as I have expounded at length elsewhere, to continue to operate on the assumption that the normal metal in the high-T_c superconducting cuprates is a Fermi liquid in the face of mounting experimental evidence is to be either very unsophisticated, very unwise, or very stubborn. At the same time, almost equally convincing evidence is available that it does have a Fermi surface. Our task for a couple of years, then, has been to produce a theory of a substance which is not a Fermi liquid in the Landau sense but does have a Fermi surface; and, at first, just for a name to call it, I borrowed Duncan Haldane's soubriquet of "Luttinger Liquid" as a description of a more general liquid which obeys Luttinger's theorem. Now I think I know a great deal more about it. The plan of my talk, then, is as follows:

I. A discussion of general approaches to the quantum liquid
    (1) Via Renormalization-Poor-Man's, Gallivotti-Shankar approach
    (2) Low Density approach
    (3) Luttinger Liquid approach

II.
    (1) Some generalities about Luttinger Liquids: related to Jastrow
    (2) Basic conditions for them
    (3) Tomography

III. Explicit calculation of anomaly

IV. Relation to Bethe Ansatz
    1 Dimension
    Holons and Spinons
    Nature of Spinons

*This work was supported by the NSF Grant # DMR-8518163 and AFOSR # 87-0392.

I. General approaches

(1) *Poor-Man's RNG.*   A little study of the derivations and the history quickly convinces one that the Landau Fermi liquid is not purely a perturbation approach, in that at the very start of the theory two non-perturbative steps are taken: (1) Certain of the interactions are singled out for special attention and called "Fermi liquid parameters"; and only in the low-density theory has a rigorous attempt ever been made to connect those parameters or the residual interactions with the fundamental interactions. (2) The perturbation theory starts from an assumed vertex renormalized by ladders, in the form of an "effective range" a which, it turns out, must then be recalculated in each order as part of the perturbation series in n or U. Since, as we shall see, this is an infinite renormalization, such a procedure is rather circular.

Gallivotti and Benfatto, I in some unpublished notes, and Shankar have all concluded that the real RNG on which Landau theory is based is a "Poor Man's theory" in which one envisages a cutoff momentum-energy Λ which defines a shell around the Fermi surface and Λ is continually decreased by Poor-Man's rescaling, $\Lambda \to \Lambda - d\Lambda$. One then studies the resulting effective Hamiltonian in terms of orders in Λ. The appropriate rescaling of this effective Hamiltonian is $\mathcal{H}_{\text{eff}}/\Lambda$. Clearly, the original kinetic energy $\sum_k \epsilon_k n_k$ is relevant; it receives only finite modifications leading to an m^* but $\mu(\Lambda)$ changes in order to maintain constant density, so that the Fermi sea is still effectively a free FS with N particles. These terms are *relevant*: of order $1/\Lambda$.

$$\mathcal{H}_0 - \mu N \to \mathcal{H}^* - \mu^* N \times 1/\Lambda.$$

But, as Landau points out, there are a set of marginal terms, of order $\Lambda/\Lambda = 1$, which do not vanish but do not change the nature of $\mathcal{H}^*$. These are the forward scattering or *Landau parameter* terms

$$\sum_{kk'} f_{kk'\sigma\sigma'} n_{k\sigma} n_{k'\sigma'}$$

which, because they do not violate the exclusion principle, remain finite. They must be treated at mean field level in all responses.

Finally, ordinary scattering, because of the limitation on intermediate states due to the exclusion principle, is of order $\Lambda^2/\Lambda \sim \Lambda$, irrelevant. All of the diagrams of conventional diagrammatic perturbation theory are renormalization-irrelevant, and we cannot expect that they will necessarily signal any failure of the renormalization procedure; if they do—as Fukuyama has shown they do—we are lucky only. One term remains which does not drop out, the Cooper scattering $k, -k \to k', -k'$: Which renormalizes logarithmically away for repulsive U. It is responsible for BCS.

(2) *Low Density Approach.* Is the above really justified? Can it fail? This is approached, so far as I know, only in certain low-density calculations of Galitskii, Bloom and others, and for truly low density it appears to work. The reason it works is *recoil*: It does *not* work in one dimension because the forward scattering phase shift remains finite at lower momentum, with little or no effect of recoil: Leading to effective Landau parameters which diverge as $1/k - k'$; we will discuss these divergences at length very shortly. But it is important to realize that the convergence of the calculation of Landau parameters rests on these few studies. Many people have noticed—e.g., Hamann—that the problems we have with Fermi surface singularities and/or one dimension are only converged by recoil to an effective range theory. But we will see that in general they are not.

(3) *Luttinger Liquid Theory.* Haldane showed—following ideas of Luther and others—that in a variety of non-FL one-dimensional models one nevertheless has a much more general category of liquid, the Luttinger liquid, where, again, all dynamical properties are determined by a low-energy limit, and certain very general theorems apply—such as, specifically, Luttinger's theorem. The models to which it applies are at least—but more general than—all which are solved by Bethe Ansatz.

II. Luttinger liquids in $D > 1$

(1) Let me then continue with some of the key points about Luttinger liquids. Let us first make a very important remark of Haldane: That not only Landau but also Luttinger liquids are what he calls "Gaussian": They are characterized at least after "poor-man's scaling" by noncritical *Gaussian* fluctuations about some non-interacting Fermi gas (or in some cases other system). This is equivalent to saying that none of the two-particle correlations of Fermion operators show giant eigenvalues, i.e., ODLRO. This allows us to assume that there is *some* set of harmonic bosonic variables in which the Hamiltonian may be diagonalized, although these are never simply the RPA density operators.

(2) A second feature of Luttinger liquids is that they are closely related to Jastrow approximations. Chester showed long ago that the boson operators in superfluid Helium give the long-range part of the Jastrow correlation function in a Jastrow product approximation, and this is easy to show for the RPA. I believe that this is more generally intrinsic to the "Gaussian" assumption.

(3) Finally, it is very important that Luther long ago showed that the bosonization procedure is a perfectly good, complete description of the free Fermi gas in any dimension; in other words, he generalized the Tomonaga method to general dimensions. This was done by defining a new kind of bosonization quite distinct

from the RPA bosonization of the usual Green's function approach. Density operators are defined only for excitations very near the Fermi surface—again, the Poor-Man's shell idea—but there are density operators for every direction Ω around the Fermi surface

$$\mathcal{H}_{\text{eff}} \rightarrow \sum_{\substack{Q,\hat{\Omega} \\ \sigma}} v_F Q \rho^{\Omega}_{Q,\sigma} \, \rho^{\Omega}_{-Q\sigma}$$

and this is a complete description of any Fermi sea.

Luther was disappointed when he found that, using conventional effective range theory, interactions added nothing new. But of course what we now understand he found is that, indeed, the fixed point Hamiltonian under Fermi Liquid conditions is the free Fermi Gas.

III. Explicit calculation of anomaly

Now, finally, we are ready to calculate what really does happen in the two-dimensional Hubbard model. When does it turn into a Luttinger Liquid? First let us see what happens in 1D. In 1D whenever two particles interact, they scatter with a finite phase shift:

$$e^{iQ\cdot(r_1-r_2)} \rightarrow e^{-iQ\cdot(r_1-r_2)+2i\eta(Q)},$$

where η is finite. When we put this together with boundary conditions, we find that $\delta Q = \eta/\pi L$. In particular, when $Q \rightarrow 0$ there is a minimum Q of $\eta/\pi L$: no two momenta can be the same. This has the effect of a *fractional exclusion principle*. Haldane has recently shown that this fractional exclusion principle idea is the most generalizable definition of fractional statistics.

We may put this result either into the bosonization formalism or into the Landau parameter formalism for the 1D Hubbard model. We find that the effective Landau parameter $\delta E_{k'}/\delta n_k$ diverges as $Q \rightarrow 0$ as $1/L$. L the size of the sample; and bosonization indeed gives one a non-Fermi liquid solution in which charge and spin have separate Fermi velocities.

How and why does the system continue to obey Luttinger's theorem? Why is easy to answer: The total density is controlled by the number of nodes crossing the boundary—effectively, one treats the system as though it was non-interacting outside the boundary. Thus a finite η actually modifies the density. It is necessary then, in order to maintain charge neutrality, to shift η relative to the mean potential—essentially, to add a weak long-range potential to bring $\bar{\eta}$ back to zero. But the *true* exclusion principle tells us $\eta_{\uparrow\uparrow} = 0$ if we had only a short-range potential; so the correction merely brings $\eta_{\uparrow\uparrow} + \eta_{\uparrow\downarrow} = 0$, and when we add a singlet pair the effects cancel out. Thus the *separation of charge and spin* plays an essential role, *independently* of the less profound exponent shifts which occur in the spinless models and which are renormalizable.

In two dimensions the calculation of the phase-shift is much trickier. First we must define explicitly what is required because this is the most difficult part of the discussion.

We are discussing fluctuations around an occupied Fermi sea. The relevant fluctuation involves a pair of particles of opposite spin, and we may neglect lifetime effects because they are of higher order in Λ. What we are interested in, if we want to study occupancies, is the S-matrix element for elastic scattering of two real excitations in the states $k_\uparrow$ and $k'_\downarrow$. The S-matrix connects the real, physical incoming states to outgoing states, and it is strictly the S-matrix which, through the Wigner-Friedel Theorem,

$$\delta n = 1/i\pi \, \mathrm{Tr} \log S$$

determines the change in occupancies caused by the shift in boundary conditions at the origin. What we are looking for is a finite eigenvalue of $\log S$, i.e., a finite phase shift η in some single channel. If a finite phase shift occurs in any channel, that channel will act the same as a one-dimensional system and be in essence a one-dimensional Luttinger liquid.

The actual calculation is much more a matter of why the forward scattering phase shift for $\uparrow$ and $\downarrow$ spins should ever be zero than why it should not be. The key is, as remarked years ago by many people, the calculation of recoil: The summation of all possible intermediate scatterings of $Q_0 \to Q' \to Q''$, etc., except those returning precisely to Q_0. That is, the S-matrix element $e^{i\eta(Q)}$ comes from the vertex for scattering of a pair of particles, irreducible for returning precisely to the energy shell. This is nothing like the prescription for calculating the conventional T-matrix element, where the energy is treated as a continuous variable: $\arg T(Q, \omega)$ is a continuous function of energy ω, η is a *number* for each Q. η represents the boundary condition, in other words, for scattering of a real particle.

Let us do this little calculation. What we want eventually to do is to solve the two-particle wave equation *with boundary condition at* ∞ for a pair of particles in momentum states $\uparrow$, $(K + Q)/2$ and $\downarrow$, $(k - Q)/2$. The wave equation itself will take care of repeated scatterings into the actual asymptotic states $(K \pm Q)/2$, because we will write explicit wave functions in the asymptotic region $e^{iQ(r_1 - r_2)} + e^{iQ(r_1, r_2)}$. What we need is to take into account all *virtual* scatterings into other states $Q' \neq Q$, giving the wave function modification in the interaction region. These are included by doing a K-matrix calculation

$$K = \tan \eta(E_{Q_0}) = \frac{n(E_Q)U}{1 - U \sum_{Q \neq Q_0} \frac{F(Q)}{E - E_Q}},$$

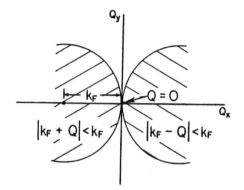

Figure C.1. Region of recoil momentum Q excluded by Pauli principle.

where $F(Q)$ is zero when *either* state is occupied. This exclusion is seen in the diagram, (fig. 1) and we see that although the sum (in two dimensions) is divergent as $\ln L$ without $F(Q)$, it is perfectly convergent *with* it in two dimensions and we get

$$\tan \eta \sim \frac{1}{\frac{1}{N(0)U} + \ln k_F a}.$$

To understand the meaning of this calculation, I would like to use an analogy I call "rocks in a box." We have to recognize that $\uparrow$ and $\downarrow$ spins are total strangers to each other, so if we insert an $\uparrow$ spin, as far as the $\downarrow$ spins are concerned it might as well be a dynamical pebble and not an electron. What we want to know is—how does inserting the pebble change the wave functions of the $\downarrow$ spin? And what is the resulting self-energy *of the pebble*? The pebble analogy tells us that such an object can change the boundary conditions on *all* the $\downarrow$ spin wave functions, an effect which leads to a *kinetic* energy shift which is not easily calculated perturbatively; rather, it is the result of solving for the mutual wave function.

This η essentially tells us 3 things:

(1)   The nature of the boundary condition which determines the Jastrow hole which surrounds every particle. As we have seen, the Jastrow hole is like $\log r/a$ in 2 dimensions, but must terminate at r_s, where it has a finite slope.

(2)   The effective fractional exclusion in the $k + k' = K$ channel which leads to statistical, exclusionary—i.e., infinitely long-range—interactions.

(3)   And finally, it tells us the nature of the divergence as $k \to k'$ in the Landau parameters. This part of the calculation has only been done sketchily elsewhere and may be worth repeating in more detail. The actual shifts in energies depend

on the boundary conditions at ∞: periodic in x and y, circular symmetric, outgoing, etc., but the energy picture does not. With circular symmetry only one eigenvalue is shifted outwards, that in the most isotropic channel, but by a constant amount $\eta/(\pi\rho(E))$. With periodic boundary conditions, we instead see an incompressible dilation of the lattice of k' values around k, so that

$$\delta k' = \frac{\eta}{\pi} \frac{k' - k}{(k' - k)^2} \left(\frac{\pi}{a}\right)^2.$$

A second important point is that we are inserting $k\uparrow$ at a *fixed* k, and asking for the resulting shift in k'; i.e., $\delta k' = \delta Q$; total momentum K is not conserved, because we must take *both* Hilbert spaces as incompressible, and k is inserted at a specific point in the k lattice. This correlates with the idea that we must shift μ to keep the average density in momentum space constant, so initially, before adding a given spin, both lattices are correctly occupied.

Adding an irrelevant constant

$$\frac{1}{2} \frac{(k - k')^2}{(k - k')^2},$$

we see that the energy shift of k' is

$$\frac{\eta}{2\pi} \frac{k'^2 - k^2}{(k - k')^2} \times \left(\frac{\pi}{a}\right)^2$$

and the effective Landau parameter is of the form—in the interesting region near the Fermi surface, it diverges as $(\epsilon_k - \epsilon_{k'})/(k - k')^2$. It falls off rapidly (as $1/Q^2$) with angle around the Fermi surface, essentially being singular only between states at the same point on the Fermi surface. It is for this reason, in fact, that statistical interactions such as the exclusion principle itself could be accurately treated by Luther in the tomographic model.

By far the simplest way to deal with this interaction is to recognize that it is indeed of exactly the same form that an interaction which embodied the exclusion principle would take, but is simply a *fractional* exclusion principle, a *fractional* statistics of amount η/π, and therefore it leads to the identical type of interaction given by Luther but between *opposite* spins.

$$\mathcal{H}_{\text{int}} = \sum_{Q,\Omega} \rho^{\Omega}_{Q\uparrow} \rho^{\Omega}_{-Q\downarrow} \frac{\eta}{\pi} v_F Q.$$

It is exactly this combination of bosonized interactions which Haldane and others have rediagonalized to obtain the fixed-point level solution of the 1D Hubbard model, and we refer the reader to these fundamental papers for the details of the Bogoliubov transformation which leads to the bosonic charge and spin variables with separate Fermi velocities $v_c(\Omega)$ and $v_s(\Omega)$. As observed by Haldane following Luther and Peschel and others, the backward-scattering exchange terms are only weakly singular in the 1D problem and lead only to small logarithmic renormalization, which we will ignore.

IV. Properties of the Luttinger liquid

As we see, as far as singular terms are concerned, 2D is 1D, so we simply take the properties of the 2D Luttinger liquid as angular averages over those in 1D. These are actually simplest to derive from the original Lieb-Wu solution, as rethought by Ogata, Shiba, Ren, and others.

Spinons are known to be essentially identical in properties with the spin 1/2 excitations of the "squeezed" Heisenberg model with N sites, and are algebraically *unaware* of the holes moving through the substrate except for an overall ("gauge" or "backflow") long-range coupling. They are, thus, true semions in Haldane's sense, excluding half a state and having only half a spectrum and chiral symmetry, $s_{i\uparrow}^+$ not being distinct from $s_{i\downarrow}$. Their $2k_F$ spin correlations are enhanced, like

$$(x^2 - v_s^2 t^2)^{-1/2}(x^2 - v_c^2 t^2)^{-p}$$

($p \sim 1/4$ for strong coupling): Smeared out in real space by holon motion but not in spin-order space. The spinons are the residue of the quasiparticles, do not change statistics continuously (non-Abelian group—Korepin) and determine the Fermi surface. It is a key fact that every electron or hole excitation carries *exactly one* spinon.

Holons are collective, fractionalized objects, like Haldane's original Luttinger liquid. They can be thought of as an electron or hole with its spin screened by the spinon condensate—the electron carrying $\sim k_F$, the screening also $\sim k_F$, to give $2k_F$ as the essential momentum. Key points to notice are

$$G_{1D} = \frac{e^{ik_F x}}{(x - v_s t)^{1/2}(x - v_c t)^{1/2}(x^2 - v_c^2 t^2)^a} + \propto e^{-ik_F x}.$$

From this can be easily derived the most important property: The response functions are *dead* to time-invariant, $\omega = 0$ perturbations. In particular, the gas *confines* and is superconducting at $T = 0$. A very important sketch suggested by Haldane describes this behavior: There is a critical V_{scatt} or $t_\perp$ for any U below which V_{scatt} or $t_\perp$ have no effect. (Fig. 2)

This flows from spin-charge separation and the non-renormalizability of spinons. Second, it is important to recognize that the spinons have a hidden Girvin-Read order parameter: "Extended s" pairing with a zero of their

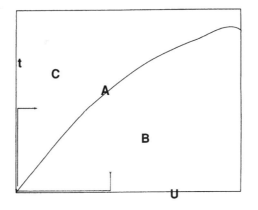

Figure C.2.

gap at E_F, just like the half-filled case. I believe this gapless condensation *in spin space* is the best physical way to understand the properties. The long-range transverse gauge fields are screened away, and chirality is very strongly *confined* in the Luttinger liquid state.

Conclusions

(1) The Luttinger Liquid in 2D exists.
(2) It has all the properties of the "Normal" Cuprates.
(3) Superconductivity is a trivial consequence, as seen 4 years ago—the mechanism is a fertile source of understanding of the *harder* normal problem.
(4) Fractionalization is easiest without a gap (it has been with us as the "X-ray edge effect" and the "Kondo effect"for decades).
(5) It may recur in organics (not 1D—usually unstable—but may show excess conductivity) and in 3D models at least as an unstable fixed point—as e.g., in heavy electron and high T_c conventional superconductors.
(6) Beauty is in the eye of the beholder; anyon superconductivity can be, depending on your point of view, either "too beautiful not to be true" or too untrue to be beautiful. Myself, I prefer not to second-guess Nature.

D

Breaking the Log-Jam in Many-Body Physics: Fermi Surfaces without Fermi Liquids

P. W. Anderson

(Received 4 June 1992; accepted 10 June 1992)

1 Introduction

In this article I want to put forward three general themes, all of which may at first appear somewhat controversial but which reinforce each other in sending the same general message: that the conventional perturbative renormalized theory of the Fermi surface has been unsuccessful in handling much of the interesting physics which has engaged the attention of quantum many-body theorists for the past decade or two, and must be supplemented. At least one promising variant, the Luttinger liquid, can be identified. This revolution is forced upon us by the failure of conventional theory in the high-T_c cuprate superconductors, but appears to be much more general.

The first proposition is that the classic theory had been failing us more often than it has been succeeding for two decades or so. The second is that the conventional perturbative derivation of Fermi liquid theory has a basic flaw, and that a variant, the Luttinger liquid theory, can be demonstrated to hold in several important cases; and the third is that the Luttinger liquid's properties correspond to many observed anomalies.

2 Successes and failures of conventional theory: history and a list of unsolved problems

The present theory of metals was put together in the 1950s as a further elaboration of the successes of Feynman diagram theory and renormalization in quantum field theory. As a matter of fact, except for QED it was more effective in this field than in its area of origin. Initial successes were the correlation energy problem, the Yang-Lee theories of hard core bosons and fermions, the Landau-Fermi liquid theory of collective modes and mean field responses, and the Fröhlich-Migdal theory of electron-phonon interactions, but the crowning glory was the application in the years 1957–65 to the theory of conventional BCS phono-mediated superconductivity [1, 2].

It was perhaps more evident at the origin of this process than it is now that the structure of the theory of metals is not a single, unitary, rigorous whole. Two basic insights—which I, for one, do not question—inform all of it: the Luttinger theorem of invariance of the volume of the Fermi surface to interactions [3], and the exclusion principle restriction on phase space for inelastic scattering which requires eigenexcitation energies to become real rapidly as one approaches zero frequency [4]. These seemed to lead unequivocally to the quasiparticle concept, which is the central idea of the whole present structure: that the eigenexcitation is a renormalized electron. It seems to have escaped peoples' notice that aside from the problems mentioned above, and a very incomplete and heuristic success in the ^{3}He superfluidity problem in 1972–74 [5], the perturbation theory has not helped us much with any of the many new physical phenomena which have come up since its heyday. The elegance of our methodology, both theoretical and experimental, seems to have concealed from us its essential sterility. Let me list what I consider to be wholly or partially unsolved problems. I happen to have a listing by dimensionality but any order would to.

2.1 Low-dimensional physics

1. Classic "organic superconductors"—BEDT and Bechgaard salts: juxtaposition of antiferromagnetism and superconductivity, low electron density for superconductivity, interesting diamagnetic responses, anomalous transverse conductivity, etc. [6].

2. Simple 1D chains, polypropylene, e.g., Conductivity is too large.

3. Sliding CDWs. Were Bardeen's misgivings about the classical nature of this process justified? Odd hysteretic behavior. Why is $\sigma_{\text{sliding}} = \sigma_{\text{free}}$? [8].

4. "Nesting" CDWs in $NbSe_2$, TaS_2, etc. T_c too low, gap too large, and the nesting of Fermi surfaces is too implausible an explanation. Superconductivity: density of electrons is too low, dependence on structure strange [9].

5. High-T_c normal state is not Fermi liquid [10].

6. Too much surface reconstruction, no free Fermi liquids, in surfaces of semiconductors.

7. Polyacetylene: solved by solitons? Outside perturbation theory [11].

8. Quantum Hall effects: solved outside perturbation theory.

2.2 Restricted-dimensional physics in 3D matrix

1. Buckey Ball superconductivity.

2. Chevrel superconductors—many mysteries. Why so disconnected from f-bands, mechanism of superconductivity in narrow bands? [12].

3. The high-T_c superconducting state.

4. The still mysterious chain and cluster compounds of "old" high T_cs: "A15s" (Nb_3Sn, V_3Si) CDWs at too low a temperature: strange transport behavior and magnetic susceptibility [13].

5. $LiTi_2O_4$, BKB superconductors [14].

2.3 On to 3D

1. Localization and interactions; hundreds of papers, no solutions of the many mysteries.

2. Heavy-electrons and mixed valence

(a) Low T transport—unsolved [15].

(b) Mechanism of superconductivity: "spin fluctuations" too vague, does not provide convincing heuristics [16].

(c) Mixed and interfering magnetic-superconducting phase transitions [17].

(d) Animal (superconductors); mineral (magnets); vegetable (nothing?), none of the above (mixed valence insulators); who decides which state will win? [18].

3. Many old problems (V_2O_3, ^3He microscopics, ferromagnetism, etc.).

2.4 Conclusion

The science behind much of materials science is missing. Much of what is done is futile or worse: futile repeats of failed techniques, band calculations as an end in themselves, etc. We propose (at least) three new mechanisms or possibilities which will follow from going to a new point of view.

(a) Possibility of decoupling of different bands or Fermi surface points, and of recoupling by pairing as a superconductivity mechanism.

(b) Fractionalization and multiple soliton formulation: novel and multiple transport mechanisms (note: this is the secret of the FQHE, and of polyacetylene, the two successful theories of the 1980s).

(c) Spin-charge separation frees both spin and charge susceptibility to have stronger Fermi surface singularities.

3 Landau and Luttinger liquids: Unified framework

The second point is deriving the Landau and Luttinger liquids on a common basis. This part is heavily indebted to F.D.M. Haldane and A. Tsvelik [19].

This part of the theory is based on a single postulate: the existence and meaningfulness of a Fermi surface. It is not sufficiently appreciated what a spectacular generalization the Fermi surface is, relative to, for instance, the Fermi point of a massless relativistic quantum field theory. The classic Dirac sea is inert and even "chiral anomalies" involve only a single degree of freedom; whereas the Fermi surface is a (D-1)-dimensional collection of Fermi points and has a very significant dynamics of its own.

The basic theorem on which the original one-dimensional Luttinger liquid theory was based, which is still valid in any dimension, is that the interior of the Fermi sea has no low-frequency dynamics: all of the low-frequency dynamics may be described in terms of quantities at the Fermi surface. This is simply the Luttinger–Friedel theorem in dynamical guise, exploiting the fact that the electron is a conserved quantum field obeying the exclusion principle whose local phase counts eigenstates and thus counts the total particle number. Thus we choose to describe the physics not by the set of individual particle positions, momenta, or field variables, most of which are redundant in a filled Fermi sea, but by the local displacements of the Fermi surface $\Delta k_F(\widehat{\Omega}, \sigma, x)$. Here $\widehat{\Omega}$ is a direction around the Fermi surface, σ is spin and x is a coarse-grained position variable.

The above is not really an Ansatz but an analytic result. The basic Ansatz of Landau–Luttinger liquid theory is that the ground state and all low-energy fluctuations may be described in terms of relatively small quadratic fluctuations of the relevant variables Δk_F. This is by no means to restrict ourself to perturbation theory or to RPA; this includes the renormalizations of all low-frequency fluctuations such as zero sound and spin waves, as well as of plasmons, and was shown by Haldane to include many "anomalous" 1D systems which don't even show a sharp Fermi surface. If we assume a preliminary "poor-man's" shell or renormalization it also includes most of the virtual excitations to higher energies.

Let us then do a sketch of the Haldane theory (soon to be published) based on this Ansatz. It is not appropriate to go into too much detail here but a few of the "buzz-words" may be introduced.

The basic quantities Δk_F obey the commutation relations of a "Kac-Moody" algebra:

$$
\begin{aligned}
[\Delta k_F(\Omega, x, \sigma), &\Delta k_F(\Omega', x', \sigma';)] \\
&= (2\pi)^3 i D\{\delta^D(x - x')\delta^{D-1}[k_F(\Omega), k_F(\omega')]\delta(\sigma, \sigma')\}
\end{aligned}
\tag{1}
$$

where D is an operator defined by

$$
Df(x, \Omega) = \hat{n}(\Omega) \cdot \nabla f(x, \Omega)
\tag{2}
$$

where $\hat{n}(\Omega)$ is the normal to the Fermi surface at Ω. We can Fourier transform $\Delta k_F(x)$ in terms of plane waves e^{iQx} and we find that it is effectively the Tomonaga density wave operator

$$
\Delta k_F(Q) \sim \sum_k^{\Omega} c_{k+Q}^+ c_k
\tag{3}
$$

where k is summed only near the point Ω and one thinks of this sum as basically radial to the local Fermi surface. In terms of the Δks, the kinetic energy ε_k transforms into

$$E_k = \int dx \int d\Omega \frac{v_F(\Omega)}{2}[\Delta k_F(x, \Omega, \sigma)]^2 \qquad (4)$$

so that the exclusion principle and the kinetic energy indeed depend only on the local Fermi surface variables. v_F is the obvious local Fermi velocity perpendicular to the Fermi surface.

Because of the exclusion principle restriction on two-particle scattering around the Fermi surface, to quadratic order the only effects of interactions appear as Landau forward scattering terms:

$$\mathcal{H}_{int} = \int d^D(x)\, d^D(x') \int d\Omega \int d\Omega'$$
$$\times \Gamma_{\sigma\sigma'}(\Omega, \Omega'; x - x')\Delta k_F)x, \Omega, \sigma)\Delta k_F(x', \Omega', \sigma') \qquad (5)$$

Γ represents simply forward local scattering, in a finite number of partial waves, of particles at Ω by those at Ω' and acts exactly like—in fact, is—the Landau shift of local one-particle energies depending on the overall Fermi distribution. Its dynamic effects are confined to producing a small, finite number of collective oscillations like zero sound; but the enormous (D-1)-dimensional manifold of single-particle excitations is unaffected unless Γ has a singularity as $\Omega \to \Omega'$.

The reader will note that in this order quasiparticle lifetimes and the "residual interactions" of Landau play no role at all: in fact, they come from higher-order, anharmonic terms in the basic variables. These irrelevant, high order terms are the only ones which normal perturbation theory calculates directly.

Another way of expressing the above theory, which I have advocated, is in terms of a "poor-man's" or shell renormalization, eliminating states outside a shell of energy Λ above the Fermi level. The two terms given here are order Λ^0 and Λ^1; lifetimes are of order Λ^2 (ω^2 or T^2) and become irrelevant as $\Lambda \to 0$; we must consider the order Λ as marginally relevant.

One remarkable fact about the above theory is that in the absence of T-breaking terms the Kramers degeneracy at each point of the Fermi surface gives us an extra symmetry: symmetry $= U_2(\Omega)$, a local U_2 gauge group with conservation of both spins separately at each Ω. This is not a necessary consequence of the overall symmetry group, which is $SU_2 \times U_1(\Omega)$, conserving charge and spin independently for each Ω. The symmetry-breaking $U_2 \to SU_2 \times U_1$ is thus not unexpected, and is known to happen always in the one-dimensional case. There is no symmetry reason, in other words, why v_F should be the same for charge and spin fluctuations.

In one dimension Γ necessarily contains a piece which is of the same form as the unperturbed Hamiltonian, but connects $\Delta k_F(\uparrow)$ and $\Delta k_F(\downarrow)$ at each of the two Fermi points. This then leads to a Hamiltonian which is of the form

$$\mathcal{H} = v_{FC}(\Delta k_F\uparrow + \Delta k_F\downarrow)^2 + v_{F\,spin}(\Delta k_F\uparrow - \Delta k_F\downarrow)^2 \tag{6}$$

with spin-charge separation. A simple transformation *a la* Bogoliubov to more conventional bosons (bosonization) leads to the usual Luttinger liquid solutions of the interacting Fermi gas, which have been discussed elsewhere. These bosons now have a complex relationship to the original Fermion operators, which gives these 1D solutions several very distinctive features.

Most important is the separation of charge and spin, most clearly expressed in the asymptotic Green's function for electrons:

$$G_1(x, t) = \frac{e^{ih_F x}}{[(x - v_s t)(x - v_c t)]^{1/2}} \frac{1}{(x^2 - v_c^2 t^2)^\alpha} \tag{7}$$

where the charge and spin parts of the electron propagate separately, at different velocities. Both charge and spin have strongly singular correlation functions, at the $2k_F$ spanning vector of the Fermi surface, behaving like

$$\begin{aligned}\langle S(0,0)S(x,t)\rangle &\propto \\ \langle \rho(0,0)\rho(x,t)\rangle &\propto \end{aligned} \left\{ e^{2ik_f x} \frac{1}{(x^2 - v_s^2 t^2)^{1/2}} \frac{1}{(x^2 - v_c^2 t^2)^\beta} \right\} \tag{8}$$

$\propto$ is a variable power between 0 and $\frac{1}{16}$, β between $\frac{1}{4}$ and $\frac{1}{2}$.

The "backward-scattering" part of Γ connecting $\pm k_F$ has a much weaker, logarithmically singular effect. The small exponent α is responsible for the well-known, but experimentally almost irrelevant, absence of a sharp Fermi surface in $n(k)$, which behaves near k_F like

$$n(k) \simeq \frac{1}{2} \left[1 \pm \left(\frac{k - k_F}{k_F} \right)^{2\alpha} \right] \tag{9}$$

A common misapprehension is that this is the major effect of interactions; physically, however, it has little effect and G_1 is probably best approximated by its first two factors alone. Many calculations have neglected $v_c - v_s$, which is like neglecting the broken symmetry in superconductivity: it leads to inconsistent nonsense.

4 Cause of Fermi surface anomalies

The universality of these anomalous effects in one dimension is a consequence of the fact that a simple short-range scattering interaction is indistinguishable from an exclusionary statistical interaction, since it has the effect of modifying the boundary condition on the relative wave function so that this cannot have relative momentum $\Delta Q \equiv 0$ but instead $\Delta Q \sim \eta/L$ (L is the sample size and η the scattering phase shift) which is of the same order as the spacing of the individual momentum space wave functions.

The new realization which has been forced upon us by the behavior of the high-T_c cuprates is that this same statistical type of interaction is common in higher dimensional systems too. The derivation in two dimensions is given elsewhere,[20] and that in three dimensions—which requires specially strong interactions and tight-binding bands—will be presented in a forthcoming paper. A summary of the arguments follows.

What we must first ask ourselves is not "why does perturbation theory ever fail?" but "why does it ever work, in the presence of a free Fermi surface?" The reason it should not work is the well-known "infrared catastrophe" which leads to such phenomena as the "X-ray edge" singularity. It was shown several decades ago that the introduction of a scattering center into a Fermi gas leads to a singular response because of the excitation of an infinite number of soft electron–hole pair excitations. The new equilibrium state is orthogonal to the original one in the large N, macroscopic limit by a factor

$$N^{-1/2} \sum_l (\eta_l/\pi)^2 \cdot 2l + 1 \text{ const}$$

where η_l is the scattering phase shift of the lth partial wave at the Fermi surface. Note that if $\eta_0 = \pi$, this overlap corresponds precisely to the $N^{-1/2}$ overlap between the extra localized state implied by such a phase shift and the extra continuum state. This simply typifies the physical meaning of the non-orthogonality: that the wave function of every electron is modified by the scattering.

When we introduce an extra electron into an interacting Fermi gas *a priori* it must act as a scattering center and modify all the other wave functions, hence the overlap between the new and old states of the total Fermi liquid should be zero; hence $Z \equiv 0$ in perturbation theory terms.

The standard perturbation theory avoids this dilemma by maintaining constant density and treating the effects of interactions as perturbations of the chemical potential. In crude physical terms, one subtracts from the interaction a very long-range potential whose effect is to reduce η to zero, and adds that potential back in as a contribution to the chemical potential. Thus, apparently, the Hartree self-consistent field can be cancelled by a change in chemical potential and therefore never causes a change in the quasiparticle wave functions.

This turns out to be successful if and only if the effect of the particle–particle scattering is sufficiently short-range as to be quite separable from these long-range corrections. The criterion in three dimensions is that one can use an "effective range" theory in which, as the relative momentum of two scattered particles approaches zero, the phase shift is equivalent to a finite displacement a of the relative wave function

$$\psi(r_1 - r_2) \to \frac{\sin Q(r_1 - r_2 + a)}{r_1 - r_2} \tag{10}$$

which clearly is a phase shift vanishing in the long wave limit:

$$\eta = Qa \to 0 \quad \text{as } Q \to 0$$

Two free particles with a short-range repulsion colliding in 3D do indeed satisfy an effective range theory of this sort. In two dimensions they do not, but the phase shift still vanishes as $Q \to 0$ as

$$\mu_0 \sim \frac{1}{|\ln Qb|}$$

where b is the range of the potential.

What has been missed in previous work is that this renormalization is a recoil effect depending on the ability of both particles to readjust their wave-functions in the interaction region. But if each particle is in the presence of a Fermi sea, the exclusion principle removes the available recoil momenta Q according to a factor

$$F_Q = [1 - f(k + Q)][1 - f(k' - Q)]$$

which, for $k \simeq k' \simeq k_F$ removes almost all recoil momenta out of k_F. When this effect is inserted into the two-dimensional Schrödinger equation, it turns out that the boundary condition on the relative wave function corresponds to a finite phase shift

$$\eta_0 \approx \frac{1}{(1/U) + \ln k_F^2 b^2} \tag{11}$$

as $Q \to 0$. (Let me emphasize that this is unrelated to the phase of a conventional scattering vertex as defined in perturbation theory, a misunderstanding which has led to incorrect "refutations" in the literature. The phase shift appropriate to the Hartree terms, which involves the precisely on energy shell S-matrix, is singularly ambiguous in perturbation theory.)

In three dimensions the situation is more complex. The simple calculation of each total momentum channel independently of every other leaves us with wave-functions for the interior of the Fermi sea which are not orthonormal. In this case we must rely on more general arguments of the Friedel type which show that if a finite amplitude is projected out of the interior of the Fermi sea, that amplitude must appear as an appropriate shift of the phases at the Fermi surface. This is the same theorem of dynamical irrelevancy of the interior of the Fermi sea on which the Haldane theory we discussed above is based. The detailed argument will be given elsewhere but the conclusion is that any interaction which is sufficiently strong to project "anti-bound states" out of the top of a tight-binding band (i.e., to produce an "upper Hubbard band") is going to cause a finite phase shift as $Q \to 0$ leading to Luttinger liquid anomalies and, in particular, charge–spin separation.

The argument that a finite phase shift as $Q \to 0$ automatically gives a singular Γ is the same as in one dimension. A finite phase shift η implies that $\Delta Q \sim (\eta/L)$ for the scattering of $\uparrow$ and $\downarrow$ particles at relative momentum $Q \simeq 0$. This in turn means that the two asymptotic states $k\uparrow$ and $k'\downarrow$ cannot have the same momentum, on the scale of the granularity of states in phase space: they obey a fractional exclusion principle or, equivalently, have a statistical, infinite-range effective interaction. This interaction is like the longitudinal Coulomb-like (but unscreened) part of a non-Abelian gauge field; the transverse part (which is the core of the anyon theory) is, as far as can be seen, absent. I think of this as a Higgs–Anderson screening of these transverse parts by the spinon gas.

The two ways of thinking of this effect are equivalent. Simplest is just to think of state $k\uparrow$ as introducing a dilation of size $(\eta/\pi) \times (\pi/L)^D$ into the incompressible manifold of k' values, thus producing a k' shift of the form

$$\delta k' = \frac{k' - k}{(k - k')^D} \frac{\eta}{\pi} \tag{12}$$

which clearly is singular at $k - k' = 0$. The effect of kinetic energy and the fractional exclusion of opposite-spin electrons is to modify the H_0 terms just as in one dimension.

5 Consequences of Luttinger liquid behavior: Physical effects

There are three ways in which Luttinger liquids give usefully different behavior from conventional Fermi liquids.

5.1 Enhanced Fermi surface responses

The intuitive assumption is that the $2k_F$ "nesting" responses of the Luttinger liquid should be weaker rather than stronger than those of a Fermi liquid, since the Fermi surface is less sharp. Quite the opposite is the case: in several ways the Luttinger liquid has enhanced physical effects of the Fermi surface. When charge and spin Fermi surface fluctuations become independent, each becomes more responsive to external perturbations. As we have already remarked, $2k_F$ fluctuations of both charge and spin have up to $1/2$ power more singular responses in one dimension, even in the doped case, while the half-filled case is a whole power more singular [21]. That is, if β in eq. (8) is $1/4$, as it will be in strongly interacting systems, the $2k_F$ correlation function falls off as $x^{-3/2}$; this corresponds to a singularity in the Q-space response of $(Q - 2k_F)^{-1/2}$. When averaged over angles in two dimensions, this gives a logarithmic BCS-like susceptibility—i.e., interacting theory in 2D imitates mean field theory in 1D. For the insulating, half-filled Hubbard model the singularities are half a power stronger, such that even in 3D there would be a $2k_F$ antiferromagnetic singularity.

We see that the susceptibility for a 2D metal can be logarithmically singular at $2k_F$ without any Fermi surface nesting at all. The existence of incommensurate SDWs in so many systems, agreeing with Fermi surface spanning vectors, has been a great mystery to which this seems to be the answer.

One hesitates to carry the argument even into 3D but the "Martensitic" transitions of the A15 metals V_3Si and Nb_3Sn are very suggestive of anomalous $2k_F$ nesting.

A second type of Fermi response, the de Haas van Alphen type, is better described under the next heading.

5.2 Anomalous transport in Luttinger liquids

I should emphasize that there can be a number of types of Luttinger liquids in higher dimensions, as there are in one, depending on signs of interactions, type of band—degenerate or non-degenerate, for instance—etc. So far the transport properties have only been worked out in the one case of the 2D Hubbard model appropriate to high-temperature cuprate superconductors. In this case the transport properties are dominated by charge–spin separation and the long mean free time of the spinon. When the electron is accelerated by an electric field, it breaks up immediately into charge and spin excitations with different velocities. This rapid rate of decay, with its rate proportional to the available phase space $\propto \omega$, is the dominant dissipative process and causes d.c. resistance $\propto T$ and the infrared relaxation rate $\propto \omega$, as well as the breadth of the one-particle G_1 also $\propto \omega$ as observed in photoemission.

On the other hand, acceleration by a magnetic field does not cause breakup into charge and spin because the acceleration is only parallel to the Fermi surface. Thus no anomalous extra dissipative process is excited and one sees only the intrinsic decay rate of the excitations, the longest being the spinon mean free time which is controlled by spinon–spinon scattering $\propto T^2$ or ω^2. Thus the Hall angle $\theta_H \sim \omega_c \tau$ sees a spinon mean free time $\tau \sim (1/T^2)$ as is observed [22].

This is true in the CuO_2 planes, but $\perp$ to the planes the particles which are moving are electrons, not spinons, since spinons cannot be transferred from plane to plane, and the Hall effect in this direction is given by the simple metallic formula $R_H \propto 1/ne$ [23].

Finally, the effective sharpness of the Fermi surface is that of the spinons and the Fermi surface is much better defined than would be indicated by the short transport mean free time. Hence De Haas-van Alphen measurements of the Fermi surface might be surprisingly easy.

The properties of Fermi surface particles measured by magnetic means, NMR and χ, will depend on whether it is $q \simeq 2k_F$ or general q which is relevant. χ at general q and particularly $q = 0$ will behave like a free Fermi liquid, as will the Korringa relaxation: this is the spinon response. But at $q = 2k_F$ there is enhancement, which seems to be relevant for the Cu nuclear resonance.

Where may we see other anomalous transport properties? Possibilities are:

(a) Anomalously large σ, absence of localization in many 1D polymers [7].

(b) Puzzling transport and saturation effects in some strong coupling superconductors [24].

(c) Puzzling transport properties below T_F in many heavy electron systems.

One of the difficulties in transport theory has been the excessive reliance on diagrammatic perturbation theory and the Kubo formula. In localization theory this can be barely—and possibly incorrectly—justified by the scaling technique, but in problems with charge–spin separation the situation is too complex for simple perturbation theory, especially in that the response functions are not uniquely determined by time-ordered Green's functions. We are finding that whole new forms of transport theory are required.

5.3 "Confinement"

This is the most striking effect of all—it is in a sense a transport effect but has other consequences. It is easily shown that spin–charge separation (or in fact any form of soliton formation) can block hybridization between bands or sublattice entities when the hybridizing matrix elements are small [25].

In particular, the chains and layers of the cuprate superconductors may have no coherent motion of electrons between them at all. This is a consequence, like the conductivity process, of the breakup of electrons—which are the entities carried by the hybridizing matrix elements—into incoherent sets of holons and spinons. "Coherent" means $t \rightarrow \infty$, but at $t = \infty$ the holons and spinons are infinitely far apart! This effect is:

(a) the cause of high c-axis resistivity in cuprates and other 1- and 2-dimensional plane and chain systems (e.g., $4\,h\,TaS_2$, intercalated $NbSe_2$, TTF–TCNQ, etc.).

(b) This also may lead to the isolation of different parts of the Fermi surface in mixed valence systems and/or in chevrels and other cluster compounds.

(c) This is the key to the superconductivity mechanism at least in the cuprates, where pair tunneling is not subject to confinement so self-consistently provides the energetic motivation for superconductivity [26].

6 Conclusion

Many of the problems of the quantum physics of metals have been endlessly discussed in the past two decades or so without any conclusive resolution in most cases. Many phenomena seemed to lack even qualitative understanding, and many others required quantitatively implausible assumptions.

The strong possibility is that the new physics opened up by the abandonment of the straightjacket of the Fermi liquid theory and its renormalized perturbation theory will lead to resolution of many of these problems. We see that this straightjacket has been incorrectly enforced in too many important systems— even in one dimension, for instance, where it surely fails. There are many systems—the 2D layer materials, for instance—where we can prove that the conventional methods fail, and many others where we may hope to resolve problems by taking the Luttinger liquid point of view.

Thus a new physics as well as a new technology has been born with the discovery of high-T_c superconductivity. The importance of the latter should not obscure that the former can also be revolutionary.

REFERENCES

1. Schrieffer, J. R., "Superconductivity" (Benjamin, New York 1963), and references therein; see also McMillan, W. L. and Rowell, J. M., in "Superconductivity" (edited by R. D. Parks) (Dekkar, New York 1969), p. 561.
2. An excellent modern text of many-body theory is Mahan, G. D., "Many Body Theory" (Plenum, New York 1981).
3. Luttinger, J. M., Phys. Rev. **121**, 942 (1961).
4. See, e.g., Anderson, P. W., "Concepts in Solids" (Benjamin, New York 1963).
5. See Brinkman, W. F. and Anderson, P. W., in "Physics of Liquid and Solid Helium" (edited by K. H. Bennemann and J. B. Ketterson) (Wiley, New York 1978), vol. II, p. 177.
6. For a review, see Jérome, D., Creuzet, F., and Bourbonnais, C., Physica Scripta **T27**, 130 (1989).
7. See Phillips, P. and Wu, H. L., Science **252**, 1805 (1991).
8. Bardeen, J., Physics Today **43** (12), 25 (Dec. 1990).
9. DiSalvo, F. J. and Rice, T. M., Physics Today **34** (4), 32 (Apr. 1979); Phys. Rev. **B20**, 4883 (1989), etc.

10. Anderson, Science (in proof).

11. See: Prange, E. and Girvin, S. M. (editors), "The Quantum Hall Effect" (Springer, Berlin 1987).

12. Shenoy, G. K., Dunlap, B. D., and Fradin, F. Y. (editors), "Ternary Superconductors" (North-Holland, New York 1981); see review articles by Anderson, P. W., p. 309; Ö. Fischer, p. 303.

13. Weger, M., in "Solid State Physics: Research and Applications" (edited by H. Ehrenreich, F. Seitz, and D. Turnbull) (Academic Press, New York 1973), vol. 28, p. 2; Testardi, L. R., Rev. Mod. Phys. **47**, 637 (1975).

14. Bosovic, L., et al., preprint.

15. Levin, K., Qimiao Si., Yuyao Zha, Kim, J. H., and Lu, J. P., J. Phys. Chem. Solids **52**, 1337 (1991).

16. Anderson, P. W., in "Windsurfing the Fermi Sea" (edited by T. Kuo and J. Speth) (North-Holland, New York 1987), p. 61; see also Varma, C. M., p. 69.

17. Steglich, F., in "Springer Series on Solid State Sciences" (edited by T. Kasuyai and T. Saso) (Springer, Berlin 1985), vol. 62, p. 23.

18. Tajima, K., Endoh, Y., Fischer, J. E., and Shirane, G., Phys. Rev. **B38**, 6955 (1988); Hundley, M. F., Canfield, P. C., Thompson, J. D., Fisk, Z., and Lawrence, J. M., Phys. Rev. **B42**, 6842 (1990); Takabatake, T., Teshima, F., Fujii, H., Nishigori, S., Suzuki, T., Fujita, T., Yamaguchi, Y., Sakurai, J., and Jaccard, D., Phys. Rev. **B41**, 9607 (1990); see also various contributions to "Physical Phenomena at High Magnetic Fields" (edited by E. Manousakis, P. Schlottman, P. Kumar, K. S. Bedell, and F. M. Mueller) (Addison Wesley, New York 1992), (Tallahasee, May 1991).

19. Private communication.

20. Anderson, P. W., Phys. Rev. Lett. **65**, 2306 (1990).

21. Ogata, Y. and Shiba, H., Phys. Rev. **B41**, 2326 (1990); Ren, Y. and Anderson, P. W., Phys. Rev. (submitted).

22. Anderson, P. W., Phys. Rev. Lett. **67**, 2092 (1991).

23. Ong, N. P., et al., Phys. Rev. Lett. (submitted).

24. Anderson, P. W. and Yu, C., in "Varenna '85: Proceedings of International School of Physics '89" (edited by F. Bassani, F. Fumi, and T. Tosi) (North-Holland, New York 1985), p. 767.

25. Anderson, P. W., Phys. Rev. Lett. **67**, 3544 (1991).

26. Wheatley, J., Hsu, T. C., and Anderson, P. W., Phys. Rev. **B37**, 5897 (1988).

E

The "Infrared Catastrophe": When Does It Trash Fermi Liquid Theory?*

P. W. Anderson

(Received 9 February 1993)

ABSTRACT

We give a historical discussion of the "infrared catastrophe" and the "x-ray edge anomalies" of Mahan associated with scatterers in a Fermi sea of electrons. The infrared catastrophe provides a perspicuous way into understanding the difficulties with many-body perturbation theory which have recently been discovered as a result of a study of high-T_c superconductivity, and we show how this "catastrophe" is avoided in some cases, but cannot be avoided in the 1- and 2-dimensional electron gas systems. Finally, we indicate the new type of theory which is necessary in the event of such a breakdown.

Nearly 30 years ago G. D. Mahan [1] conjectured a fundamental anomaly associated with a Fermi sea of electrons interacting with a local potential, which he supposed to be responsible for experimentally observed power law anomalies in the x-ray absorption and emission edges of metallic systems.

This conjecture, based on extrapolating the results of a perturbation treatment, was turned into a proof in Ref. (2) and into a full theory of the x-ray edge effect by Nozieres and de Dominicis in 1969 [3].

Mahan's original conjecture encountered a great deal of resistance at the time, in fact his paper was delayed in the refereeing process; and the general community did not accept the reality of the effect until NDD's formal theory appeared (or even after). This is somewhat understandable: it is easy, and was done from time to time, to produce false Fermi surface anomalies, such as false "bound states," caused by the sharp edge in the density of states at the Fermi level; and Kohn and Majumdar [4] had apparently given a general refutation of such efforts a year or two earlier. The proofs of Refs. (2) and (3) showed that

*This work was supported by the NSF, Grant #DMR–9104873

the effect did exist and did not contradict any fundamental principles, however, including those discussed in Ref. 4.

The effect is so simple in the form described in Ref. (2) that it is worth demonstrating here. We suppose ourselves to have introduced a local, finite scattering potential (representing the effect of the inner-shell electron in the x-ray process) $V(r - r_0)$ which is so short-range that its matrix element for scattering a free electron from state k to state k' is essentially a constant for all states near the Fermi surface:

$$V_{kk'} \simeq V \tag{1}$$

We calculate the probability that state k below the Fermi surface is scattered into state k' above it ($\epsilon_{k'} > 0$) which is, to lowest order

$$P_{k \to k'} = \frac{|V|^2}{|\epsilon_{k'} - \epsilon_k|^2}.$$

The sum of all such scattering probabilities diverges logarithmically:

$$\sum_{k,k'} P_{k \to k'} = |V|^2 \sum_{\epsilon_{k'} < 0} \sum_{\epsilon_k > 0} \frac{1}{|\epsilon_{k'} - \epsilon_k|^2} \tag{2}$$

$$\sim |N(E_F)V|^2 \ln \frac{(\epsilon_{k'} - \epsilon_k)_{max}}{(\epsilon_{k'} - \epsilon_k)_{min}} \tag{3}$$

(where $N(\epsilon_k)$ is the density of states at the Fermi level) and the minimum possible $\epsilon_{k'} - \epsilon_k$ is of order (size of system $\Omega)^{-\frac{1}{2}}$, so the divergence is as $\ln \Omega$.

Such a divergence was shown, in Ref. (2), to mean that the ground state of the system *with* the scattering potential V (a Slater determinant of scattered wave functions) is orthogonal to the original free electron Slater determinant in the limit $\Omega \to \infty$. The overlap contains a singular power law

$$(\psi_0(V), \psi_0(0)) \propto (\Omega)^{-\frac{1}{2}|N(E_F)V|^2} \tag{4}$$

which can also be shown to appear, as a function of ω, in various response functions as well, and causes the "x-ray edge anomalies."

The property of V which is crucial is that it leads to at least one finite scattering phase shift η (in this case, in the isotropic $\ell = 0$ channel). The wave functions in the asymptotic region, far from the scatterer, are, for free electrons,

$$\varphi_{\ell=0}^{\text{free}} \sim N \frac{\sin kr}{r} \tag{5}$$

and when the scatterer is introduced,

$$\varphi_{\ell=0}^{\text{scatt}} \sim N \frac{\sin(\tilde{k}r + \eta)}{r} \tag{6}$$

where N is a normalization constant and $\tilde{k}$ must be modified to fit boundary conditions at the edge R of the sample. The expansion of the scattered wave functions (6) in terms of free wave functions (5), when we take into account the boundary condition which makes k different from k, gives exactly the overlaps which one obtains as transition probabilities from perturbation theory:

$$S_{kk'} \propto \frac{\eta}{k - k' + \frac{\eta}{R}}.$$

This way of thinking of the effect of the scatterer is, however, more precise and generalizable: one must find the scattered wave functions for a given set of boundary conditions and calculate the overlap of the appropriate Slater determinants.

This effect, apparently simple in old-fashioned Brillouin-Wigner perturbation theory or in conventional scattering theory, is not easy to deal with in modern many-body theory. Among many other reasons, boundary conditions, which play a vital role, are absent in many-body theory; and it is assumed in that theory that the set of wave vectors $[k]$ cannot change. Nozieres and coworkers [5] labored through two long, difficult papers before getting the effect, and Ref. (3) uses a radically different method from the usual techniques. Many-body perturbation theory is based on the Feynman diagram technique, which in turn depends crucially on the fact that a hole (a "positron") is simply the time-reverse of an electron, with the Dirac Sea having no dynamical consequences (this is the famous "Z" diagram of Feynman, describing pair creation as simply reversing the space-time path of the electron.) In these phenomena which we treat here, the "Sea" is no longer a dead, meaningless object but—as we see from the fact that the logarithm in (3) contains both the lower *and upper* cutoffs in energy—plays a vital role. In metals, we may have to distinguish holes from electrons—we paint the "hole" branch of the space-time path a different color from the "electron" one, in a manner of speaking.

The "orthogonality catastrophe" is closely tied in with two other important Fermi Sea concepts, again not natural to diagram techniques: the Friedel theorem and bosonization. The Friedel theorem is that in the presence of a scattering potential, the region near the scatterer contains an excess of particles given by

$$\delta n = \sum_{\ell} \delta n_{\ell} = \sum_{\ell} (2\ell + 1) \frac{\eta_{\ell}(E_F)}{\pi}. \tag{7}$$

(This is related to certain theorems of scattering theory due to Wigner: the object on the right is the trace of the imaginary part of $\ln S$, S the general scattering matrix at the Fermi energy.) In fact, the exponent in the overlap is proportional to the sum of the squares of the δn_{ℓ}'s. This demonstrates that a phase shift of π is exactly equivalent to the formation of a locally bound state, 2π is 2 bound states, etc., as far as the Ω-dependence of the overlaps is concerned. This fact that $\eta = n\pi$—which corresponds to a zero scattering cross-section—has this profound effect, more than anything else demonstrates that the properties of a Fermi sea cannot really be understood without taking into account the "counting theorems" of Friedel and Luttinger, in addition to the simple diagrammatic perturbation theory of electrons at the Fermi surface.

Schotte and Schotte [6] showed that the overlap could be calculated by "bosonizing" the variables in the $\ell = 0$ channel, which is basically equivalent to treating the position of the Fermi surface in the relevant channel in momentum space as the appropriate dynamical variable to describe the Fermi sea, rather than using particle coordinates and momenta. This very important and useful concept has been recently generalized by Haldane [7]; but the rather formidable mathematics involved is not necessary to our story.

The existence of the orthogonality catastrophe of course called into question immediately the whole structure of many-body perturbation theory as it is applied to metals. If an electron added to the Fermi sea acted as a scatterer for the other electrons, with a finite scattering phase shift in some channel, the new state of the "sea" electrons would be orthogonal to the original state. The obvious consequence is that the quasiparticle renormalization constant "Z" would vanish since this is just the overlap between the relaxed and "bare" particle states. This would mean, among other things, that the Fermi surface discontinuity of occupancy of states in momentum space would vanish. The basic point of a vanishing "Z" is even deeper than that: Z represents the connection between the "bare" particle states and the exact low-energy eigen-excitations of the interacting system. A finite Z means that these remain in one-to-one correspondence with each other, while a vanishing Z means that the exact excitations can—and do—have totally different character from the original excitations. If "Z" were zero in quantum electrodynamics, that would mean that the underlying theory contained no "bare" entities resembling the physical electrons which we see in nature. Such a phenomenon actually occurs in the transformation from bare quarks to physical nucleons.

Even for electrons this is not impossible at all: in fact, the bewildering complexity of the various exact or asymptotically exact theories of one-dimensional interacting electrons (for a review see Solyom [8]) can be traced to just this fact: that in one dimension an added electron has a finite scattering phase-shift for the electrons near the Fermi surface. As a result these theories have no discontinuity at the Fermi surface and the eigen-excitations are not electrons but bosons representing vibrations of the local Fermi surface in k-space [9]. Metzner and de Castro [10], as well as Haldane [9], have recently emphasized that for real electron systems there is really only a single free parameter—representing this phase shift—in the known solutions of the one-dimensional interacting electron gas, except at special points where commensurability with the lattice plays a role, and aside from a trivial mean field (RPA) response.

But many Fermi systems of ordinary dimensions (2 or 3) behave more or less like "Fermi liquids," which is the shorthand for a system described by quasiparticle excitations, presumably the result of applying a convergent perturbative renormalization to the bare Hamiltonian, leading to quasiparticles with effective masses and Fermi velocities, and restricted types of interaction, but not to totally new entities. D. R. Hamann [11], to my knowledge, was the first to think about and to solve the problem posed by the "orthogonality catastrophe" for Fermi liquid theory. He pointed out that in dimensionality ≥ 2, the Fermi surface scattering is much reduced by the dynamic recoil of the electron, i.e., simply by the fact that when the particle of momentum k' is scattered to a momentum $k' + q$, the original particle of momentum k is scattered to $k - q$. In general, this removes the vanishing energy denominators in the crucial sum (2). This argument is entirely valid for foreign light particle scatterers such as positrons or μ mesons, for which there are never any infrared problems. The inner shell electrons which are involved in the x-ray edge process are, however, so narrow-band that they cannot recoil appreciably.

The step which can avoid Fermi-edge divergences is the replacement of the scattering matrix element $V_{kk'}$ by a "pseudopotential" or "scattering vertex" $T_{kk'}$. This idea was one of the seminal ideas behind many-body theory, in its early form of "multiple-scattering theory," which was aimed at dealing with strong individual scattering potentials and was first applied (to nuclear matter) by Brueckner and collaborators [12]. One attempts to solve the problem of repeated scattering of two particles and to insert the result of the repeated scattering as a formal renormalized effective or "pseudo"-potential. The calculation of repeated scatterings is essentially the same as solving the Schrödinger equation for the two particles and deriving from this their scattering amplitude. This renormalizes the large effect of strong repulsive interactions of "hard cores" but it also has the effect of allowing the scatterer to recoil. In diagrammatic terms the idea is to replace the individual scattering vertex $V_{kk'}$ by the sum of "ladder diagrams" (see Fig. 1) in which repeated scatterings $k \rightarrow k + q \rightarrow k + q' \cdots \rightarrow k$ and $k' \rightarrow k' - q \rightarrow k' - q' \cdots \rightarrow k'$ are allowed. The other particles are assumed not to be involved in this process. This procedure is therefore exact in the limit of low density, and in fact was the procedure used in the treatment of the low density Bose and Fermi systems by Yang [13] and coworkers and by Gor'kov and Galitskii [14], and later on by Bloom [15] for the 2D electron gas. But it is actually essentially correct even at finite densities for the purposes of our problem, because for particles close enough to the Fermi surface or within it real scatterings by other particles are restricted severely by the exclusion principle, and virtual scatterings by other particles can be renormalized away. The only important correction is that the ladder diagram sum must take into account the fact that states $k' - q$ etc., may be excluded because of being already occupied. So it is literally exact, in principle, to consider an electron of up spin as a foreign particle scatterer for down spins

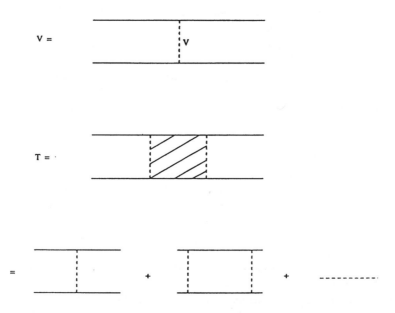

Figure E.1. The ladder sum for the effective scattering potential.

exactly like the inner shell hole of the x-ray edge problem. Like that hole, it acquires a singular self-energy due to the resulting modification of all of the energies and wave-functions of the down-spin electrons—this is exactly the method of calculation used by Nozieres and de Dominicis, in fact, for the x-ray problem.

In many cases this "recoil" calculation actually works to reduce "Z" to a finite value. These are the cases in which the scattering can be described, for small scattering vectors Q, by an "effective range" or "scattering length" theory, as was done in the works by Yang et al. [13, 14, 15], referred to above and as is assumed without proof in AGD [16]. This seems to be correct in the limit of low densities or weak scattering in 3 or more dimensions.

The "effective range" theory assumes that the scattering is like that from a hard sphere of radius a, with "a" a parameter which can be determined by solving the radial wave equation for two free particles. The wave function in relative coordinates, near the origin, becomes

$$\varphi(r) \sim \frac{\sin Q(R - a)}{r} \qquad (8)$$

so that the phase-shift η is obviously Qa, which vanishes as Q, the relative momentum, goes to zero. This is sufficient to restore Fermi liquid theory, for reasons which require a little subtlety to explain.

We consider the state $(k^\uparrow)$ to be the added particle whose self-energy Σ_k and renormalization constant $Z_k = (1 - \frac{\partial \Sigma}{\partial \omega})^{-1}$ we are seeking. The state $k'_\downarrow = (k + Q)_\downarrow$ is the scattered state whose energy and wavelength shift due to the existence of $k^\uparrow$ we want to calculate. In order to fit the new s-wave state to the boundary condition at a radius $\sim R$ ($R^3 \sim \Omega$), we must shift Q by

$$\delta QR = Qa$$

$$\delta Q = \frac{aQ}{R}. \tag{9}$$

But plane wave states k and k' are only fractionally in an s-wave state with respect to each other, the number of angular momenta possible for a given $k - k'$ being

$$N_L \simeq QR \tag{10}$$

only one or, at most, a small finite number of these angular momenta are affected by the scattering so that

$$\delta k' \propto \frac{\delta Q}{QR} \propto \frac{a}{R^2}. \tag{11}$$

The separation between k-values $\Delta k'$ is $= \frac{\pi}{R}$ so that

$$\frac{\delta k'}{\Delta k'} \simeq \frac{a}{R} \lll 1, \tag{12}$$

vanishing in the $\Omega \to \infty$ limit. This is the condition that no orthogonality catastrophe takes place; this slight shift can be compensated, as extra particles are added, by a shift in mean potential, adjusting the chemical potential so that the density of k-values in momentum space remains constant. Any higher power-law dependence of η on Q is a fortiori even more harmless: the key condition is that the forward scattering phase shift vanishes or not as $Q \to 0$.

It is often stated that Fermi liquid theory is based on a convergent perturbation theory in which no resummation of divergent terms has taken place. This is obviously not the case in view of the above derivation: the "ladder" summation which reduced η ($Q = 0$) to zero and gave us effective range theory is precisely such a summation. The rather indirect way in which this summation is argued around in the standard derivations is the subject of a short paper which has been submitted [17], but that argument is not necessary to the present discussion. Suffice it to say that the summation of formally divergent ladder diagrams as $Q \to 0$ is an integral part of the theory.

In dimensions lower than three the recoil argument becomes more difficult. In fact, in one dimension η ($Q = 0$) does not vanish even for free particles and, as we already pointed out, no interacting system has a finite Z. In two dimensions a free scatterer can be shown to obey

$$\eta_{2D}(Q) = \frac{1}{|\ln Qa|}$$

which, as Bloom showed, is just adequate to give a convergent first few terms of perturbation theory in the formal zero-density limit of the 2d hard core gas.

In 2 or 3 dimensions there are certain theorems ("Levinson theorems") which demonstrate that in the absence of bound states the scattering phase shifts go to zero for free particles at zero relative energy. However, there is no proof of such theorems for scattering of two particles which live in a Fermi sea and must obey the exclusion principle. The effect of the exclusion principle is to make it much harder for the two particles to recoil, especially at very low energies: essentially, "soft" recoils become impossible, because the states involved are occupied. Levinson's arguments have no relevance in this case.

It is very important to understand precisely the scattering problem we are considering, and why it is done in this way. We are adding a particle of $k_\uparrow$ near (but below) or on the Fermi surface: i.e., we visualize having added the particles one by one and space has been left for this one. We now consider each particle $k'_\downarrow$ in the down-spin Fermi sea and allow it to adjust its wave-function to the new scattering by $k_\uparrow$. The scattering problem for each k' is the "on-shell" problem in which we study scattering which is precisely elastic. Lifetimes, near the Fermi surface, of the holes that these particles will fill are $\propto (k - k_F)^2$ and can be made negligible by well-known arguments.

Both particles we are taking as renormalized quasiparticles as far as high-energy virtual interactions with the Fermi sea electrons are concerned. These various statements can be justified quite rigorously by lengthy, fussy arguments whose general character is quite obvious and irrelevant to the present considerations. This is a slightly unfamiliar scattering problem, not the same as the one which would be done to produce the conventional "scattering vertex" for two added particles or holes in perturbation theory, because we are concerned with the wave functions of electrons actually present in the Fermi sea, not with added excitations over the Fermi sea as vacuum. As in the x-ray problem, we are looking for the response of the "sea" considered as a physical fluid to the addition of a perturbing particle.

The diagrammatic description of this process is that we sum all ladders which scatter k, k' to $k + Q, k' - Q$, eliminating all occupied states for both particles with a factor

$$F(Q) = (1 - f_{k+Q})(1 - f_{k'-Q}) \tag{13}$$

but allowing the particles to return to their initial state $Q = 0$. The result is a simple Schrödinger equation which, using a Hubbard-model local potential U, reads

$$1 = U \sum_Q \frac{1}{E - E_Q} F^*(Q)$$

where E_Q is $\epsilon(k + Q) + \epsilon(k' - Q)$ and F^* is (13) for all states Q except those for which $E - E_{Q_0}$ is of order ΔE_{Q_0}, the spacing between levels of different Q_0.

These are the states which must be treated separately to allow the scattered wave function to satisfy the boundary condition enforced by the scattering. Essentially, these states must be allowed to move in k-space in order to let the wave function modify itself in order to satisfy Schrödinger's equation both at ∞ and at the origin. Formally, we do this by writing

$$\frac{1}{U} = \sum_Q \frac{F^*(Q)}{E - E_Q}$$

$$= \sum_{Q \simeq Q_0} \frac{1}{E - E_{Q_0}} + P \int \frac{dN}{dE_{Q'}} \frac{dE_Q}{E - E_{Q'}} F(Q')$$

$$= N(E) \cot \pi \eta + P \int \frac{dN}{dE_Q} \frac{dE_{Q'}}{E - E_{Q'}} F(Q').$$

This, in 2D, leads in general to a finite η as $Q \to 0$ and $k \to k_F$, because $F(Q)$ vanishes as Q^2 and the principal part integral is convergent as $Q \to 0$.

This computation is trivial, and in fact has been checked in much more complete calculations by Fukuyama et al. [18] and by Engelbrecht et al. [19]. Both find that *on the energy shell* the forward scattering phase shift is finite. What they find is that the phase of the scattering matrix $T(Q, \omega)$ is finite exactly along the line $E_Q = \omega$ but not in either the limit $\omega \to 0$ or $Q \to 0$. Thus less careful calculations miss this "on-shell" effect. Where we differ is in the use of this information: I point out that this phase shift implies a shift of the momentum values δk which is turn leads to the orthogonality catastrophe. These authors, however, do not draw the obvious conclusion that the orthogonality catastrophe occurs, because they are working with conventional perturbation theory, in which the orthogonality catastrophe does not appear and the consequences of a finite phase shift are not evident. As we showed earlier, the orthogonality phenomenon shows up as a divergence when a finite system is taken to $\Omega = \infty$, but perturbation theory assumes that the infinite volume limit is approached uniformly. Special stratagems are necessary to uncover the orthogonality phenomenon without recourse to taking the limit from finite systems.

One of the interesting side implications of this result is that it may not be possible in general to draw conclusions about the states of finite systems using many-body perturbation theory as it now exists. In finite systems, the modification of wave functions upon adding a particle is not small and must be taken into account. This may be one of several reasons for difficulties with the LDA approach to such problems. Genuine Hartree-Fock, which does take the change in wave functions into account, may often give qualitatively preferable answers.

It is not yet clear when or whether 3D systems also are liable to the orthogonality catastrophe and the breakdown of perturbation theory. The simple forward scattering calculation (14) does not lead to finite η because $N(E_Q) \to 0$ at $Q = 0$. General arguments suggest that two conditions may need to be satisfied in 3D:

(1) The relevant band must be isolated by an energy gap from higher, free-electron like bands.
(2) U must be stronger than a critical value.

These are the conditions for validity of the "t-J" model transformation of the interacting Hubbard model to a projective model in which the exact "double occupancy" restriction is enforced. Such a restriction can be removed with the addition of a local gauge symmetry, which leads in general to a major modification in the elementary excitation spectrum.

Conclusion

A great deal of work has been done involved in fitting this breakdown of perturbation theory into a possible scheme for dealing with the 2-dimensional electron gas. This scheme involves using the multidimensional bosonization scheme of Luther and Haldane, which they presented as a transcription of Landau's Fermi Liquid Theory. As they show, the hydrodynamics of the Fermi liquid are not adequately described by the random-phase approximation; however, in the Fermi liquid theory there are no interesting collective modes other than at $Q \simeq 0$, where RPA is satisfactory. In their picture, the hydrodynamics of the Fermi sea is that of an incompressible liquid blob in k-space, with the low-frequency modes being vibrations of this blob at each point of its Fermi surface, with conservation laws for every angle around that surface.

The singularity due to forward scattering fits neatly into this hydrodynamics and can be expressed by modifying the bosons representing the fluctuations of the Fermi surface, giving separate Fermi velocities for charge and spin bosons. This seems to be not only theoretically sound but also vindicated by a wide variety of experimental phenomena in the cuprate metals, some of which were predicted prior to observation. The mathematics of this approach is, however, beyond the scope of this article.

This approach leaves open the next order corrections: coupling of bosons on different patches of the Fermi surface, etc. It also has, so far, only been applied in the 2D case. The conventional many-body theory has been noteworthy in its failure to deal adequately with strongly interacting and low-dimensional systems; we now have the first breakthroughs which demonstrate that this is not a mere problem of difficulty in application but a failure in principle. It is likely to be essential to completely revise our conceptual structure before going further.

REFERENCES

1. G. D. Mahan, *Phys. Rev.* **163**, 612 (1967); *Phys. Rev.* **153** (1967) 882.
2. P. W. Anderson, *Phys. Rev. Lett.* **18** (1967) 1049.
3. C. J. DeDominicis and P. Nozieres, *Phys. Rev.* **178** (1969) 1097.
4. W. Kohn and C. Majumdar, *Phys. Rev. A.* **138** (1965) 1617.
5. P. Nozieres, O. Raulet, and J. Gavoret, *Phys. Rev.* **178** (1969) 1072, 1084.
6. U. Schotte and D. Schotte, *Phys. Rev.* **182** (1969) 429.
7. F.D.M. Haldane, Unpublished Notes, Les Houches Summer School, 1992.
8. J. Solyom, *Adv. in Physics*, **28** (1979) 201.
9. F.D.M. Haldane, *J. Phys C* **14** (1981) 2585.
10. W. Metzner and C. di Castro, preprint.
11. D. R. Hamann, private communication.
12. K. Brueckner et al., *Phys. Rev.* **92** (1953) 1023; *Phys. Rev.* **95** (1954) 217, etc.
13. T. D. Lee, K. Huang, and C. N. Yang, *Phys. Rev.* **106** (1958) 1135.
14. L. P. Gorkov and G. Galitskii, as quoted in Ref. 17.
15. P. Bloom, *Phys. Rev.* **B12** (1975) 125.
16. A. A. Abrikosov, L. P. Gorkov, and I. Dzialoshinskii, *Methods of Quantum Field Theory*, Prentice Hall, N.J., 1963.
17. P. W. Anderson, submitted to *Phys. Rev. Lett.*
18. H. Fukuyama, O. Narikiyo, *J. Phys. Soc. Japan* **60** (1991) 372; **60** (1991) 1032.
19. J. R. Engelbrecht and M. Randeria, *Phys. Rev. Lett.* **65** (1990) 1032.

It is well known that perturbation theory fails in the face of a phase transition. It is not possible to describe a liquid with a perturbation theory starting from the crystal, for instance. The reason is that at a phase transition the Hilbert space of the excitation spectrum becomes orthogonal to that described by $\mathcal{H}_0$. Each of $N \to \infty$ coordinates which describe the state displaces by a finite amount δx_i at such a transition, leading to an overlap $\sim e^{-N(\delta x_i^2)} \to (\infty)^{-1}$ between any of the original states and any of the final states. Basically, a phase transition is signaled by a divergence of perturbation theory, and the true different phases must be described by different $\mathcal{H}_0$'s.

A less well known, but equally obvious, case is that of a projective transition. In this case the interaction $\mathcal{H}'$ causes a subset of the states of the original Hamiltonian to break off from the original spectrum. A finite gap Eg develops, separating each of these bound—or "antibound"—states from others of similar quantum numbers. One can think of the states originating as eigenstates of $\mathcal{H}_0$ as dividing up into two subspaces, with all of the states in one subspace separated by a finite gap Eg from those in the other (the "upper" and "lower" Hubbard bands, for instance). Then the physical subspace is only a projection of the original space.

This situation violates the requirement of a continuous, adiabatic, unique unitary equivalence U between unperturbed and perturbed states, hence cannot be described by a perturbation theory. The physical subspace does not have the same dimension as the unperturbed space, and U has no inverse.

The simplest example is the half-filled Hubbard model. The interacting system is described by a spin Hamiltonian, with 2^N states; the unperturbed system by the Hubbard model with 4^N states. No continuous unitary transformation carries one into the other.

There are many different ways of saying this obvious fact. Each projective constraint imposes a conservation law on the system, so that there are $\sim N$ conserved quantities. An appropriate $\mathcal{H}_0$ must contain those constraints. This can be managed by introducing a Lagrange multiplier for each constraint: hence the many attempts to introduce massless gauge bosons into $\mathcal{H}_0$, which are correct in principle but so far not very useful in practice. But any attempt to do perturbation theory without enlarging $\mathcal{H}_0$ is foredoomed. A possible resolution has been proposed: it is suggested that somehow, in the course of the solution, the gauge variables become massive, hence have little effect; but then one can easily see that the space of $\mathcal{H}_0$ would remain unmodified and the original problem returns.

In the case of the half-filled Hubbard model, the solution in 1D is identical with that for the Heisenberg model: spinons. Clearly J of the Heisenberg model is, for large U, $\propto \frac{1}{U}$ where U is the Hubbard interaction; the appropriate transformation is the "t-J" transformation in which doubly occupied states are projected out of the space (as first shown in [Anderson, '59]). This is clearly not perturbative in U in any sense. It is probable that most or all 2D half-filled Hubbard models, and all large U Hubbard models, are similar, although the Heisenberg model solutions are ordered in these cases. Note that this *non*perturbative mechanism is behind most (if not all) physical examples of antiferromagenetism or of antiferromagnetic "spin fluctuations." The $t - J$ transformation can be thought of as a renormalization to the neighborhood of the $U \to \infty$ fixed point. The half-filled case is thus nonperturbative. In an RNG diagram, the flow is towards $U = \infty$, not $U = 0$, and the only discontinuity in the space of solutions occurs at $U = 0$. There are many arguments which show that this is also the case for many incommensurately filled Hubbard models:

(1) Certainly for all 1D models;
(2) Certainly for all large U: there is a critical U;
(3) Probably for all 2D models

(1) In 1D it is enough to appeal to the Lieb-Wu exact solution and its more recent reinterpretations. As shown clearly by Ogata and Shiba, for large U and all n, or for any U and n close to 1, the solutions explicitly break up into a product of a charge and a spin sector. The spin sector is identical with the $n = 1$ case, i.e., the Heisenberg model: one has a Heisenberg model on the "squeezed" lattice with holes omitted, which exhibits a spinon spectrum. But the spinon spectrum has the same strange property that it has on the Heisenberg model: there are exactly N spinon momenta, i.e., the spectrum ends at k_F, not at the edge of the zone. This implies the existence of a hidden order parameter which connects a spinon of momentum k with a formal antispinon of momentum $k - 2k_F$, hence the latter is not an independent state and can be said not to exist.

It is possible to show that the spinon system has a broken symmetry and is hence intrinsically nonperturbative, as it must be because it is projective.

That this system is projective has also been shown by Gulasci, who has demonstrated that all n and U there is a branch of "complex k" solutions at high energy. They constitute a concrete demonstration of the existence of the "anti-bound states" which Anderson and Hsu proposed. The low-energy subspace must have these anti-bound states projected out.

For sufficiently large U at any finite dimension, but not at $D = \infty$, anti-bound states develop. The band of one-electron states, in the Hubbard model, has an upper bound in energy so two-quasiparticle states do also. If U is large enough, there will be one state separated from the continuum for pairs of opposite-spin particles, for each total momentum. Such a set of states forms a discrete subspace which must be projected out of the low-energy dynamics. Thus that low-energy subspace does not have the right dimension to be described by the algebra of N fermions.

This holds a fortiori in 2 dimensions, where any interaction no matter how weak causes a bound state to form at one edge of the band or the other. Thus it seems almost certain that in 2D $U_{\text{crit}} \equiv 0$.

The reader should note that these phenomena depend crucially on spin, especially in the Hubbard model. As I have argued elsewhere more directly, spin is vital to the breakdown of perturbation theory and all discussion of spinless models in this regard is completely irrelevant. It is not possible to "generalize trivially" to a model with spin, as has been claimed. The spin sector exhibits the broken symmetry in the 1D system, and the anti-bound state is a spin singlet. The power laws which appear in the simple Luttinger model and in the "Luttinger liquid" of Haldane seem to result from a relatively simple logarithmic divergence of perturbation theory which can be resummed, since the Hilbert space remains unaltered. The spin sector, while not exhibiting incommensurate power laws, is where intrinsically nonperturbative effects happen. Independently of U, the spinon remains an exact semion: a result which is manifestly nonperturbative.

In conclusion, then, I am pointing out that a very large class of Fermi systems of physical interest are not treatable with conventional perturbation methods (for which a rather inappropriate shorthand is "Fermi Liquid Theory"). This class includes at least all 1D and quasi-1D systems, and the cuprate superconductors. It is likely to include many mixed valence systems, buckeyballs, and $(BaK)BiO_3$, and may extend to the transition metals.

G

Charge-Spin Separation in 2D Fermi Systems: Singular Interactions as Modified Commutators, and Solution of the 2D Hubbard Model in the Bosonized Approximation

P. W. Anderson and D. Khveshchenko

Joseph Henry Laboratories of Physics, Jadwin Hall, Princeton University
(Received 22 June 1995)

ABSTRACT

The general two-dimensional fermion system with repulsive interactions (typified by the Hubbard model) is bosonized, taking into account the finite on-shell forward-scattering phase shift derived in earlier papers. By taking this phase shift into account in the bosonic commutation relations, a consistent picture emerges showing the charge-spin separation and anomalous exponents of the Luttinger liquid.

The proper description of the effect of finite forward-scattering on-shell phase shifts on Fermi systems for $D > 1$ (Ref. 1) has been the subject of a number of papers.[2–4] The existence of such scattering, leading to on-shell singularities in the T matrix, was confirmed by Fukuyama and Narikiyo,[5] Metzner and di Castro,[6] and others. The discovery which these papers confirmed followed from the fact that when two particles are embedded in their respective Fermi seas, effectively all soft recoils are forbidden to them by the exclusion principle. Under these conditions, the logarithm of the S matrix for relative motion retains a finite eigenvalue η_0 in two dimensions (2D),

$$S_0(Q) = \exp i[\eta_0(Q)],$$

in the limit that the relative momentum $Q = k_\uparrow - k'_\downarrow \to 0$, and that both states are below the Fermi level. (On general grounds, it seems likely that this can happen in 3D as well, but this has not been proven.) It is important to be clear that S is not the conventional T matrix defined for a hole at k scattering against another at k', whose imaginary part represents an incoherent decay process, vanishing at the Fermi surface; S is the on-shell scattering matrix of particles embedded in the Fermi sea, and its phase determines the boundary condition for their asymptotic wave functions at the origin of relative coordinates $r - r' = 0$. Addition of a particle modifies the wave functions of all other particles, and our endeavor is to investigate the consequences of this fact.

Our initial description[7] was heuristic, merely pointing out that the effect of such a phase shift mimics that of a change in statistics by enforcing a partial exclusion principle between electrons of opposite spins. We also described the type of singular interaction which would give energy shifts similar to those which take place, and emphasized that these would "trash" Fermi-liquid theory. Some papers have focused on this singular interaction (specifically Engelbrecht and Randeria[2] and Stamp[3]). The treatment in terms of an interaction is in several respects unsatisfactory, as clarified by Baskaran.[4] But even Baskaran's discussion does not give us a clear insight into clean formal ways to deal with the situation.

The problem is that the effect is best thought of as a constraint on the wave functions, not as an interaction. This is most clearly seen in the Hubbard model, where the effect of a strong enough repulsive potential $U \to \infty$ is to enforce a projective constraint, expressed as the Gutzwiller projector acting on the kinetic energy in the t-J model, for instance. Since the exchange term also is expressible purely in terms of the projected operators, the t-J system is confined to the subspace defined by projected operators.

It is worth emphasizing that renormalization-group derivations of Fermi-liquid theory (FLT) as a theory of the low-energy states, such as that of Shankar, implicitly assume a free Fermion starting Hamiltonian. If the starting problem itself is projected onto a subspace, this property will remain after renormalization and FLT changes into the theory we shall derive.

In general (in 2D) the constraint appears as a phase shift, which is a boundary condition for the asymptotic wave function in the relative coordinates of a pair of particles. Such a wave function is indeterminate unless it has a boundary condition both at $|r - r'| \to \infty$ and at $r - r' \to 0$. Arguments in several of the original papers show that the rest of the particles may be satisfactorily dealt with by taking the exclusion principle into account; the multiparticle encounters are not crucial.

This local boundary condition on the asymptotic wave function at $r - r' \to 0$ is a kinematic, rather than a dynamic, effect: there is a change in the wave functions of the particles, not directly in their energy. We are used to this with hard-core potentials: the effect is best expressed as one purely on the kinetic energy, not on the potential. This kinematic effect dominates here because the scattering region where the potential acts is small, of order N^{-1} compared to the asymptotic region in which the kinetic energy is modified. The way to make this point is that such a boundary condition can actually change the dimensionality of the Hilbert space of allowed wave functions. In simple terms, such a boundary condition forces the wave function's nodes to shift in such a way that a particle moves into or out of the distant boundary, so that the same volume contains $N \pm \eta/\pi$ particle states rather than N. This is what is meant by a change in the dimensionality of Hilbert space. This change of Hilbert space occurs in 1D even as a consequence of an ordinary interaction potential (hence the flexibility of statistics in 1D), but in all other dimensions it is distinct from the kind of interaction effects which can be treated perturbatively.

The conclusion we came to is therefore that the effect of a finite phase shift is best modeled as a modification of the algebra of the particles, expressed in their commutation relations. Projected fermions

$$(c_{i\sigma}^+)_{\text{proj}} = P_G^i c_{i\sigma}^+ = (1 - n_{i-\sigma})c_{i\sigma}^+ \tag{1}$$

do not have the same commutation relations as ordinary fermions, obviously, but we have not found the fermion representation convenient to work with. It is much simpler to use the bosonized representation in terms of the Fermi-surface fluctuations.[8,9] The bosonized version of Fermi-liquid theory can be equivalently thought of as the appropriate gauge theory in the presence of a Fermi surface, since the bosonic variable is essentially the phase of the Fermi-surface wave function.

Haldane, particularly, has emphasized that the most useful description of the dynamics of a Fermi system is via the operators Δk_F describing the position of the Fermi surface in k space, taken to be dynamical variables, functions of a coarse-grained space, and time. That is, he argues that Luttinger's theorem holds exactly during sufficiently long-wavelength and low-frequency fluctuations. (Parenthetically, even the conventional derivations of Luttinger's theorem[10] depend not on the convergence of perturbation theory but merely on the assumption that excitations precisely at the Fermi surface (FS) do not decay; hence the Green's function is real.) We define operators

$$\Delta k_{F\sigma}(\Omega, r, t),$$

giving the Fermi-surface fluctuations of spin σ at a point on the FS parameterized by Ω, and at coarse-grained r, t. These are the bosonic variables: they commute for different Ω and r, and, for noninteracting electrons, for different σ. We can introduce a phase variable θ_σ of the wave function at the Fermi surface, which is a function of Ω, r, and t, and then $\Delta k_{F\sigma}$ is

$$\Delta k_{F\sigma} = \hat{\mathbf{n}}_\Omega \cdot \nabla \theta_\sigma, \tag{2}$$

where $\hat{n}_\Omega$ is the local normal to the fiduciary Fermi surface. θ and Δk_F, which is equivalent to the particle density at Ω, $\rho(\Omega)$, are conjugate variables, and for free fermions have canonical commutation relations:

$$[\theta, \rho] = i\pi\delta(r - r')\delta(\hat{\Omega} - \hat{\Omega}'). \tag{3}$$

As Haldane has pointed out, this representation can be motivated by the idea of expressing the fermion field in terms of two real operators ρ and θ:

$$\psi(x) = \rho e^{i\theta(x)} \tag{4}$$

rather than by the earlier "Tomonaga" definition of $\rho(q)$ as a density of fermions $\sum_k c_{k+q}^+ c_k$. This latter representation is not possible when the fermions are projected operators. But we can still speak of a Fermi surface and a Fermi-surface phase for each spin which satisfies Luttinger's theorem, and hence determines the density of particles at each point on the Fermi surface. In this transcription of the original idea of bosonization we follow Khveshchenko.[11] If a Fermi surface exists, this implies zero-frequency modes at each point on it, hence separate, independent conservation of particle and spin currents at the Fermi surface at each Ω, even allowing for Fermi-surface fluctuations which may be integrably singular at low frequencies.

However, this does not imply that, in the presence of interactions, θ_σ and ρ_σ (or $\Delta k_{F\sigma}$) remain the appropriate canonically conjugate variables. These are variables which measure, respectively, the particle number at a particular patch on the Fermi surface and a given spin and the phase of the wave function at the Fermi surface. If there is a finite phase shift for forward scattering of opposite-spin electrons, as we have shown,[1,7] the order of doing these operations matters. If we add a particle of up-spin, the phase of the down-spin wave function depends on whether the particle of up-spin was added before or after the phase was measured. The failure of commutation for opposite spins is the phase shift η/π, just as adding a particle of up-spin below the Fermi surface enforces a change in up-spin phase by the amount π. We may express this by writing the free-particle commutator in matrix form:

$$[\rho_\sigma, \theta_{\sigma'}]_{\text{bare}} = i\pi \begin{bmatrix} 1 & 0 \\ 0 & 1 \end{bmatrix} \delta(r - r')\delta(\Omega - \Omega'), \tag{5}$$

while

$$[\rho_\sigma, \theta_\sigma]_{\text{interacting}} = i\pi \begin{bmatrix} 1 & \dfrac{\eta}{\pi} \\ \dfrac{\eta}{\pi} & 1 \end{bmatrix} \times \delta(r - r')\delta(\Omega - \Omega'). \tag{6}$$

Let us explain these equations in detail. Equation (5) means in the one-dimensional model that if we insert an extra particle into the Fermi sea at a point r, because of the exclusion principle the wave function at the Fermi surface [which is the basic interpretation of Eq. (4)] must have an extra node inserted into it near r; hence the phase difference between left- and right-going (or ingoing and outgoing) waves must shift by π as a consequence. Hence after we insert one particle in ρ, θ will change by π, but not vice versa: one is the generator of displacements of the other. Equation (6) must be interpreted in exactly the same way. The insertion of an up-spin particle at r, near Ω, means that the phase of the down-spin wave at Ω is shifted by η, while the up-spin wave is shifted by π. This means that $\theta_\uparrow$, $\rho_\downarrow$ and $\theta_\downarrow$, $\rho_\downarrow$ are no longer canonically conjugate; the correct canonically conjugate variables are proportional to

$$\theta_s + \frac{\theta_\uparrow - \theta_\downarrow}{\sqrt{2}}, \quad \rho_s = \frac{\rho_\uparrow - \rho_\downarrow}{\sqrt{2}} \tag{7}$$

and

$$\theta_c = \frac{\theta_\uparrow + \theta_\downarrow}{\sqrt{2}}, \quad \rho_c = \frac{\rho_\uparrow + \rho_\downarrow}{\sqrt{2}}. \tag{8}$$

The equations of motion of the charge and spin bosons follow from the commutation relations and the Hamiltonian, which as we explained is simply the original kinetic energy, the interaction terms being completely subsumed in the commutation relations. The Hamiltonian is, as for free particles, the one given by Haldane:

$$\mathcal{H} = \frac{1}{2} \int d\Omega \int d^D r \, v_F(\Omega)[\Delta k_F(\Omega), r, t)]^2$$

$$= \frac{1}{2} \int d\Omega \sum_q v_F[\rho^2(q, \Omega) + q^2\theta^2(q, \Omega)]. \tag{9}$$

Then

$$[H, \theta_{c,s}(q, \Omega)] = v_{c,s}q\theta_{c,s}(q, \Omega), \tag{10}$$

with

$$v_s = v_F(\Omega)\left[1 - \frac{\eta(\Omega)}{\pi}\right],$$

$$v_c = v_F(\Omega)\left[1 + \frac{\eta(\Omega)}{\pi}\right], \tag{11}$$

and bosons are left as harmonic-oscillator variables with frequencies

$$\begin{Bmatrix} qv_c & (\Omega) \\ qv_s & (\Omega) \end{Bmatrix}.$$

For free particles the Fermion operator is made up from bosons via the formula

$$\text{Free: } \psi_\sigma^\Omega(r) = \rho_0 e^{-i\sqrt{2}(\theta_c + \sigma\theta_s)}, \tag{12}$$

which gives the Green's function

$$G_{\text{free}} = \frac{1}{\sqrt{r - v_s t + i/\Lambda}} \frac{1}{\sqrt{r - v_c t + \frac{i}{\Lambda}}} \quad (v_c = v_s). \tag{13}$$

But we cannot assume that the connection between interacting electrons and the modified bosons obeys (12). The coupling of the two Fermi surfaces which leads to the modified CR means that (12) creates an object which can be thought of as a pseudoelectron, with the suitable backflow caused by the fractional opposite-spin hole which accompanies it, so it describes an exact eigenexcitation of electronlike character moving in the exact ground state. These excitations are analogous to bosonized versions of the exact eigenexcitations of charge (I_i) and spin (J_α) of the Lieb-Wu solution of the 1D Hubbard model. (The discussion here was foreshadowed in Ren's thesis.[12]) These ladders of excitations can be described in terms of appropriate bosons since they have linear energy-momentum relations near zero energy, and these are the bosons which we have derived. But the actual electron operator creates a physical electron, not the pseudoelectrons described by these bosons, and hence must have the backflow compensated out. This leads to the fractional exponents in the Green's function and other correlation functions characteristic of the Luttinger liquid. As in the 1D case (as shown in Ren's thesis) the coefficients may be deduced from conservation laws and from the Luttinger theorem of incompressibility of the Fermi sea in momentum space.

Note that the pseudoelectron has the quantum numbers of a true electron, and in fact it is one of the packet of exact eigenstates created when a true electron is inserted at the appropriate momentum, though with vanishing amplitude as $L \to \infty$. When a real electron is added, a cloud of particle-hole excitations in addition to the two semions is excited, analogous to the cloud of particle-hole excitations which causes the x-ray edge anomaly. This is the backflow. The modified commutation relations of the charge and spin bosons still leave them as a bosonic description of particles which are semions in the sense that two of them make an electron. The transformation which diagonalizes the CR is not modified from the free-particle case, i.e., it is independent of η.

This is essentially because we maintain Luttinger's theorem of incompressibility as a constraint, so that no net down-spin particles are removed by the scattering process: they are merely redistributed in momentum space, which is the backflow we must now calculate. η/π particles are displaced from the neighborhood of the scatterer particle at $k\uparrow$, and we must find how they displace the Fermi surface bosons, i.e., how the phases are shifted at the Fermi surface. But first we must take into account some consequences of the non-Abelian spin symmetry which we have been ignoring so far.

A key theorem of the bosonization technique follows from the symmetry properties of the states at the Fermi surface. As we stated above, the existence of a Fermi surface implies separate conservation of each component of spin at each point on the Fermi surface. But spin conservation must remain independent of the choice of axes, and we must be able to choose the axes at each point independently. A related requirement is that Kramers' degeneracy of the spin at each point of the fermi surface independently must be maintained. This is not possible if spin at different Fermi points is coupled relevantly as $\omega \to 0$. As is seen in the 1D Hubbard model, this implies that the spin bosons cannot acquire an anomalous dimension, and must retain the same semionic character that they have for free fermions. In our situation, this expresses itself by the observation that our scattering calculation is slightly incomplete. We have not required formal spin rotation invariance [SU(2) symmetry] of the S matrix for scattering, which requires that the phase shift have the form

$$\eta = \eta_c + \eta_s(\sigma \cdot \sigma') \qquad (14)$$

and allows for a spin-flip scattering, which we have so far ignored, of half the magnitude η of the potential term. This requires the scattering to take place entirely in the singlet channel, rather than the up-down channel, as we have implied in our discussion so far. Our previous picture left us with one spin $k_\uparrow$ plus a hole of magnitude η/π in $k_\downarrow$. This left $1 + \eta/\pi \uparrow$ spins, but now we have spin-flip scattering of $\eta/2$ giving $\eta/2\pi$ missing down-spins and $1 - \eta/2\pi$ up-spins or one net spin. Correspondingly, this gives matching currents of up-spin in the scattered channels which leaves us with displacements only of charge, not spin, bosons in the backflow. The comoving hole of magnitude η/π is now in the charge channel.

In the actual 1D Hubbard model, this theorem is satisfied only to logarithmic accuracy, leading to $(\ln \omega)^{-1}$ and $(\ln q)^{-1}$ corrections to power laws; the relevant coupling constant goes to zero only logarithmically. We expect the same pathology in 2D. But dominant power laws will be correctly determined by bosonization. (All of this was foreshadowed in Haldane's "Luttinger liquid" treatment of the 1D Hubbard model.)[13] When the spin-flip component is taken into account, we now can determine how the phases at the Fermi surface are shifted, specifically when we insert an electron at Ω, q in order to calculate the one-particle Green's function. The rule is very simple: we calculate the phase shifts we would have expected using naive up-spin down-spin scattering, and replace these by phase shifts in the pure charge channel. Let us first discuss the 1D case, which was worked out by Ren.[12]

In 1D, the amount of charge η/π which is displaced from the state $k = k_F - q$ appears, half at the left-hand Fermi point and half at the right, i.e., $\eta/2\pi$ at each. These components multiply the Green's function by the factor

$$e^{i\theta_c^r(\eta/2\pi)(1/\sqrt{2})} e^{i\theta_c'(\eta/2\pi)(1/\sqrt{2})},$$

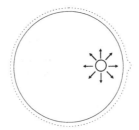

Figure G.1. Dilation of the Fermi surface due to the comoving hole.

which gives, in space-time representation, a factor

$$\left[\frac{1}{(x^2 - v_c^2 t^2)} \right]^{1/4(\eta/2\pi)^2} , \tag{15}$$

which has the maximum exponent $(\frac{1}{2})^2 \times \frac{1}{4} = \frac{1}{16}$, as pointed out by Ren. This gives the famous Fermi-surface smearing exponent $2 \times \frac{1}{16} = \frac{1}{8}$ in the strong-coupling case, and with the strictly local interaction appropriate to the Hubbard model.

These two displacements are the total backflow. The net momentum of the backflow is zero, and the net charge η/π, as it must be.

The situation in 2D is not quite so simple. Again, we recognize that η/π worth of charge boson—i.e., η/π enclosed by an internal Fermi surface—has been displaced from the region of momentum k. We may calculate the displacement of a circular Fermi surface which would result from elastic incompressible deformation of the lattice of k values. (We use a circular FS for illustrative purposes.) This would give us

$$\delta \mathbf{k}' = \frac{\mathbf{k}' - \mathbf{k}}{(k' - k)^2} \frac{\eta}{2\pi^2}$$

and

$$\delta k_F^{(\Omega)} = \frac{\mathbf{k}_F(\Omega) \cdot (\mathbf{k}_F - \mathbf{k})}{(k_F - k)^2} \frac{\eta}{2\pi^2}. \tag{16}$$

See fig. 1. If k is chosen at $\theta = 0$, and $k = k_F - \epsilon$,

$$\delta k_F(\Omega) \simeq \frac{\epsilon k_F}{\epsilon^2 + k_F^2 \theta^2} \frac{\eta}{2\pi^2} + \frac{\eta}{2\pi^2} \frac{1 - \cos \theta}{2(1 - \cos \theta)}$$

$$= \frac{\eta}{2\pi} \delta(\theta) + \frac{\eta}{(2\pi)^2}. \tag{17}$$

In this case, half of the displacement is in the forward direction, and half is a uniform displacement of the Fermi level—essentially an s wave, equivalent to isotropic potential scattering. This, however, is not quite the whole story. In one dimension the backflow compensated the charge and momentum exactly, since the left- and right-moving pieces were identical. Here, however, we have an uncompensated momentum of the forward-moving wave, $\eta/2\pi \times \mathbf{k}_F$. The correct displacement satisfying the Luttinger-Ward theorems is not merely a dilation of the momentum lattice, but a rigid displacement of $-\eta/2\pi k_F$ as well.

The simple uncompressible dilation of the Fermi surface which we postulated in (16) is too simple: the interactions must satisfy momentum as well as particle conservation, and so the backflow must carry no net momentum, as in 1D. The relative s-wave channel must carry momentum $-(\eta/2\pi)k_F$, which compensates the extra momentum of the δ-function peak at $\mathbf{k}_F$. This is equivalent to a uniform translation of the Fermi surface, which is a simple unitary transformation (multiplication of all states by a common factor) and does not lead to any anomalous dimensions. On the other hand, the s-wave does do so, and the anomalous dimension of the Green's function is, as in 1D, $(\eta/2\pi)^2 \times \frac{1}{2} = \alpha$, $0 \leq \alpha \leq \frac{1}{8}$. Another way of describing this part of the backflow is as a Fermi-surface shift proportional to $(1 - 2\cos\theta)$ rather than simply to 1. This is not a scattering in the p-wave channel; rather it is more like a "Mossbauer" zero-phonon, coherent recoil of the Fermi sea as a whole.

The form of the Green's function is quite different from 1D: it will look something like

$$G(r, t) \propto \int d\Omega e^{ik_F(\Omega)\cdot r} \left\{ \frac{1}{[\mathbf{r} \cdot \hat{\mathbf{n}}(\Omega) - v_s t]^{1/2}} \frac{1}{[\mathbf{r} \cdot \hat{\mathbf{n}}(\Omega) - v_c t]^{1/2 + 1/4(\eta/2\pi)^2}} \right\}$$
$$\times \left[\frac{1}{r^2 - v_c^2 t^2} \right]^{(\eta/2\pi)^2 \times 1/4}. \tag{18}$$

$\hat{n}(\Omega)$ is the Fermi surface normal unit vector at Ω, and $\cos\theta = \hat{\mathbf{n}}(\Omega) \cdot \hat{\mathbf{r}}$. The stationary phase will ensure that $G(r, t)$ comes almost entirely from the "patch" $n(\Omega) \| r$.

Experimentally, several hints suggest that $\alpha > \frac{1}{8}$ in fact, in some of the cuprates. We must not be surprised by the parallel-spin interaction also being finite and repulsive, which will enhance the charge-channel backflow without affecting spin properties except to lower v_s further, and make the electrons even less Fermi liquid. For the Hubbard model there is a fixed relation between η_c and η_s in (14), but in the physical case η can be larger.

Most of the physical phenomena which depend on G and other correlation functions can be calculated using the simple homogeneity property

$$G = \frac{1}{t^{1+\alpha}} F\left[\frac{r}{t}\right]. \tag{19}$$

This determines the infrared spectrum in parallel and perpendicular polarizations,[14,15] and the Fermi-surface smearing; a similar property will give the exponent for $1/T_1$. Only angle-resolved photoemission spectroscopy (ARPES) requires the full G. This will depend critically on details of the single-particle dispersion and Fermi surface, and so will require a separate investigation.

With (18) we have a principle the asymptotic solution of the 2D electron gas with a local, repulsive interaction. This is expected to be valid in the regions of the phase diagram of the Hubbard model reasonably far from half-filling (where umklapp terms are important and can pin down the charge bosons) and $U \to \infty$ at high density, where ferromagnetic coupling of Landau mean-field type will possibly be important, and lead to Nagaoka ferromagnetism. Finally, we exclude strong magnetic fields, strong being enough to allow interference after a full cyclotron orbit; i.e., we require $\omega_c \tau \ll 1$ where ω_c is the cyclotron frequency. Under this condition transverse gauge transformations are simple reparameterizations of the Fermi surface and meaningless; i.e., the Fermi surface and anyons are mutually incompatible. $\omega_c \tau \gg 1$ destroys the symmetries implicit in the Fermi surface, and causes gaps in the spectrum which are incompatible with bosonization. With $\omega_c \tau < 1$ bosonization is the only gauge theory of the interacting Fermi system; there is no meaningful other.

Khveshchenko has argued that in ≥ 2 dimensions the equations of motions of the bosons are a very crude approximation valid only for very small q and ω. This is clearly so in our approach, since the δ function in Eq. (17) is actually of width q. We have argued that Chern-Simons types of terms are not important if $\omega_c \tau < 1$, but insofar as charge and spin velocities differ, there can be effects such as those we have postulated in the past caused by mixing of bosons over a finite area of the Fermi surface, when electrons of finite $q = k - k_F$ are excited. Thus the above is a first approximation to a much more complex theory which we do not yet have under control. Nonetheless it seems the only way to proceed.

We would like to acknowledge valuable discussions with Y. Ren, F.D.M. Haldane, and especially G. Baskaran; also the hospitality of the Dept. of Theoretical Physics, Oxford. This work was funded by the NSF under Grant No. DMR 9104873.

REFERENCES

1. P. W. Anderson, Phys. Rev. Lett. **64**, 1839 (1990).
2. G. Engelbrecht and M. Randeria, Phys. Rev. Lett. **65**, 1032 (1990); **66**, 3325 (1991); Phys. Rev. B **45**, 12 419 (1992).
3. P.C.E. Stamp, Phys. Rev. Lett. **68**, 2180 (1992); J. Phys. (France) I **3**, 625 (1993).
4. G. Baskaran (unpublished).

5. H. Fukuyama, O. Narikiyo, and Y. Hasegawa, J. Phys. Soc. Jpn. **60**, 372 (1991); **60**, 1032 (1991).

6. C. di Castro, C. Castellani, and W. Metzner, Phys. Rev. Lett. **72**, 316 (1994).

7. P. W. Anderson, in *The State of Matter: A Volume Dedicated to E. H. Lieb*, edited by M. Aizenman and H. Araki, Advanced Series in Mathematical Physics Vol. 20 (World Scientific, Singapore, 1994), pp. 279–92; also in Rev. Math. Phys. **6**, 1085 (1994).

8. F.D.M. Haldane, in *Varenna Lectures, 1992*, edited by R. Schrieffer and R. A. Broglia (North-Holland, Amsterdam, 1994).

9. A. Houghton and J. B. Marston, Phys. Rev. B **48**, 7790 (1993).

10. A. A. Abrikosov, L. P. Gorkov, and I. E. Dzyaloshinskii, *Methods of Quantum Field Theory in Statistical Mechanics* (Dover, New York, 1974).

11. D. Khveshchenko, Phys. Rev. B **52**, 4833 (1995).

12. Y. Ren, Ph.D. thesis, Princeton, 1991.

13. F.D.M. Haldane, J. Phys. C **14**, 2585 (1981).

14. D. G. Clarke, S. Strong, and P. W. Anderson, Phys. Rev. Lett. **74**, 4499 (1995).

15. P. W. Anderson (unpublished); A. El Azrak, R. Nahoon, A. C. Boccara, N. Bontemps, M. Guilloux-Viry, C. Thiret, A. Perrin, Z. Z. Li, and H. Raffy, J. Alloys Compounds **195**, 663 (1993).

H

Interlayer Tunneling and Gap Anisotropy in High-Temperature Superconductors

Sudip Chakravarty, Asle Sudbø, Philip W. Anderson, Steven Strong*

(Received 2 April 1993; accepted 28 May 1993)

ABSTRACT

A quantitative analysis of a recent model of high-temperature superconductors based on an interlayer tunneling mechanism is presented. This model can account well for the observed magnitudes of the high transition temperatures in these materials and implies a gap that does not change sign, can be substantially anisotropic, and has the same symmetry as the crystal. The experimental consequences explored so far are consistent with the observations.

The gap in the electronic spectrum at the Fermi energy is a distinguishing feature of a superconductor; it is also the order parameter that describes the broken symmetry of the superconducting phase. The macroscopic properties of a superconductor follow once the gap is known. Here we explore a gap equation that was recently proposed by one of us *(1)* to explain the properties of the high-temperature cuprate superconductors. The underlying mechanism that leads to this gap equation is an interlayer tunneling phenomenon. We show (i) that the high transition temperatures of these materials can be natural consequences of this mechanism, (ii) that the superconducting gap can exhibit substantial anisotropy similar to that observed in recent photoemission measurements *(2)* but does not change sign (it is not possible to detect the sign of the gap in photoemission experiments), and (iii) that the superconducting state does not exhibit a Hebel-Slichter peak in the nuclear magnetic relaxation rate. Because the gap does not change sign and has the same symmetry as the crystal, even a moderate amount of nonmagnetic impurity scattering is likely to have little

*S. Chakravarty and A. Sudbø, Department of Physics, University of California–Los Angeles. S. Strong, Department of Physics, Princeton University.

effect on the superconducting transition temperature T_c. We refer to this gap as an anisotropic s-wave gap.

The interlayer tunneling mechanism *(1)* is based on the presence of well-defined CuO layers in these materials. The idea is to amplify the pairing mechanism within a given layer by allowing the Cooper pairs to tunnel to an adjacent layer by the Josephson mechanism. This delocalization process of the pairs gives rise to a substantial enhancement of pairing only if the coherent single particle tunneling between the layers is blocked, which we argue to be the case on phenomenological as well as theoretical grounds. In this sense the presence of bilayers or triple layers in these materials is important. The principle of amplification, however, is indifferent to the specific mechanism within a given CuO layer; the pairing can be due to electron-phonon interaction or spin fluctuations (3). We shall assume that the electron-phonon interaction is the dominant mechanism. As shown below, this leads to a natural explanation of the anomalous isotope effect seen in these materials. It is known that the isotope effect is negligibly small for materials with the highest T_c's and reverts to near normal for very low ones.

The normal state of these materials exhibits properties that require us to go beyond the conventional Fermi liquid theory *(4)*. Here we assume that, owing to strong electronic correlation effects, coherent single particle tunneling is not possible between the adjacent layers of a given bilayer even though the bare hopping rate, as obtained from electronic structure calculations, is substantial, of the order of 0.1 eV—a phenomenon that has been termed confinement (1). To justify this assumption, we briefly recall the phenomenology of the normal state of the high-temperature superconductors; for a more complete discussion, see *(4)*.

Suppose for the moment that the normal state is described by a Fermi liquid. Then, the two CuO layers hybridized by the single particle tunneling matrix element, $t_\perp(\mathbf{k})$, would lead to a symmetric and an antisymmetric combination of the quasiparticle states for each value of the wave vector $\mathbf{k}$ in the plane. (Throughout this report, $\mathbf{k}$ sill refer to a two-dimensional wave vector.) Because these states are dipole-active, the transition between them should lead to a prominent signature in the frequency-dependent c-axis conductivity, which, to date, has not been observed. [Because of the $\mathbf{k}$ dependence of $t_\perp(\mathbf{k})$, the signature is likely to be broad, but should still be observable in the range 500 to 1000 cm^{-1}.] In contrast, if the one-particle Green's function does not exhibit a quasiparticle pole but a power-law relaxation for asymptotically long times, the coherent quasiparticle tunneling can be blocked as a result of the orthogonality catastrophe *(5)*. The observation of a rapidly growing c-axis resistivity with decreasing temperature in these materials *(6)* is also consistent with the confinement idea. Note that the observed Fermi surface in photoemission experiments does not establish the Fermi liquid behavior of the normal state; as a counter example, it is only necessary to recall the well-established Luttinger liquid behavior of one-dimensional interacting Fermi systems *(7)*. Thus, the Fermi surface defined as the surface of low-energy excitations in **k**-space that encloses a volume appropriate to the density of electrons can exist in a non-Fermi liquid; in this case, the singularity in the derivative of the electronic occupation number at the Fermi surface will be a power-law instead of a δ function.

To treat the superconducting state, we note that the quasiparticle peak, a δ function broadened by the thermal as well as the experimental resolution factors, is recovered in this state *(8)*. In contrast, in our picture, the observed normal-state spectral function reflects only a similarly broadened power-law singularity. However, the c-axis infrared conductivity continues to be small in spite of the development of c-axis superconductivity *(9)*. Therefore, the quasiparticles are still not mobile in the c direction, but only in the ab plane. Thus, we ignore coherent single particle tunneling in the c direction and consider coherent Josephson pair tunneling which does not suffer the orthogonality catastrophe. We therefore approximate the motion in the motion in the ab plane by the dispersion relation, $\varepsilon(\mathbf{k}) = -2t[\cos(k_x a) + \cos(k_y a)] + 4t' \cos(k_x a) \cos(k_y a)$, where $t = 0.25$ eV and $t'/t = 0.45$ *(10)*. We have used the electron picture rather than the hole picture.

The gap equation can be derived by considering two close CuO layers described by the BCS (Bardeen-Cooper-Schreiffer) reduced Hamiltonian and coupled by the momentum conserving Josephson pair tunneling term, H_J, which we write as

$$H_J = -\sum_{\mathbf{k}} T_J(\mathbf{k})(c_{\mathbf{k}\uparrow}^{(1)\dagger} c_{-\mathbf{k}\downarrow}^{(1)\dagger} c_{-\mathbf{k}\downarrow}^{(2)} c_{\mathbf{k}\uparrow}^{(2)} + \text{h.c.}) \tag{1}$$

where $c_{\mathbf{k}\uparrow}^{(i)\dagger}$ is the electron creation operator, pertaining to the layer (i), of wave vector **k** and spin $\uparrow$ and h.c. is the Hermitian conjugate. This description is applicable to two-layer compounds, such as YBCO, two-layer Bi2212, or Tl2212; the extension to other types of layered materials is straightforward.

The effective Hamiltonian H_J is to be understood in the context of renormalization group theory. At high energies, of the order of the bandwidth, the microscopic Hamiltonian should be well described by the band theory that couples the layers by a single particle matrix element, $t_\perp(\mathbf{k})$, and clearly conserves the wave vector **k**. As the degrees of freedom are integrated out starting at the band edge, the term H_J is generated. One also generates, in addition, a particle-hole pair hopping term that would be important if we were to describe the magnetic ordering in these systems. Thus, the symmetry of the coefficient in Eq. 1, $T_J(\mathbf{k})$, is dictated by the symmetry of $t_\perp(\mathbf{k})$ calculated from band theory. For simplicity, we choose $T_J(\mathbf{k}) = t_\perp(\mathbf{k})^2/t$. We believe that this is approximately correct in magnitude as well and is certainly what one gets from a second-order perturbative renormalization group analysis. This is also supported by the fact that H_J reproduces the Josephson energy in conventional superconductors. To see this, it is necessary to evaluate it only in first-order perturbation theory, for simplicity at $T = 0$. The result differs from the conventional answer by a numerical factor close to unity. We do not, however, attempt to continue this perturbative renormalization group analysis all the way to the Fermi energy because the perturbative analysis must eventually break down. Instead, we consider the effective Hamiltonian, H_J, generated from a partial integration of the degrees of freedom as the starting point of our mean-field theory. Finally, we note that at long as the Coulomb interaction is well-screened, the vertex corrections to H_J are unimportant. All other effects involving Coulomb interactions are included in the Coulomb pseudopotential parameter, μ^*, as in BCS theory.

From the BCS mean-field theory the gap in the quasiparticle spectrum is derived to be

$$\Delta(\mathbf{k}) = T_J(\mathbf{k})b_k - \sum_{\mathbf{k'}} U_{\mathbf{k},\mathbf{k'}} b_{\mathbf{k'}} \tag{2}$$

where $b_\mathbf{k}^* = \langle c_{\mathbf{k}\uparrow}^\dagger c_{\mathbf{k}\downarrow}^\dagger \rangle$. We have considered the solution $b_\mathbf{k}^{(1)} = b_\mathbf{k}^{(2)} \equiv b_\mathbf{k}$. The attractive kernel $U_{\mathbf{k},\mathbf{k'}}$ is assumed to be due to electron-phonon interaction but could be due to other mechanisms. The sum over $\mathbf{k'}$ is cut off at an energy ω_D from the Fermi energy, where ω_D is of the order of the Debye energy. It is crucial, however, that the first term on the right-hand wide of Eq. 2 is local in $\mathbf{k}$. In fact, the striking properties of this gap equation would not follow if this term were smoothed out in the momentum space over a range comparable to the second term. As stated above, the locality follows from the conservation of the parallel momentum. The role of disorder and inelastic scattering clearly deserves further study, but we believe that $\mathbf{k}$ is a good quantum number, at least close to T_c.

The self-consistent equation for $b_\mathbf{k}$ is

$$b_\mathbf{k} = -\frac{\chi(\mathbf{k})}{1 - T_J(\mathbf{k})\chi(\mathbf{k})} \sum_{\mathbf{k'}} U_{\mathbf{k},\mathbf{k'}} b_{\mathbf{k'}} \tag{3}$$

where $b_\mathbf{k} = \Delta(\mathbf{k})\chi(\mathbf{k})$, and the pair susceptibility, $\chi(\mathbf{k})$, is

$$(1/2E_\mathbf{k}) \tanh(E_\mathbf{k}/2T).$$

The quasiparticle spectrum is given by $E_\mathbf{k} = \sqrt{\Delta(\mathbf{k})^2 + \varepsilon(\mathbf{k})^2}$, where, from now on, $\varepsilon(\mathbf{k})$ will be measured from the chemical potential ε_F.

Consider first a $T_J(\mathbf{k}) = T_J$, independent of $\mathbf{k}$ (1). With the simplification that $U_{\mathbf{k},\mathbf{k'}} = -V$, $(|\varepsilon(\mathbf{k})|, |\varepsilon(\mathbf{k'})| < \omega_D)$, we see that as $\lambda = N_n(0)V \to 0$ [$N_n(0)$ is the normal state density of states at the Fermi energy], $T_c - T_J/4 \approx (\lambda/1 - \lambda)(T_J/4)$. The transition temperature T_c tends to $T_J/4$ even when $\lambda \to 0$. Moreover, the phonon enhancement is linearly proportional to λ, in contrast to $T_c \approx \omega_D e^{-1/\lambda}$ in the weak coupling BCS theory. Of course, in the same limit, we expect strong fluctuation effect in **k**-space, which will act to reduce the mean field transition temperature. These **k**-space fluctuations are easily seen in the pseudospin formalism (12). However, when T_J is small, we recover the BCS result. As in BCS, the repulsive screened Coulomb interaction modifies this simple result. The expression for the Coulomb pseudopotential, μ^* is different, however, and is $\mu^* \approx \mu[1 + \mu \ln |\varepsilon_F/(\omega_D - T_J/2)|]^{-1}$, where μ is the Coulomb matrix element times the density of states.

Consider the anisotropic equation. The principal symmetry of $t_\perp(\mathbf{k})$ is evident from the electronic structure calculations. The two-dimensional bands of the two layers touch along the line joining ΓM. The point M corresponds to $(\pi/a, \pi/a)$. The largest splitting of the hybridized bands of the two layers is seen to be at the point X, which is $(\pi/a, 0)$. For clarity, we have chosen the special lattice, instead of the face-centered tetragonal notation. (Small orthorhombic distortions in these materials are unimportant for our discussion.) Thus, if we choose

$$t_\perp(\mathbf{k}) = \frac{t_\perp}{4}[\cos(k_x a) - \cos(k_y a)]^2 \tag{4}$$

we can reproduce the qualitative features of the Fermi surface quite well; the magnitude $t_\perp$ is in the range 0.1 to 0.15 eV *(11)* which is also consistent with the magnetic neutron scattering experiments *(13)*. This leads to $T_J(\mathbf{k}) = (T_J/16)[\cos(k_x a) - \cos(k_y a)]^4$, where $T_J = t_\perp^2/t$. Note that the choice $[\cos(k_x a) - \cos(k_y a)]$ will not agree with the calculated band structure, nor would the choice $|\cos(k_x a) - \cos(k_y a)|$, which would produce unwanted cusps in the spectrum. We have also verified the correctness of our choice from a tight-binding calculation.

Solving Eq. 3 at $T = 0$ gives

$$\Delta(\mathbf{k}) \left[1 - \frac{T_J(\mathbf{k})}{2\sqrt{\varepsilon(\mathbf{k})^2 + \Delta(\mathbf{k})^2}} \right] = \Delta_0 \theta(\omega_D - |\varepsilon(\mathbf{k})|) \tag{5}$$

where $\Delta_0 = V \sum_{\mathbf{k'}} \Delta(\mathbf{k'}) \chi(\mathbf{k'}) \theta[\omega_D - \varepsilon(\mathbf{k'})|]$ must be calculated self-consistently from the full solution $\Delta(\mathbf{k})$. To gain some insight, consider the solution on the Fermi surface

$$\Delta(\mathbf{k_F}) = \frac{T_J(\mathbf{k_F})}{2} + \Delta_0 \tag{6}$$

We see several important features: (i) the gap is highly anisotropic, but has the full symmetry of the crystal. Note that Δ_0 is not a BCS gap; it contains the enhancing effect of $T_J(\mathbf{k})$. (ii) The anisotropy is quadratic in $t_\perp$, and therefore the gap rapidly acquires pure s-wave character as $t_\perp$ decreases. (iii) Here, unlike d-wave, the gap never changes sign. (iv) At the Fermi surface points

$\pm k_{xF} = \pm k_{yF}$, $T_J(\mathbf{k_F}) = 0$, and the gap attains its smallest value Δ_0. (v) For a Fermi surface closed around the point Γ ($\varepsilon_F \leq -4t'$), the maxima of the gap are in the directions $(1, 0)$ and $(0, 1)$. The maximum value of the gap is obtained when the Fermi surface includes the points $(\pi/a, 0)$, $(0, \pi/a)$ and is $T_J/2 + \Delta_0$. For larger electron fillings, when the Fermi surface is not closed around the point Γ and does not include the maxima in $T_J(\mathbf{k})$, a more complicated structure develops, with the maxima in off-symmetry directions.

We now focus on an electron Fermi surface closed around the Γ point which corresponds to recent experiments on Bi2212 (2). Along the directions $(0, 1)$ or $(1, 0)$, the gap on the Fermi surface is given by

$$\Delta(\mathbf{k_F}) = \Delta_0 + \frac{T_J}{32}\left(1 - \frac{2t + \varepsilon_F}{4t' - 2t}\right)^4 \tag{7}$$

Thus, it is not difficult to obtain anisotropies similar to those seen in experiments. Clearly, the gap anisotropy on the Fermi surface is sensitively dependent on doping. From numerical solutions we have explored the structure of the gap for a range of parameters; a typical example is shown in fig. 1. Because it is not possible to detect the sign of the gap in angle-resolved photoemission spectroscopy, it is necessary to explore the structure of the gap in more detail and precision, if it is to be distinguished from a d-wave gap.

We next turn to a brief discussion of the superconducting density of states, $N_s(\omega) = \frac{1}{2}\Sigma_{\mathbf{k}}\delta(\omega - E_{\mathbf{k}})$, obtained from the quasiparticle spectrum $E_{\mathbf{k}}$ at $T = 0$. We have calculated $N_s(\omega)$ for a variety of parameters; an example is shown in fig. 2. The low-energy features are as follows. At the lower gap edge, Δ_0, $N_s(\omega)$ has a step discontinuity if $T_J \neq 0$, which can be proven analytically. There are also logarithmic singularities (proven analytically) at higher energies (see fig. 2) similar to the logarithmic singularities in the d-wave density of states. Clearly, these integrable singularities smoothed further by the observed inelasticity (14) cannot give rise to the Hebel-Slichter peak in the nuclear magnetic relaxation rates. The absence of the Hebel-Slichter peak is an important distinguishing feature of the high-temperature superconductors.

The critical temperature T_c can be determined from Eq. 3. Note that one can obtain a lower bound on T_c by neglecting the phonon enhancement term in the gap equation. We find $T_c \geq [T_J(\mathbf{k})/4]_{\max}$. When the Fermi surface is closed around Γ (that is, $\varepsilon_F \leq -4t'$), this leads to

$$T_c = \frac{T_J}{64}\left(1 - \frac{2t + \varepsilon_F}{4t' - 2t}\right)^4 \tag{8}$$

However, when the Fermi surface is open around Γ (that is, $\varepsilon_F \geq -4t'$), we find

$$T_c = \frac{T_J}{64}\left(1 + \frac{2t - \varepsilon_F}{4t' + 2t}\right)^4 \tag{9}$$

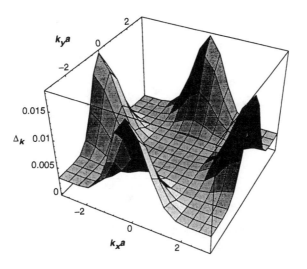

Figure H.1. Anisotropic gap $\Delta(\mathbf{k})$ in electron volts as a function of $k_x a$ and $k_y a$ at $T = 0$, with $\lambda = 0.31$ ($\Delta_0 = 0.003$ eV), $T_J = 0.03$ eV $= 10\Delta_0$, and $\varepsilon_F = -4t' = -0.45$ eV, corresponding to an electron Fermi surface closed around Γ. For clarity of visualization we have omitted the factor $\theta[\omega_D - |\varepsilon(\mathbf{k})|]$, which multiplies the gap.

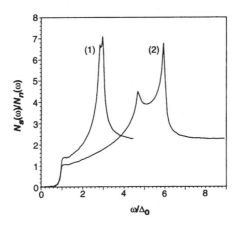

Figure H.2. Density of states $N_s(\omega)/N_n(\omega)$ at $T = 0$ for $\varepsilon_F = -4t' = -0.45$ eV, for two sets of parameters: (1) $T_J = 0.012$ eV, $\lambda = 0.29$ ($\Delta_0 = 0.003$ eV); (2) $T_J = 0.03$ eV, $\lambda = 0.25$ ($\Delta_0 = 0.003$ eV). The numerical evaluation of the density of states was carried out by replacing the δ function by a Lorentzian of width 10^{-4} eV.

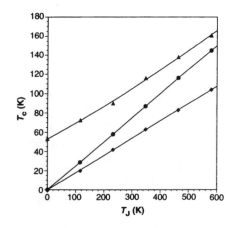

Figure H.3. Critical temperature T_c as a function of T_J; $\omega_D = 0.02$ eV; ($\triangle$) $\varepsilon_F = -4t' = -0.45$ eV, $N_n = 2.5$/eV-spin, $\lambda = 0.625$; ($\circ$) $\varepsilon_F = -4t' = -0.45$ eV, $N_n = 2.5$/eV-spin, $\lambda = 0.0625$; ($\diamond$) $\varepsilon_F = -2.67t' = -0.3$ eV, $N_n = 0.8$/eV-spin, $\lambda = 0.2$.

Remarkably, the lower bound, $T_c = T_J/4$, when $\varepsilon_F = -4t'$, is identical to the isotropic estimate when $T_J \gg V$. The numerical evaluation of T_c is shown in fig. 3 for typical sets of parameters.

It is clear from above that T_c is highest (not only the lower bound) when $\varepsilon_F = -4t'$ but falls away from this doping. A high value of T_c is obtained when the maxima of $T_J(\mathbf{k})$ are included within the shell $2\omega_D$ of the Fermi surface, and has nothing to do with the existence of van Hove singularities in the density of states. Note that the lower bound does not even involve the density of states at the Fermi surface. For lower T_c materials, T_J plays a less important role and T_c begins to be increasingly controlled by the phonon coupling, thereby reaching the standard BCS value. Therefore, for materials with highest T_c's, T_J dominates and the isotope effect is expected to be small. Conversely, materials with low T_c's should exhibit a normal isotope effect. This is in agreement with our earlier isotropic estimate which showed that when T_J dominated, even the phonon contribution, linearly proportional to λ, is independent of isotopic substitution. On the other hand, when T_J was small, we had the BCS result corresponding to normal isotope effect. It is now evident that the dispersion of $t_\perp(\mathbf{k})$ is essential in producing this crossover in the isotope effect.

From the gap equation, it is possible to prove that as long as $\Delta_0 \neq 0$, all the Fourier modes of $\Delta(\mathbf{k})$ vanish at the same temperature T_c. It is also important to note that a value of $T_J \approx 400$K used to obtain a T_c of the order of 100 K is an underestimate from what we know about $t_\perp$; T_J could be as large as 1000 K. Of course, fluctuation effects, not included in our mean field theory, will reduce T_c.

In the present report we have tried to be as quantitative as possible for two-layer materials. The pairing mechanism within a given layer was assumed to be BCS-type electron-phonon interaction because we believe this to be the most realistic possibility. In previous work *(1, 15)* we gave a simple formula for T_J which can be used to generalize to other materials with variable layer numbers (including single-layer compounds for which T_J characterizes the coupling between the CuO layers in adjacent unit cells), and have shown that it yields a good heuristic fit to T_c versus number of layers for Bi and Tl one-, two-, and three-layer materials as well as a prediction for ∞-layer materials. This leads to a near zero T_c for Bi one-layer materials, but to a $T_c \sim 60$ to 70 K for Tl one-layer materials [T_J is considerably larger in this case (16)]. The fit to the recently discovered Hg-based materials *(17)* is also satisfactory. A prediction, based on the present ideas, is that $La_{2-x}Sr_xCuO_4$ as well as $Nd_{2-x}Ce_xCuO_4$ should exhibit roughly isotropic gaps because $t_\perp(\mathbf{k})$ is not very anisotropic within tight-binding approximation.

REFERENCES AND NOTES

1. P. W. Anderson, in *Superconductivity, Proceedings of the ICTP Spring College in 1992*, P. Butcher and Y. Lu, Eds. (World Scientific, Singapore, in press).
2. Z.-X. Shen et al., *Phys. Rev. Lett.* **70**, 1553 (1993).
3. V. J. Emery, *Ann. Phys. N.Y.* **28**, 1 (1964).
4. P. W. Anderson, *Science* **256**, 1526 (1992); C. M. Varma et al., *Phys. Rev. Lett.* **63**, 1996 (1989); ibid., **64**, 497 (1990), erratum.
5. G. D. Mahan, *Many Particle Physics* (Plenum, New York, 1990). For a recent discussion of the orthogonality catastrophe, see P. W. Anderson (*Math. Rev.*, in press) and references therein. If we adiabatically decouple the motion of an electron perpendicular to the layers, for each **k**, one can construct a two-state system coupled to the bath of in-plane excitations. In a non-Fermi liquid any attempt to tunnel between the layers will significantly disturb all the electrons in the system. Thus, the tunneling matrix element will be multiplied by a overlap matrix element involving $N(N \to \infty)$ *electrons, which can vanish depending on the spectral density of the bath of excitations; see* S. Chakravarty, *Phys. Rev. Lett.* **49**, 681 (1982), and A. J. Leggett et al., *Rev. Mod. Phys.* **59**, 1 (1987). Here, this would happen if the spectral function corresponding to the one particle Green's function satisfies the homogeneity relation $A(\Lambda\omega, \Lambda k) = \Lambda^{\alpha-1}A(\omega, k)$ and the exponent $\alpha \geq 1$. There are, however, logarithmic corrections as $\alpha \to 1$. The value $\alpha = 1$ will lead to a precise linear T dependence of the in-plane resistivity in these materials. We emphasize that the physical effect will not be very different even if $\alpha < 1$ provided that T is larger than the crossover temperature $t_\perp (t_\perp/W)^{\alpha/(1-\alpha)}$, where W is of the order of the bandwidth. For another derivation, see P. W. Anderson, in preparation.
6. See, for example, B. Batlogg in *High Temperature Superconductivity*, K. S. Bedell et al., eds. (Addison-Wesley, Reading, MA, 1990). The observation, in some samples, of a c-axis resistivity which follows a linear (or more complex) temperature

dependence does not imply coherent tunneling of quasiparticles between the planes. This behavior can arise if the hopping rate between the layers, renormalized by the coupling to the intralayer excitations, is smaller than the temperature: a crucial point is that the magnitude of the c-axis resistivity seems to be considerably larger than the ab plane resistivity, generally much above the Mott limit at T_c; see N. Kumar and A. M. Jayanavar, *Phys. Rev. B* **45**, 5001 (1992).

7. That k_F remains unshifted in the Luttinger model was first pointed out by D. C. Mattis and E. H. Lieb. *J. Math. Phys.* **6**, 304 (1965). This follows from a counting argument because there are no level crossings. Such a counting argument is independent of the dimensionality of the electron gas.

8. C. G. Olson et al., *Science* **245**, 731 (1989); D. S. Dessau et al., *Phys. Rev. Lett.* **66**, 2160 (1991).

9. K. Tamasaku, Y. Nakamura, S. Uchida, *Phys. Rev. Lett.* **69**, 1455 (1992).

10. See, for example, J. Yu and A. J. Freeman, *J. Phys. Chem. Solids* **52**, 1351 (1991).

11. See, for example, S. Massida, J. Yu, A. J. Freeman, *Physica C* **152**, 251 (1988); M. S. Hybertsen and L. F. Mattheiss, *Phys. Rev. Lett.* **60**, 1661 (1988); H. Krakauer and W. E. Pickett, *ibid.*, p. 1665.

12. P. W. Anderson, *Phys. Rev.* **112**, 1900 (1958).

13. J. Traquada et al., *Phys. Rev. B* **40** 4503 (1989); J. Traquada *et al., ibid.* **46**, 5561 (1992). Note that for O_6, the optic mode in the spin wave spectrum due to bilayers was not observed up to 60 meV. For O_{66}, some indication of the optic mode has been observed for energies greater than 40 meV.

14. D. A. Bonn et al., *Phys. Rev. Lett.* **68**, 2390 (1991).

15. J. M. Wheatley, T. C. Hsu, P. W. Anderson, *Phys. Rev. B* **37**, 5897 (1988).

16. We argue this on the basis of the band structure calculations of D. J. Singh and W. E. Pickett [*Physica C* **302**, 193 (1992)] and D. R. Hamann and L. F. Mattheiss [*Phys. Rev. B* **38**, 5138 (1988)].

17. A. Schilling et al., *Nature* **363**, 56 (1993); S. N. Putilin et al., ibid. **362**, 226 (1993).

18. We thank E. Abrahams, W. E. Pickett, and Z.-X. Shen for useful discussions. This work was supported in part by the National Science Foundation, DMR-9104873 (P.W.A. and S.S.), DMR-9220416 (S.C.), and in part by the Office of Naval Research, ONR-92J1101 (S.C. and A.S.). S.C. also thanks the John Simon Guggenheim Memorial Foundation for support through a fellowship and S.S. the AT&T Bell Laboratories for a Ph.D. scholarship.

I

Interlayer Josephson Tunneling and Breakdown of Fermi Liquid Theory

Sudip Chakravarty and Philip W. Anderson

Department of Physics, University of California, Los Angeles
(Received 10 January 1994)

ABSTRACT

From the single assumption that the spectral density of the single particle excitations in high-temperature superconductors obeys a scaling property characterized by a nonvanishing exponent α, an interlayer tunneling Hamiltonian is obtained. A nonvanishing value of the exponent α implies breakdown of the Fermi liquid theory at sufficiently low energies and low temperatures, consistent with experiments.

PACS numbers: 74.20.−z, 71.27.+a, 74.50.+r

Recently, we have explored a gap equation based on an interlayer Josephson tunneling Hamiltonian of high-temperature superconductors and have found the results to be satisfactory [1–3]. To many, however, this Hamiltonian has appeared to be enigmatic, to say the least. In the present paper, we wish to demonstrate that this Hamiltonian is a consequence of the non-Fermi liquid behavior of the normal state; however, we are unable to provide a rigorous proof. We characterize the non-Fermi liquid behavior by a spectral density of the single particle excitations that satisfies a homogeneity relation with a nontrivial exponent [4], at sufficiently low temperatures and low energies. Although considerable progress is being made, the derivation of this spectral density from a realistic microscopic Hamiltonian remains an open problem. To be more specific, we show the following: (1) For the assumed non-Fermi liquid Green's function, and for a range of α, the Josephson critical current I_c is proportional to Δ^2/ω_c, where Δ is the gap and ω_c is the frequency scale of the order of the in-plane bandwidth. This allows us to reformulate the Josephson effect in terms of an instantaneous Hamiltonian. In contrast, for a Fermi liquid, this is not possible because I_c is proportional to $|\Delta|$. (2) The same assumptions lead to incoherent single particle tunneling between the layers for a range of α. Thus, one can formulate a simplified effective Hamiltonian in which single particle tunneling between the layers is missing, but pair tunneling is possible [1–3].

To set the stage, we revisit the Josephson effect. The well-known expression for the critical current for two conventional superconductors with equal gaps [5],

$$I_c = \frac{\pi \Delta(T)}{2eR_N} \tanh\left(\frac{\Delta(T)}{2T}\right), \tag{1}$$

where R_N is the normal state tunneling resistance, is unusual in two respects. As $T \to 0$, I_c is a nonanalytic function of the gap ($\propto |\Delta|$), but is analytic close to T_c ($\propto \Delta^2$). It is also remarkable that I_c is independent of the cutoff, ω_D, the Debye energy. The dependence on the Fermi energy is buried in the definition of R_N. We claim that this is a subtle manifestation of Fermi liquid theory. Consider the expression for the critical current at $T = 0$, given by

$$I_c \propto \sum_{k,q} |T_{k,q}|^2 \frac{\Delta_k}{E_k} \frac{|\Delta_q|}{E_q} \frac{1}{E_k + E_q}, \tag{2}$$

where $E_k = \sqrt{\varepsilon(k)^2 + |\Delta_k|^2}$, Δ_k is the energy gap, and the quasiparticle energy $\varepsilon(\mathbf{k})$ in the normal state is measured from the chemical potential. The integration over the momenta can be converted to energy integrations which extend only to the Debye energy. If the energy integrations can be extended to infinity with impunity, it can be seen by scaling the energy variables that $I_c \propto |\Delta|$; the correction is of order Δ/ω_D. This result crucially hinges on the following asymptotic property of the single particle spectral function, $A(k, \omega)$ (k is measured from the Fermi wave vector and ω from the chemical potential): $A(\Lambda k, \Lambda \omega) = \Lambda^{\alpha-1} A(k, \omega)$, where the exponent α is zero. Note that the spectral function of a Fermi liquid, ignoring the incoherent multiparticle excitations, $Z\delta(\omega - v_F k)$, where v_F is the Fermi velocity, clearly satisfies the homogeneity relation with $\alpha = 0$. Any nonzero value of the exponent α will result in a cutoff dependence of the critical current I_c, and if this cutoff is much larger than the gap, the critical current can be an analytic function of the gap, proportional to $\Delta(\Delta/\omega_D)$. This will allow us to reformulate the Josephson effect in terms of an instantaneous pair tunneling Hamiltonian on energy scales smaller than the cutoff. [Note that, close to T_c, the magnitude of the Josephson effect, derived from Ginzburg-Landau theory [6], is proportional to $\Delta(\Delta/T)$ and the analytic result is due to the thermal smearing.]

We now calculate the critical current I_c for an assumed form of the non-Fermi liquid Green's function. Consider first the Nambu matrix Green's function, G, with Bardeen-Cooper-Schrieffer (BCS) theory [7], and examine the simplest possible generalization to include a non-Fermi liquid spectral function, i.e.,

$\alpha \neq 0$. For BCS, we can write

$$G_{\text{BCS}}^{-1} = \begin{pmatrix} G_e^0(k, \omega)^{-1} & -\Delta_k^* \\ -\Delta_k & G_h^0(k, \omega)^{-1} \end{pmatrix}. \tag{3}$$

Note that the off-diagonal terms are simply the Hartree-Fock self-consistent pairing potentials, while the normal state Green's functions are $G_e^0(k, \omega)^{-1} = \omega - \varepsilon(\mathbf{k})$ and $G_h^0(k, \omega)^{-1} = \omega + \varepsilon(\mathbf{k})$. Here the energies are measured from the chemical potential, and we are interested in low energies where we can linearize and write $\omega - v_F k$ instead of $\omega - \varepsilon(\mathbf{k})$, etc. The analyticity properties must conform to the general properties of the Lehmann representation. The simplest possible generalization to include non-Fermi liquid spectral function is

$$G_s^{-1} = \begin{pmatrix} [\omega - \varepsilon(\mathbf{k})]^{1-\alpha} \omega_c^\alpha & \Delta_k^* \\ -\Delta_k & [\omega + \varepsilon(\mathbf{k})]^{1-\alpha} \omega_c^\alpha \end{pmatrix}. \tag{4}$$

Here, the cutoff ω_c is introduced to give the Green's functions their proper dimensions. The self-consistent Hartree-Fock potential must be calculated in the usual manner [7]. This Green's function can be diagrammatically motivated as follows. Consider, for example, the element G_{s11} and write it as $G_{s11}^{-1} = (G_e^0)^{-1} - \Sigma_n - \Sigma_a$, where Σ_n is the normal part and Σ_a is the anomalous part of the proper self-energy, such that Σ_n does not contain Σ_a as an internal part. This approximate decomposition is possible because even if we included Σ_a as an internal part in Σ_n, Σ_n would not have changed appreciably; being an internal block, the singularities of Σ_a would have been integrated over. Besides, Σ_a is only expected to make a change over a small range of energies of order $|\Delta_k|$ around the Fermi energy and its effect on Σ_n should be proportional to $|\Delta_k|/\omega_c \ll 1$. Similarly, Σ_a is assumed to contain only internal hole lines. Thus, we can immediately write down Dyson's equations, $G_e = G_e^0 + G_e^0 \Sigma_n G_e$ and $G_{s11} = G_e + G_e \Sigma_a G_{s11}$. This adiabatic decoupling allows us to use the essentially exact Green's function of the normal state, which in the present problem we have assumed to be of the non-Fermi liquid form.

In addition to the branch point at $\pm \varepsilon(\mathbf{k})$, the Green's function has poles at $\omega(\mathbf{k})^2 = \varepsilon(\mathbf{k})^2 + \exp[2\pi i n/(1 - a)]|\Delta_{\mathbf{k}}^{\text{eff}}|^2$, where n are arbitrary integers, positive or negative, and $|\Delta_{\mathbf{k}}^{\text{eff}}| = |\Delta_{\mathbf{k}}|(|\Delta_{\mathbf{k}}|/\omega_c)^{\alpha/(1-\alpha)}$. These poles surround the point $\varepsilon(\mathbf{k})^2$ in the complex plane densely and uniformly on a circle of radius $|\Delta_{\mathbf{k}}^{\text{eff}}|^2$ for any nonzero irrational α. For rational α, we have a discrete set of poles. Because of the presence of the branch points at $\pm \varepsilon(\mathbf{k})$, the Green's function is not single valued. We make it single valued by defining the cut plane by $0 \leq \theta < 2\pi$; then, the only physically relevant pole corresponds to $n = 0$, and the excitation spectrum is given by $\omega(\mathbf{k})^2 = \varepsilon(\mathbf{k})^2 + |\Delta_{\mathbf{k}}^{\text{eff}}|^2$, as in BCS theory, with no imaginary part, although the reactive nature of the coupling to the non-Fermi liquid spectral function is manifest in $\Delta_{\mathbf{k}}^{\text{eff}}$. Note that the gap collapses as $\alpha \to 1$ (cf. Balatsky [8]).

Using Eq. (4), the Josephson critical current can be shown to be [9]

$$I_c \propto \langle |T_{\mathbf{k},\mathbf{q}}|^2 \rangle v_F^{-2} \Delta(\Delta/\omega_c) f(\alpha, \Delta/\omega_c), \tag{5}$$

where the angular brackets imply an energy average and $f(\alpha, \Delta/\omega_c)$ is a complicated function. We have, however, been able to check the following limiting cases: (a) $f(0, \Delta/\omega_c) = \omega_c/\Delta$, for $\Delta/\omega_c \ll 1$; (b)

$$f(\alpha, \Delta/\omega_c) = \left[\frac{4^\alpha \Gamma(\alpha)\Gamma(1 - 2\alpha) \sin \pi(1 - \alpha)}{\pi \Gamma(1 - \alpha)} \right]^2$$

$$\times \int_0^\infty \frac{ds}{s^{4\alpha}} \gamma^2(2\alpha, s), \tag{6}$$

for $\frac{1}{2} > \alpha > \frac{1}{4}$, $\Delta/\omega_c \ll 1$, where $\gamma(2\alpha, s)$ is the incomplete gamma function. As is clear from Eq. (6), the result can be analytically continued across $\alpha = \frac{1}{2}$.

Thus, for $\alpha > \frac{1}{4}$, the dependence of $f(\alpha, \Delta/\omega_c)$ on Δ/ω_c disappears in the limit $\Delta/\omega_c \ll 1$. Although we have not been able to evaluate $f(\alpha, \Delta/\omega_c)$ for $\alpha < \frac{1}{4}$ in a useful form, this is no longer true. We suspect, however, that for α not too small, the dependence on Δ/ω_c is weak. The difference between the present result and the conventional result is a striking manifestation of the cut spectrum of the normal state. We emphasize that it is ω_c that sets the scale, not Δ as in the conventional case.

We now obtain an instantaneous pairing Hamiltonian that describes the Josephson effect. This Hamiltonian is certainly justified for $\alpha > \frac{1}{4}$, but should be approximately true for α not too small as well, for reasons stated above. Consider a Josephson pair tunneling Hamiltonian H_J for two superconductors (1) and (2), given by

$$H_J = -\sum_{\mathbf{k},\mathbf{q}} \frac{|T_{\mathbf{k},\mathbf{q}}|^2}{\omega_c} (c_{\uparrow\mathbf{k}}^{(1)\dagger} c_{\downarrow-\mathbf{k}}^{(1)\dagger} c_{\downarrow-\mathbf{q}}^{(2)} c_{\uparrow\mathbf{q}}^{(2)} + \text{H.c.}), \tag{7}$$

and evaluate the ground state energy in first order perturbation theory. The resulting critical current is essentially the same as that obtained above. The same result can be obtained in second order degenerate perturbation theory, used to derive the conventional Josephson effect [7], provided we recognize that the weight is concentrated at high frequencies due to the non-Fermi liquid behavior of the normal state and the energy denominator can be replaced by the high frequency cutoff. Note that the wave vector dependence of H_J is dictated by the wave vector dependence of the single particle tunneling matrix element. Thus, the mapping to H_J should be qualitatively correct, and we believe that its magnitude will be approximately correct with a proper definition of the electronic scale ω_c. We emphasize, once again, that such a mapping is *not* possible for the conventional Josephson effect with a Fermi liquid spectrum of the normal state because $I_c \propto |\Delta|$ in that case. If we specialize to the situation where the tunneling takes place between two layers, such that the parallel momentum is conserved during single particle tunneling and the electron operators do not depend on the wave vector perpendicular to the layers, then we can immediately write [10]

$$H_J = -\sum_{\mathbf{k}} \frac{|T_{\mathbf{k}}|^2}{\omega_c}(c_{\uparrow\mathbf{k}}^{(1)\dagger} c_{\downarrow-\mathbf{k}}^{(1)\dagger} c_{\downarrow-\mathbf{k}}^{(2)} c_{\uparrow\mathbf{k}}^{(2)} + \text{H.c.}). \tag{8}$$

Now the electron operators $c_{\sigma\mathbf{k}}^{(i)}$ refer to the electrons of the ith layer, of *parallel* wave vector $\mathbf{k}$ and spin σ. Previous attempts [11] to derive this Hamiltonian by coupling the single particle tunneling Hamiltonian to the spinon pair field does not lead to a momentum conserving H_J, as above, without further assumptions.

We now examine the conditions under which *coherent* tunneling of single electrons is not possible. This does not necessarily imply, however, that the single particle tunneling is irrelevant in the renormalization group sense, but that the single particle *dynamics* is incoherent. A similar effect is fairly well understood in the context of a two-state system coupled to a dissipative heat bath (see below). First, imagine a single particle Hamiltonian, which is

$$H_0 = \sum_{\sigma\mathbf{k}} \varepsilon(\mathbf{k}) c_{\sigma\mathbf{k}}^{(1)\dagger} c_{\sigma\mathbf{k}}^{(1)} + \sum_{\sigma\mathbf{k}} \varepsilon(\mathbf{k}) c_{\sigma\mathbf{k}}^{(2)\dagger} c_{\sigma\mathbf{k}}^{(2)} + \sum_{\sigma\mathbf{k}} T_{\mathbf{k}}(c_{\sigma\mathbf{k}}^{(1)\dagger} c_{\sigma\mathbf{k}}^{(2)} + \text{H.c.}); \tag{9}$$

for each $\mathbf{k}$, we have a two-state system. Imagine now that a given $\mathbf{k}$ state is coupled to a bath of bosons; in our case these are the Tomonaga-Luttinger bosons, presumed to exist due to strong electronic correlations. In fact, the very same electrons that are coupled to the bath of bosons are also creating this bath, so the whole problem must be solved self-consistently. Let us set aside the self-consistency problem for the moment and ask what would happen to the coherence of the single particle tunneling due to the coupling to the bath. We show below, however, from an entirely different approach, that identical results can be obtained. Thus, we strongly suspect that, to the level of sophistication considered here, self-consistency is not an issue. We take the coupling to the bath to be

$$H_{eb} = \sum_{\nu,\sigma\mathbf{k}} g_\nu (c_{\sigma\mathbf{k}}^{(1)\dagger} c_{\sigma\mathbf{k}}^{(1)} - c_{\sigma\mathbf{k}}^{(2)\dagger} c_{\sigma\mathbf{k}}^{(2)})(a_\nu^\dagger + a_\nu). \tag{10}$$

In a non-Fermi liquid, the $\mathbf{k}$ dependence of the coupling to the boson operators, a_ν, is likely to be unimportant. Multiparticle excitations will be spread over such a great range of energies, and hence momenta, that only an average coupling needs to be considered. This is now a two-state system coupled to a heat bath and we can simply take over the known results [12]. If the bath of oscillators is Ohmic, i.e.,

$$J(\omega) = (\pi/2) \sum_{\nu} g_{\nu}^2 \delta(\omega - \omega_{\nu}) = \alpha' \omega, \quad \omega \to 0,$$

and equal to 0 for $\omega \geq \omega_c$, then the salient features are that for $\alpha' \geq 1$, tunneling is completely quenched at $T = 0$—orthogonality catastrophe—and, for $\frac{1}{2} \leq \alpha' \leq 1$, the relaxation between the two sets is incoherent. At finite temperatures and, for $\alpha' > 1$, or for $\alpha' < 1$ and $\alpha' T \geq T_{\mathbf{k}}^{\text{eff}}$, the relaxation is an exponential, with a rate proportional to $T^{2\alpha'-1}$, where $T_{\mathbf{k}}^{\text{eff}} = T_{\mathbf{k}}(T_{\mathbf{k}}/\omega_c)^{\alpha'/(1-\alpha')}$. For $0 < \alpha' < \frac{1}{2}, T < \alpha' T_{\mathbf{k}}^{\text{eff}}$, the relaxation consists of damped oscillation at short times, but incoherent power law at long times, the magnitude of which becomes quite substantial as α' approaches $\frac{1}{2}$. For α' not too small, the oscillations are hardly important for more than a cycle; thus, the power-law relaxation is the dominant feature even in this so-called coherent regime. The Q factor of the oscillation at short times is given by $\cot[(\pi/2)\alpha'/(1 - \alpha')]$, which tends to ∞ as $\alpha' \to 0$, but tends to 0 as $\alpha' \to \frac{1}{2}$. The Ohmic oscillator bath is precisely what is relevant for our discussion, because for both Fermi and non-Fermi liquids there are excitations arbitrarily close to the Fermi surface, but we need to relate the exponent α' to α.

There is another way [13] to obtain the same result which also serves to relate α' to α. The Green's function G_F corresponding to the Hamiltonian in Eq. (10) is

$$G_F^{-1} = \begin{pmatrix} G_1^0(k, \omega)^{-1} & T_{\mathbf{k}} \\ T_{\mathbf{k}} & G_2^0(k, \omega)^{-1} \end{pmatrix}, \tag{11}$$

where $G_1^0(k, \omega)^{-1} = \omega - \varepsilon(\mathbf{k})$ and similarly $G_2^0(k, \omega)^{-1} = \omega - \varepsilon(\mathbf{k})$. This gives rise to the usual result for the hybridization and G_F has poles at $\omega = \varepsilon(\mathbf{k}) \pm T_{\mathbf{k}}$. Once again, pole prescriptions have to be added according to the Lehmann representation. Now, consider, once again, replacing G_1^0 and G_2^0 by the non-Fermi liquid Green's functions of the individual layers. Thus, we obtain the generalization G_{LL}, which is

$$G_{LL}^{-1} = \begin{pmatrix} [\omega - \varepsilon(\mathbf{k})]^{1-\alpha}\omega_c^{\alpha} & T_{\mathbf{k}} \\ T_{\mathbf{k}} & [\omega - \varepsilon(\mathbf{k})]^{1-\alpha}\omega_c^{\alpha} \end{pmatrix}. \tag{12}$$

A diagrammatic derivation of this is identical to the superconducting case discussed above, with one exception. Now the role of the anomalous self-energy is played by the self-energy that converts a $c_{\sigma\mathbf{k}}^{(1)\dagger}$ electron to a $c_{\sigma\mathbf{k}}^{(2)\dagger}$ electron, i.e., proportional to the matrix element $T_{\mathbf{k}}$; therefore, the propagator in the anomalous block is a particle propagator and not a hole propagator as in the superconducting case. From this Green's function one immediately finds that,

in addition to the branch point, it has poles at

$$\omega_k = \varepsilon(\mathbf{k}) + \exp\left(\frac{\pi i n}{1-\alpha}\right) T_\mathbf{k}(T_\mathbf{k}/\omega_c)^{\alpha/(1-\alpha)},$$

where n is an integer. As before, because of the presence of the branch points, the Green's function is not single valued. Thus, we consider the cut plane, $0 \leq \theta < 2\pi$. There is, however, a striking difference between the present case and the previous superconducting case. In this cut plane, there are two poles, corresponding to $n = 0$ and $n = 1$. The $n = 0$ pole is the analytic continuation of the larger eigenvalue $\varepsilon(\mathbf{k}) + T_\mathbf{k}$ for $\alpha = 0$. As α increases from 0, this pole undergoes a purely reactive shift and marches down to $\varepsilon(\mathbf{k})$ as $\alpha \to 1$. The pole corresponding to $n = 1$ is the analytic continuation of the smaller eigenvalue $\varepsilon(\mathbf{k}) - T_\mathbf{k}$ for $\alpha = 0$. As α increases, it immediately acquires an imaginary part which is largest when $\alpha = \frac{1}{3}$ [at this point, on the Fermi surface, i.e., $\varepsilon(\mathbf{k}) = 0$, the relaxation is totally incoherent]. The imaginary part now decreases as α increases further, but both of the poles come together and become degenerate at $\alpha = \frac{1}{2}$. As α increases, this pole escapes the principal Reimann sheet. If we follow its path through all the Riemann sheets, it spirals into $\varepsilon(\mathbf{k})$ at $\alpha = 1$. Thus, in contrast to the superconducting case where the pole does not acquire an imaginary part, the single particle tunneling is very differently affected by the coupling to the multiparticle excitations. It is important to note that our treatment depends crucially on the adiabatic principle; i.e., the time scale set by $T(\mathbf{k})$ is much larger than the typical time scale of the in-plane processes set by the strong interactions between electrons and is not expected to hold in the opposite limit.

If we identify α' to be $2\alpha/(\alpha+1)$, many of the results obtained by this simple method are identical to those given above. Note that in terms of the exponent α, the pure incoherent relaxation begins for values greater than or equal to $\frac{1}{3}$, instead of $\frac{1}{2}$ for α'. As emphasized above, even in the so-called coherent regime, $\alpha < \frac{1}{3}$, the power-law relaxation dominates for α not too small and the system is effectively incoherent.

In the present paper we have considered a model non-Fermi liquid spectral function to test the effect of the anomalous Green's function and have shown that an approximate instantaneous Josephson pair tunneling Hamiltonian can be obtained. This is certainly a simplification in the sense that the instantaneous BCS reduced Hamiltonian is a simplification of the actual coupled electron-phonon problem. However, as in BCS theory, an instantaneous Hamiltonian allows us to explore a number of properties in a simple manner; the proper extension awaits further work. The model is also simplified because we have ignored the spin-charge separation. Preliminary calculations of the normal tunneling problem with a Green's function of the same form as in one dimension [14] have been carried out [15], and the results have been found to be equivalent

to that without spin-charge separation, but with $\alpha > \frac{1}{2}$ for any finite $v_s - v_c$, where v_s and v_c are the spin and the charge velocities. The superconducting case is more difficult, but the interaction kernel is concentrated at high frequencies and may be taken to be effectively instantaneous.

We are indebted to Elihu Abrahams for pressing us hard to clarify some of the issues discussed in the present paper. The encouragement of and comments by Steven Kivelson have been invaluable to one of us (S.C.). This work was supported in part by the National Science Foundation, DMR-9104873 (P.W.A.) and DMR-9220416 (S.C.), and in part by the Office of Naval Research, ONR-N00014-92-J-1101 (S.C.). S.C. also thanks the John Simon Guggenheim Memorial Foundation for support and the Aspen Center for Physics for their hospitality.

REFERENCES

1. P. W. Anderson, in "Superconductivity, Proceedings of the ICTP Spring College in 1992," edited by P. Butcher and Y. Lu (World Scientific, Singapore, to be published).
2. S. Chakravarty, A. Sudbø, P. W. Anderson, and S. Strong, Science **261**, 337 (1993).
3. A. Sudbø, S. Chakravarty, S. Strong, and P. W. Anderson, Phys. Rev. B **49**, 12245 (1994).
4. This characterization was anticipated by X. G. Wen, Phys. Rev. B **42**, 6623 (1990). S. C. is grateful to S. Khlebnikov for emphasizing the importance of this paper to him.
5. V. Ambegaokar and A. Baratoff, Phys. Rev. Lett. **10**, 486 (1963); **11**, 104(E) (1963).
6. L. G. Aslamazov and A. I. Larkin, Pis'ma Zh. Eksp. Teor. Fiz. **9**, 150 (1969) [JETP Lett. **9**, 87 (1969)].
7. J. R. Schrieffer, *Superconductivity* (Benjamin/Cummings, Reading, MA, 1983).
8. A. V. Balatsky, Philos. Mag. Lett. **68**, 251 (1993).
9. The derivation is similar to Ref. [5], with one exception. We have found it more tractable to carry out the integrals in the time domain.
10. In the tunneling Hamiltonian, the real potential barrier is replaced by the transfer of electrons from one superconductor to the other and a proper choice of the normalization of $T_{p,q}$ is necessary; for a careful discussion of this point, see A. Abrikosov, *Fundamentals of the Theory of Metals* (North-Holland, Amsterdam, 1988). When two conventional superconductors are connected by $T_{p,q}$, $T_{p,q}$ must carry the normalization factor $(S/V_1 V_2)^{1/2}$, where S is the area of the junction and V_1 and V_2 are the volumes of the two superconductors. From the same argument, it can be seen that, for the present momentum conserving interlayer tunneling, the normalization factor carried by T_k is precisely unity.
11. V. N. Muthukumar, Phys. Rev. B **46**, 5769 (1992).
12. S. Chakravarty, Phys. Rev. Lett. **49**, 681 (1982); S. Chakravarty and A. J. Leggett, Phys. Rev. Lett. **52**, 5 (1984); A. J. Leggett *et al.*, Rev. Mod. Phys. **59**, 1 (1987).
13. This method is similar to that of Ref. [4].

14. I. E. Dzyaloshinskii and A. I. Larkin, Zh. Eksp. Teor. Fiz. **65**, 411 (1973) [Sov. Phys. JETP **38**, 202 (1974)].

15. D. C. Clarke and P. W. Anderson (unpublished).

J

"Confinement" in the One-Dimensional Hubbard Model: Irrelevance of Single-Particle Hopping

P. W. Anderson

(Received 29 July 1991)

ABSTRACT

We demonstrate, by direct use of the asymptotic Green's functions of Korepin and Ren, that the one-dimensional repulsive U Hubbard model has the property of "confinement," in that weak interchain coupling does not cause coherent particle motion in the transverse direction at absolute zero. Implications for real one-dimensional systems and for the two-dimensional Hubbard model are discussed.

PACS numbers: 75.10.Jm

It is often assumed that an array of one-dimensional chains, weakly coupled by one-electron tunneling between chains, crosses over to two-dimensional behavior. We study this question for the repulsive Hubbard model and come to the conclusion that in this, and probably many other cases when the chain is not a Fermi liquid, the opposite is true: A finite interchain hopping is required. We call this behavior "confinement" because it has a close analogy to the confinement of quarks to the interior of the hadron. It is not quite equivalent to the usual Wilson definition of confinement.

The magnetic analogy of Heisenberg or Ising chains with weak interchain J has been of course well known to have the standard crossover behavior, in the Ising case since the time of Onsager. The interchain hopping case is suggested to be different by the existence of very large resistivity anisotropy in some materials, an observation which cannot be explained by anisotropic localization as it sometimes was in the past, since that is contrary to the theory of localization: An electron delocalized in one or two directions is simply delocalized.

This problem has not been very seriously attacked. A discussion by Wen [1] does not deal with the important issue of spin-charge separation, and one by Schulz [2] removes that separation by setting spin and charge velocities equal. Both Wen and Schulz also use a questionable criterion for relevance within the

Fermi-liquid theory [3]. But actually, the main difficulty with these treatments is that the very concept of "relevance" may be itself irrelevant. At issue, as in the similar problem of localization, is coherence: the coherence or not of the tunneling process between the chains. In localization, there is no energy effect of the mobility edge, and conventional finite-temperature theory does not pick up the effect; here there are both energy and transport effects but the key is coherence.

Recent calculations [4–7] of the asymptotic one-particle Green's function at $T = 0$ for the Hubbard model are sufficient to settle this question precisely. The inter-chain "hopping" is a simple static one-particle perturbation. The lowest-order response to it is given by a diagram involving creating a particle in one chain and a hole of equal momentum in a second, which do not interact except to recombine via the perturbation itself, so the response function is simply the product of the interacting electron and hole propagators, which to lowest order can be calculated exactly. We calculate the energy response but a simple identity would relate this to the conductivity.

It is instructive to do the noninteracting case first. The problem is not as simple as might appear at first. The relevant diagram (fig. 1) leads to the response function

$$R = t_\perp^2 \sum_{k,\omega,\sigma} G_R^{1(e)}(k, \omega) G_R^{2(h)}(k, -\omega),$$

where 1 and 2 are separate Hubbard chains and we are calculating the response function to a perturbation connecting chains in an array of N chains i:

$$t_\perp \sum_i c_{k\sigma}^{\dagger(i)} c_{k\sigma}^{(i+1)} + \text{H.c.}$$

Note that R is also

$$2t_\perp^2 \int_{-\infty}^\infty dx \int_0^\infty dt\, e^{-\eta t} G_R^{(1)(e)}(x, t) G_R^{(2)(h)}(x, t).$$

In general we should use the *retarded* Green's functions (designated by R) since we contemplate turning on the perturbation $t_\perp$ adiabatically at rate $\sim\eta$ *after* any interactions are introduced on the separate chains. We are looking for an anomalous imaginary part to R representing real transitions, which signals a redistribution of electrons in k space.

The use of causal Green's functions will become the essential physical point when we consider the interacting chains, because we have to be very careful about the order of turning on the interactions U and $t_\perp$. We must realize that $U = 0$ is a singular point at which the nature of the spectrum changes qualitatively, so that we cannot expect that turning on $t_\perp$ and then U will have

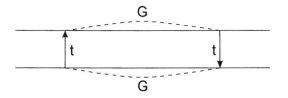

Figure J.1. Lowest-order diagram for interchain tunneling.

the same result as the opposite order. Even though some results may be obtained with renormalization methods from perturbation theory in U for the original problem with $t_\perp = 0$ we should be reasonably sure that confusing the order by using time-ordered Green's functions and not retarded ones will be incorrect. If we first go to finite, large U, where the spectrum is qualitatively different form near-zero U—having spin-charge separation, in particular—and then turn on $t_\perp$ and/or T as small perturbations, we find quite different results from those given by the standard temperature Green's functions.

Figure 2 illustrates the problem. On the vertical axis we have the perturbation $t_\perp$ (actually the same diagram holds true for a number of types of T-invariant perturbations). On the horizontal axis is U. We intend to show that a line A separates the region where $t_\perp$ is irrelevant from that where it causes finite tunneling or other effects. Clearly, to reach region B below line A, we must *first* turn on U and then $t_\perp$; if we go in the opposite order, which is implied by conventional procedures, we shall end up in region C, and will miss the effect. Thus it is essential to maintain the time ordering of events. Clearly $t_\perp = 0$, $U = 0$ must be a singular point at which Fermi liquid theory starts to fail.

If we first turn on $t_\perp$ we know easily what will happen physically. $t_\perp$ will split the unperturbed energy levels ϵ_k into a two-dimensional manifold; for a simple pair of chains 1 and 2 we will get $\epsilon_k \rightarrow \epsilon_k \pm t_\perp$, the two energies belonging to symmetric and antisymmetric linear combinations. Thus the Fermi surface will shift to $k_F \rightarrow k_F \pm t_\perp/\hbar v_F$ and, to lowest order in U, we will have no singular scatterings from $k_F + t_\perp$ to $k_F - t_\perp$ and the system will be effectively two roughly independent different Hubbard-like models with smaller effective U. If we introduce many chains there will be a two-dimensional Fermi surface and a whole new two-dimensional problem to solve. The energy response is of order $t_\perp^2$ because $\sim t_\perp$ levels have been shifted by $\sim -t_\perp$ by reoccupation; without a shift of Fermi surface there would be no response.

The most straightforward way of approaching this problem seems to lead to nonsense: no response. (The argument is foreshadowed by Kohn and Luttinger [8].) We have

$$R = t_\perp^2 \sum_k \int_0^\infty d\omega G_e(k, \omega) G_h(k, -\omega),$$

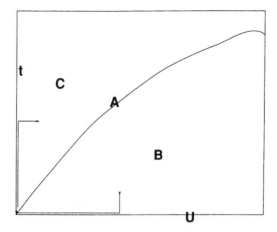

Figure J.2. Proposed $t_\perp$-U phase diagram. A represents a boundary between one- and two-dimensional behavior. One cannot reach the "confined" regime B perturbatively by the path of turning on $t_\perp$ and then U.

$$G_e(k, t) = \langle 0|c_k(t)c_k^\dagger(0)|0\rangle$$
$$= e^{i\epsilon_k t - \eta t}(1 - n_k),$$
$$G_h(k, t) = e^{-i\epsilon_k t - \eta t}(n_k),$$
$$G_e(k, \omega) = \frac{1 - n_k}{\omega - \epsilon_k - i\eta}, \quad G_h = \frac{n_k}{\omega + \epsilon_k + i\eta},$$

where $\epsilon_k = \hbar v_F(k - k_F)$ and η defines outgoing boundary conditions for the causal Green's function. Clearly, at $T = 0$, $n_k(1 - n_k) \equiv 0$ and there is zero response. This reflects correctly the fact that without readjustment of populations, there is in fact no net energy shift.

This anomaly is traditionally converged by introducing finite temperature. If

$$n_k = f(\beta\epsilon_k),$$

we get, from either the frequency or the time representation,

$$R = t_\perp^2 \sum_k n_k(1 - n_k)\frac{1}{2\eta} \sim \frac{t_\perp^2 N(0)T}{\eta}.$$

The response diverges with $1/\eta$, reflecting the coincidence of two singularities "pinching" from opposite sides of the real axis. In the time domain, this η

singularity is obvious:

$$R = t_\perp^2 \sum_k \int_0^\infty dt\,(n_k)(1 - n_k)e^{-2\eta t} = t_\perp^2 \sum_k n_k \frac{1 - n_k}{2\eta}.$$

The perturbation *for each* k acts for an infinite time and is converged only by the artificial factor η. The Matsubara perturbation technique allows us to "fuzz out" the energies ϵ_k by moving the poles off the real axis by an amount T, and one gets the right answers, at least for $t \ll T$, for a devious reason which was explored within Fermi-liquid theory by Kohn and Luttinger and by Bloch and deDominicis. But as they show, such singularities proportional to $1/\eta$ are the signal for Fermi surface shifts, for rather obvious reasons as pointed out above.

Now we are ready to approach the interacting problem correctly. After we turn on U, the space-time Green's function becomes, asymptotically,

$$G_e \sim e^{ik_F x} \frac{1}{[(x - v_s t)(x - v)_c t)]^{1/2}} \frac{1}{(x^2 - v_c^2 t^2)^\alpha}$$

$$+e^{-ik_F x}(x \to -x), \quad 0 < \alpha < \frac{1}{16};$$

for G_h, $t \to -t$. There is a multiplicative factor involving an energy cutoff $\sim \Lambda \sim \hbar v_F / a$, of order $2\alpha \Lambda^{2\alpha}$.

Fourier transforming this structure is difficult [9]. An approximation which loses none of the essential physics is to take advantage of small α and large U (for which $v_c / v_s \gg 1$) to treat the terms $(x^2 - v_c^2 t^2)^{-\alpha}$ as a slowly varying, logarithmlike term which simply multiplies by convergence factors $(i v_c t)^{-2\alpha}$ or $x^{-2\alpha} \sim (k - k_F)^{2\alpha}$, whichever is smaller.

Then we can get an adequate approximation to $G(k, t)$. Define $\Delta k = k - k_F$, $\bar{v} = \frac{1}{2}(v_c + v_s)$, and $\Delta v = v_c - v_s$; then

$$G_e(k, t) = \int dx\, e^{-ikx} G_e(x, t)$$

$$\propto e^{-\eta t} n_k e^{i\bar{v}\Delta k t} J_0(\Delta v \Delta k\, t)(i v_c t)^{-2\alpha}.$$

It is clear that now the convergence factor η is essentially irrelevant; even at k_F, $t^{-2\alpha}$ or $\Delta k^{+2\alpha}$ provides a convergence factor, and everywhere else J_0 falls off as $t^{-1/2}$. The reason is the conversion of the Fermi-liquid pole into a cut along the real axis, or physically the fact that charge and spin do not propagate with the same velocity; hence the singular perturbation has no chance to act. This is why the result is qualitatively different for the system with spin. The response to our perturbation then works out roughly as

$$R \propto t_\perp^2 \sum_k n_k(1 - n_k) \int \frac{dt\,(\Delta k)^{2\alpha}}{t^{2\alpha}\,\Delta v \Delta k\, t} e^{-2\eta t}.$$

It looks as though our convergence factor is being used twice but we believe this expression gives the right convergent behavior. There is indeed an energy response, but it is small and entirely due to virtual hopping, not real transitions, as we see by the fact that the convergence factor η does not enter the answer.

Formal theory at finite temperature seems to be difficult; the crude holon-spinon ideas we have suggested in the past may be the best way to show that at finite T or frequency there is a T-linear conductivity between chains; this could also be done by gauge methods.

This answer should not be at all unexpected. As we have repeatedly emphasized, the Luttinger liquid of the Hubbard model is an incompressible spinon liquid, if we think of spinons as the limiting case as $Z \to 0$ of electron quasiparticles (the so-called marginal Fermi-liquid concept, proposed in [10]), and we recognize that if in the Luttinger liquid, as shown by the exact Lieb-Wu solution, the spinon Fermi velocity is finite, this must mean that the self-energy as we approach this limit obeys

$$v_s = \frac{\partial \Sigma / \partial k}{\partial \Sigma / \partial \omega} = \text{finite}.$$

But if $\partial \Sigma / \partial \omega \to \infty$, this means $\partial \Sigma / \partial k \to \infty$, which implies that the spinon liquid is incompressible:

$$K = \frac{K_0}{1 + \partial \Sigma / \partial k} \to 0.$$

The spinon gas cannot change density without exciting a second type of excitation, the holon, which plays the role of a density of sites which the spins may occupy. Again, spin is vital, as is the existence of two different velocities. This means that one-particle perturbations cannot shift the Fermi level in momentum space, which prevents electrons tunneling into and out of the liquid. We hope to show in a later work that the same is true of impurity potentials: The liquid is immune to spin-independent potentials, and hence has infinite conductivity (and does not localize) even in the presence of impurities. And, of course, since we have shown that $Z \to 0$ in the 2D Hubbard model, all these properties hold in that case as well. (Confinement is also expected to hold for 2D quantum Hall systems—see the results of Boebinger [11].)

It is often argued that renormalized perturbation theory (so-called "g-ology") is an adequate method to obtain properties of one-dimensional models, in particular, the Hubbard model. Although some details differ from the more complete bosonization [6] or "Luttinger-liquid" method [12] these seemed irrelevant to many. After 20 years of g-ology, however, we now see that these responses are different from the conventional assumptions based on perturbation theory.

The immediate result is relevant for a number of "1D" systems. It seems likely that the extension to negative U, and certainly to the t-J model, is straightforward, and that if such systems take on a 1D correlated liquid state they may confine in the other two directions. The "organic superconductors," however, are in general 2D systems and we expect 2D confinement to occur in some of these, if not all. We suspect that data on transport in these systems will reward reexamination from a fresh point of view.

I acknowledge lengthy discussions with Y.-B. Xie, Y. Ren, S. Strong, X.-G. Wen, H. J. Schulz, F.D.M. Haldane, and D.-H. Lee. This work was supported by NSF Grant No. DMR-8518163 and AFOSR Grant No. 87-0392.

REFERENCES

1. X. G. Wen, Phys. Rev. B **42**, 6623 (1990).
2. H. J. Schulz, in Proceedings of the Adriatico Research Conference on the Physics of Strongly Correlated Systems, Trieste, Italy, July 1990 (to be published).
3. G. Benfatto and G. Gallivotti, Phys. Rev. B **42**, 9967 (1990); also R. Shankar (to be published). In these references it is pointed out that the Fermi liquid—and, actually, the Luttinger liquid [see P. W. Anderson, Phys. Rev. Lett. **64**, 1839 (1990)]—is an infrared stable fixed point, not a critical point, of a renormalization in which the parameter is a cutoff energy, not a length. The fixed-point Hamiltonian is the free Fermi gas for the Fermi liquid, and a modification of this for the Luttinger liquid; the only possibly relevant perturbation is a shift in the Fermi surface. The conventional critical point renormalization group is not appropriate to this problem.
4. Y. Ren and P. W. Anderson (to be published); S. Sorella, A. Parola, M. Parrinello, and E. Tosatti, Europhys. Lett. **12**, 721 (1990).
5. H. Frahm and V. E. Korepin, Phys. Rev. B **42**, 10533 (1990).
6. A. Luther and I. Peschel, Phys. Rev. B **9**, 2911 (1974); A. Luther and V. J. Emery, Phys. Rev. Lett. **33**, 589 (1974).
7. The correct Green's function, obtained by bosonization techniques as in Ref. 6. is given in Mahan's book [G. D. Mahan, *Many Particle Physics* (Plenum, New York, 1981), p. 347]. However, in this and other references the Fourier transform is specialized to $v_s = v_c$ giving incorrect physics.
8. W. Kohn and J. M. Luttinger, Phys. Rev. **118**, 41 (1969); C. Bloch and C. deDominicis, Nucl. Phys. **7**, 459 (1958).
9. We were helped with this transform by explicit formulas worked out by F.D.M. Haldane (private communication). A simpler treatment could stay in x-t space: The result is the same.
10. P. W. Anderson, in *Frontiers and Borderlines in Many Particle Physics*, Proceedings of the International School of Physics, "Enrico Fermi," Course CIV, edited by R. A. Broglia and J. R. Schrieffer (North-Holland, Amsterdam, 1987).
11. G. S. Boebinger et al., Phys. Rev. Lett. **65**, 235 (1990).
12. F.D.M. Haldane, Phys. Lett. **81A**, 153 (1981).

Incoherence of Single Particle Hopping between Luttinger Liquids

David G. Clarke,* S. P. Strong,† and P. W. Anderson

(Received 21 January 1994)

ABSTRACT

We demonstrate that for general spin-charge separated Luttinger liquids there exists a critical value of the inter-Luttinger-liquid single particle hopping, $t_\perp^c$, below which there is no coherent single particle hopping between the liquids. In the absence of coherent single particle hopping, two Luttinger liquids coupled by $t_\perp$ will not exhibit split Fermi surfaces. For many Luttinger liquids, no band dispersing between the liquids will form, and thus the system will retain a one-dimensional Fermi surface. This will have dramatic implications for the physical properties of such a system.

PACS numbers: 71.27.+a, 72.10.Bg, 74.25.−q

One of us [1] has suggested that the unusual features of the c-axis resistivities observed in the cuprate superconductors are the result of the non-Fermi-liquid nature of the in-plane ground state of these materials. This non-Fermi-liquid nature is argued to disrupt the interchain hopping of electrons so strongly that single electrons effectively do not hop between the planes, giving rise to anomalous c-axis transport properties [1] and to the anomalously large superconducting transition temperatures for these materials [1, 2]. Both of these effects hinge on the absence of normal interplane hopping and it is natural to search for theoretical models which exhibit this characteristic. An argument has previously been given that two Hubbard chains coupled by a weak interchain hopping show this effect in the character of the singularities of a particular response function [3]. Here we give a different argument which clarifies the physics considerably and establishes the conjecture of [3], that while interchain hopping is not destroyed, it is rendered completely *incoherent* for sufficiently small $t_\perp$.

*Theoretical Physics, University Oxford.
†Joseph Henry Laboratories of Physics, Princeton University.

The model we consider is that of two one-dimensional Luttinger liquids coupled with a weak interchain hopping:

$$H = H_{LL}^1 + H^2 + t_\perp \sum_{i,\sigma} [c_{1,\sigma}^\dagger(i)c_{2,\sigma}(i) + \text{H.c.}],\qquad(1)$$

where the microscopic Hamiltonians for the individual systems are such that their ground states are Luttinger liquids [4] and their single particle Green's functions take the form

$$G(x,t) \sim \frac{1}{2\pi} \frac{e^{ik_f(x-vt)}}{\sqrt{[x - v_c t + i\,\text{sgn}(t)\delta]}\sqrt{[x - v_s t + i\,\text{sgn}(t)\delta]}}$$
$$\times \left[\Lambda \left(x - v_c + \frac{i\,\text{sgn}(t)}{\Lambda} \right) \left(x + v_c t - \frac{i\,\text{sgn}(t)}{\Lambda} \right) \right]^{-\alpha}\qquad(2)$$

plus a similar term from the left Fermi point. For coupled Hubbard chains, v_c, v_s, and α are all functions of U. For our purposes we only need to know that general Luttinger-liquid arguments require that, to lowest order in the interaction, $v = \frac{1}{2}(v_c + v_s)$, $v_c - v_s \propto U$, and $\alpha \propto U^2$. For the sake of generality we will consider $v_c - v_s$ and α arbitrary and specialize to the Hubbard model as needed.

The motivation for studying this model is as follows. For Fermi liquids coupled with an interliquid hopping, the ground state for two liquids is built by making symmetric and antisymmetric combinations of the quasiparticle operators in the two liquids and then filling these new quasiparticle states. This involves constructing a ground state which is a superposition of states with different numbers of quasiparticles in a given liquid. For this to be reasonable it must be possible for such a superposition to be phase coherent. This requirement is innocuous for systems where the exact low energy eigenstates are electronlike and are hopped by $t_\perp$. However, when the system is a non-Fermi-liquid the possibility of coherence for these states must be reexamined. Previous studies of models similar to ours [5, 6] have not addressed this question. We do not necessarily disagree with the results of these works; when suitably interpreted, however, our approach is very different from that of previous work and we arrive at markedly different conclusions. Rather than examining the relevance or irrelevance of $t_\perp$ in the renormalization group sense, we ask whether or not the effect of $t_\perp$ is a coherent one. To answer this question, we consider a system prepared at time $t = 0$ such that the two liquids are separately in $t_\perp = 0$ eigenstates without any Tomonaga bosons excited. We take one liquid to have ΔN more right-moving particles of a particular spin species than the other liquid, and the two liquids otherwise identical. We then turn on the interchain hopping, $t_\perp$, and ask if the probability of the system remaining in its initial state, $P(t)$, behaves for intermediate times (the appropriate time scale will be defined later in this Letter) in a manner consistent with incoherent hopping.

One way to motivate this question is to recall the zero temperature properties of the solution of the two level system problem (TLS) [7]. In that problem one considers a system which has some variable, σ^z, which may take on two values and which is coupled to a bath of harmonic oscillators by a term $\sum_i C_i x_i \sigma^z$, where the x_i's are the oscillator coordinates, and to a biasing field, ϵ, by a term $\epsilon \sigma^z$. One turns on a tunneling matrix element, $\Delta \sigma^x$, between the two states and asks about the intermediate time behavior of $\langle \sigma^z(t) \rangle$. To see why the physics of the TLS should be related to the case of coupled Hubbard chains, consider the TLS Hamiltonian after a canonical transformation has been made to shift the x_i:

$$H_{\text{TLS}} = \tfrac{1}{2}\Delta(\sigma^+ e^{-i\Omega} + \text{H.c.}) + \tfrac{1}{2}\epsilon \sigma^z + H_{\text{oscillators}}. \tag{3}$$

Here $\Omega = \sum_i \frac{C_i}{m_i \omega_i^2} p_i$, C_i is the coupling to the ith oscillator, and m_i, ω_i, and p_i are the mass, frequency, and momentum operator. Written in a bosonized form the interchain hopping operator, $t_\perp(c_{1,\sigma}^\dagger(i)c_{2,\sigma}(i) + \text{H.c.})$, is of a form similar to the Δ term. It contains a piece which shifts the number of particles in each chain without exciting any Tomonaga bosons times exponentials of Tomonaga boson creation and annihilation operators. In place of the two states in the TLS problem we have many states labeled by the numbers of right and left movers of each spin species in each chain. In place of the bath of harmonic oscillators we have the Tomonaga bosons of the bosonized Luttinger liquids. We now briefly enumerate the zero temperature, zero bias possibilities for the TLS.

The effect of the tunneling matrix element, Δ, is essentially determined by the strength of the orthogonality catastrophe [8] among the oscillators when the system moves between the σ^z states. In the limit of strong oscillator coupling (for a definition of weak, strong, and intermediate, see [7]) the tunneling is irrelevant and, if the system is placed in one σ^z state, allowed to equilibrate without Δ, and Δ is then turned on, $\langle \sigma^z(t) \rangle$ decays to some finite constant as $t \to \infty$. In this case the system is localized at one value of σ^z. In the opposite limit, under the same circumstances, $\langle \sigma^z(t) \rangle$ will undergo damped oscillations between $+1$ and -1 at intermediate times, decaying to zero at long times. In this case, the system tunnels *coherently* between the two σ^z states. For intermediate couplings there exists a third behavior where, again with the same preparation, $\langle \sigma^z(t) \rangle$ will relax exponentially, without any oscillations, to zero. In this case there is no coherent tunneling between the states, but there is also no localization. This phase is analogous to what we find for coupled Luttinger liquids for small enough $t_\perp$: *incoherent*, finite interchain tunneling. It is important to note that in the TLS it is not claimed that $\Delta \sigma^x$ is an irrelevant perturbation in this phase. Naive renormalization group arguments show it to be a relevant perturbation, but the relevance of the tunneling term in the Hamiltonian does not guarantee that *coherent* tunneling takes place. Likewise, we do not claim that $t_\perp$ is an irrelevant perturbation, rather that its relevance is insufficient to cause coherent hopping.

In the TLS, coherence or incoherence is signaled by the intermediate time behavior of $\langle \sigma^z(t) \rangle$. In our problem, the analogous signal would come from the intermediate time behavior of $P(t)$. Unfortunately the behavior of $P(t)$ is intractable in our problem beyond the lowest order in perturbation theory. However, this order is sufficient to establish whether or not serious problems with the prediction of incoherent relaxation exist. For example, for the case of noninteracting electrons we find

$$P(t) = 1 - t_\perp^2 \, \Delta N \, t^2 + \cdots, \tag{4}$$

which we recognize as the first term in the expansion of the exact result $P(t) = \cos^{2\Delta N}(t_\perp t)$. The coherent tunneling leads to oscillations which manifest themselves as a time dependence qualitatively different from that expected from an incoherent decay.

Now we turn to the interacting case. First, we consider the case $\alpha = 0$, $v_c \neq v_s$ for which strong arguments exist in favor of coherent single particle hopping and band formation (e.g., [6]). We find, to lowest order in $t_\perp$, that

$$1 - P(t) \sim \frac{t_\perp^2 L}{4\pi^2} \mathrm{Re} \tag{5}$$

$$\times \left(\int_0^t dt' \int_0^t dt'' \int dx \frac{\exp\{-i\Delta k[x - v(t' - t'') + i\delta]\}}{[x - v_c(t' - t'') + i \, \mathrm{sgn}(t' - t'')\delta][x - v_s(t' - t'') + i \, \mathrm{sgn}(t' - t'')\delta]} \right).$$

Here Δk is $\frac{2\pi \Delta N}{L}$. The x integral can be evaluated and the result expanded for times much less than $\Delta k^{-1}(v_c - v_s)^{-1}$ to yield exactly the noninteracting time dependence. This suggests that coherent oscillation will occur so long as this time is long compared to the oscillation frequency. Here the oscillation frequency will be $t_\perp$ since that would be the frequency in the noninteracting case and we are looking precisely at those time scales where our perturbation theory shows that $P(t)$ is behaving as in that case. This leads to the requirement $t_\perp \gg \Delta k(v_c - v_s)$ for coherent interliquid tunneling. This can be satisfied trivially if Δk is not $O(1)$. The maximum Δk allowed for coherent oscillations is $O(t_\perp(v_c - v_s)^{-1})$ which is much larger than $O(t_\perp v^{-1})$, the number difference relevant to the splitting of the Fermi surfaces by an amount $\sim t_\perp$. For the infinitely many chains problem, there would be no obstacle to the formation of a coherent band of width $\sim t_\perp$. Now, however, we consider the problem with α finite.

Since all we are looking for is anomalous time dependences it is sufficient here to consider

$$1 - P(t) \sim t_\perp^2 L \mathrm{Re} \left[\int_0^t dT \int_{-\frac{T}{2}}^{\frac{T}{2}} d\tau \int dx \, G_N(x, \tau) G_{N+\Delta N}(-x, -\tau) \right].$$

This still leaves us with a complicated multiple integral to consider since the x integral now involves two branch cuts and two poles. Careful evaluation of the asymptotic time dependencies of the x integral yields, for $\alpha \ll 1$ and $t \gg \Lambda^{-1}$,

$$1 - P(t) \sim t_\perp^2 L \Lambda^{-4\alpha} \mathrm{Re}$$

$$\times \left(\int_0^t dT \int_0^T d\tau [\alpha Z_1(\Delta k, \tau) + Z_2(\Delta k, \tau) + \alpha Z_3(\Delta k, \tau)] \right), \quad (6)$$

where, for $\tau \gg \Lambda^{-1}$,

$$Z_1(\Delta k, \tau) \sim \int_0^\infty dx \left\{ e^{i\Delta k(v-v_c)\tau} + e^{-i\Delta k(v+v_c)\tau} e^{-\Delta k z} \right\}$$

$$\times z^{-2\alpha} [z + i(v_c + v_s)\tau]^{-1} (z + 2iv_c\tau)^{-1-2\alpha},$$

$Z_2(\Delta k, \tau)$ is given by a coefficient of order unity times

$$[(v_c - v_s)\tau]^{-1-2\alpha} [(v_c + v_s)\tau]^{-2\alpha} \sin \left(\frac{\Delta k(v_c - v_s)\tau}{2} \right),$$

and

$$Z_3(\Delta k, \tau) \sim e^{i\Delta k(v-v_c)\tau} \int_0^\infty dz [1 - \exp(-\Delta k\, z)]$$

$$\times z^{-1-2\alpha} [z + i(v_c - v_s)\tau]^{-1} (z + 2iv_c\tau)^{-2\alpha}.$$

Now we turn to the τ integral. For $\Delta k = 0$ there is only the Z_1 term, which behaves at long times as $Z_1(0, \tau) \sim \alpha \tau^{-1-4\alpha}$, a behavior indicative of an incoherent process, for which "golden rule" methods may be applied. As in the case of the incoherent TLS problem [7] the τ integral extended to $\tau = \infty$ is found to vanish, so that a self-consistent approach is necessary, leading to an incoherent decay rate $\Gamma \sim (\alpha t_\perp^2)^{\frac{1}{1-4\alpha}}$. For $\Delta k \neq 0$ the effect of the Z_1 term is more complicated to analyze but it still represents fundamentally incoherent processes. Further, for times large compared to $(\Delta k)^{-1}(v_c - v_s)^{-1}$, both the Z_2 term and the Z_3 terms integrate to constants as $T \to \infty$ and are therefore also incoherent. However, at times short compared to $(\Delta k)^{-1}(v_c - v_s)^{-1}$ they exhibit a dangerous, superlinear time dependence. If these terms are not severely modified by the Z_1 term then coherent single particle hopping can occur.

We now argue that, for $t_\perp < t_\perp^c$, the effects of the Z_1 term are sufficiently strong that the dangerous time behavior of the Z_2 and Z_3 terms will not survive. To show this we need to understand the effect of the incoherent transitions on the coherent ones. In a Fermi liquid, the question of incoherence is a single particle one and is straightforwardly answered by comparing the decay of the survival probabilities for a given quasiparticle due to incoherent and coherent processes at intermediate times. We believe the correct many-body generalization to a Luttinger liquid should be based on comparing, at intermediate times, the survival probabilities per unit volume of the initial many-body state due to incoherent and coherent processes. Specifically, at a time, t, our Z_1 term will have produced $N_{\text{inc}}(t) \sim \alpha t^{1-4\alpha} t_\perp^2 L$ incoherent transitions [9]. Since each involves the insertion of an extra electron into one liquid and an extra hole into the other, each incoherent operation also causes orthogonality catastrophes in each liquid leading to an overlap of the initial state with the new one vanishing like $t^{-4\alpha}$. At time t the $N_{\text{inc}}(t)$ transitions which will have occurred $O(t)$ earlier so the decay of the initial and final state overlap due to the orthogonality catastrophes initiated by the $N_{\text{inc}}(t)$ transitions are given approximately by $t^{-4\alpha N_{\text{inc}}(t)}$ or $\exp[-4L\alpha^2 t_\perp^2 t^{1-4\alpha} \ln(t)]$. The survival probability per unit volume is given by $\exp[-8\alpha^2 t_\perp^2 t^{1-4\alpha} \ln(t)]$, which is to be compared to the survival probability per unit volume coming from the coherent transitions. The Z_2 term gives, after τ and T integrations, for $t \ll (\Delta k)^{-1}(v_c - v_s)^{-1}$, a term like $Lt_\perp^2 \Delta k t^{2-4\alpha}(v_c - v_s)^{-2\alpha}$. Subject to the requirement that $t \ll (\Delta k)^{-1}(v_c - v_s)^{-1}$, this is maximized by taking $\Delta k \sim t^{-1}(v_c - v_s)^{-1}$, giving $Lt_\perp^2(v_c - v_s)^{-1-2\alpha} t^{1-4\alpha}$. We see that, in order for incoherent transitions not to dominate and destroy coherence, we need $8\alpha^2 \ln(t) \lesssim (v_c - v_s)^{-1-2\alpha}$. Provided that the appropriate time scale grows to infinity as $t_\perp \to 0$ and α and $v_c - v_s$ are finite, this cannot be satisfied for aribtrarily small $t_\perp$. If we choose to look at a time $t \sim t_\perp^{\frac{1}{2\alpha-1}}$ as suggested by renormalization group arguments as the natural scale in the problem (i.e., the inverse of the value which $t_\perp$ has renormalized to when we have scaled for long enough to have it grow comparable to the cutoff) then we find the criterion $\exp(\frac{2\alpha-1}{\alpha^{-2}(v_c-v_s)^{-1-2\alpha}}) \lesssim t_\perp$ for coherent oscillations to be possible. Specializing to coupled Hubbard chains and inserting the U dependence of $v_c - v_s$ and α we find $t_\perp \ll \exp(-\frac{\text{const}}{U^5})$ excludes the possibility of coherence [10]. Similar reasoning for the Z_3 term gives the same criterion.

Note that, independent of the more subtle form of incoherence, which we are arguing results from spin-charge separation, if $\alpha > \frac{1}{4}$ the Z_2 term and the Z_3 term will have an intermediate time behavior consistent with incoherence. Non-spin-charge separated Luttinger liquids, e.g., spinless fermions in one dimension, will also exhibit purely incoherent hopping if α is larger than this critical value. This is in spite of the renormalization group relevance of $t_\perp$ for $\alpha < \frac{1}{2}$ and is more nearly analogous to the incoherence in the TLS problem. These systems might prove a more numerically tractable laboratory for studying the physics of incoherently coupled Luttinger liquids.

It is possible that Luttinger liquids without spin-charge separation may exhibit incoherence for $\alpha < \frac{1}{4}$. This is possible because while we believe that the criterion we have used for incoherence is correct, it may be overly severe. We have chosen to restrict Δk only to values small compared to $(v_c - v_s)^{-1} t_{\perp,\mathrm{ren}}$ when one might plausibly argue that it should be no bigger than $v^{-1} t_{\perp,\mathrm{ren}}$. If this more liberal criterion is used then Luttinger liquids without spin-charge separation would also have a critical $t_\perp$ required for coherent hopping even when $\alpha < \frac{1}{4}$.

We find the destruction of the coherent single particle hopping as the result of three properties of the general Luttinger-liquid state. First, the Fermi surface is sufficiently destroyed to produce, within some time period, a finite number of the incoherent processes per unit volume when a weak interliquid single particle hopping is turned on. This is represented by our Z_1 term. Second, the velocities for spin and charge excitations are different, giving an electron spectral function whose width in energy space vanishes only linearly as we approach the Fermi surface, not quadratically as for a Fermi liquid. This results in the oscillating phase factor of our Z_2 term and the exponential decay of our Z_3 term and in general destroys coherence at sufficiently long times. Third, there is an anomalous exponent for the electron propagator, 2α, which represents the orthogonality catastrophe associated with the insertion of an extra electron or hole into a Luttinger liquid. This orthogonality causes each incoherent process to have an effect which increases with time on the coherent processes and enters our calculation in the comparison of the Z_1 term to the Z_2 and Z_3 terms. Given these three properties there will be a critical value of $t_\perp$ required to generate coherent single particle hopping between Luttinger liquids. Since all of these properties are expected to be present in higher dimensional Luttinger liquids, our results should apply directly to the cuprate superconductors.

The consequences of the absence of coherent single particle hopping should be very severe. As we have stated earlier, the whole apparatus of band theory relies on the coherence of single particle hopping. If this hopping is incoherent, then the usual band theory ground state, which involves a superposition of states with different electron number in a given liquid, cannot be an eigenstate of the interacting system. Therefore, band theory must break down completely for incoherent single particle hopping, irrespective of the relevance of $t_\perp$ in the renormalization group sense. A semiquantitative way of looking at the failure of band theory comes from recognizing that, for systems which show coherent oscillations under the circumstances we have described, the frequency of the oscillations is a measure of the energy available to the system by forming coherent superpositions of states involving different particle numbers in a given liquid. For the case of noninteracting electrons the oscillation frequency or energy available per electron is $t_\perp$. Consequently, the Fermi surface splits into symmetric and antisymmetric Fermi surfaces $t_\perp$ apart for two liquids; for infinitely many liquids a band of width $t_\perp$ forms. For purely incoherent hopping, there is no energy available and the Fermi surface will not split by any finite amount. For infinitely many liquids, no band dispersing in the perpendicular direction should form when the hopping is purely incoherent; neither should a two-dimensional Fermi surface form.

The absence of band formation will have direct and dramatic effects upon many experimentally observable quantities. We have calculated the finite frequency perpendicular conductivity for weakly coupled Luttinger liquids and find a conductivity $\sim \omega^{4\alpha}$ over a large frequency range [11]. Using a value of α close to $\frac{1}{16}$ (which is the appropriate exponent for the large-U Hubbard model, at least in one dimension) this agrees well with the conductivity obtained from optical measurements by Cooper et al. [12] on $YBa_2Cu_3O_7$. The consequences of incoherent single particle hopping for superconductivity are also particularly important. The lowering of the ground state energy coming from the kinetic energy that would be available if single particle hopping were coherent can still be achieved by the system if it goes over to a superconducting state in which Cooper pairs hop coherently [1, 2, 13].

S.P.S. acknowledges helpful discussions with S. Chakravarty. P.W.A. and S.P.S. acknowledge financial support from NSF Grant No. DMR-9104873.

REFERENCES

1. P. W. Anderson, this volume.
2. S. Chakravarty, A. Sudbø, P. W. Anderson, and S. P. Strong, Science **261**, 337 (1993).
3. P. W. Anderson, Phys. Rev. Lett. **67**, 3844 (1991).
4. F.D.M. Haldane, J. Phys. C **14**, 2585 (1981); Phys. Lett. A **176**, 363 (1993).
5. X. G. Wen, Phys. Rev. B **42**, 6623 (1990); C. Bourbonnais and L. G. Caron, Int. J. Mod. Phys. B **5**, 11 033 (1991); H. J. Schulz, Int. J. Mod. Phys. B **5**, 57 (1991); M. Fabrizio, A. Parola, and E. Tosatti, Phys. Rev. B **46**, 3159 (1992); F. V. Kusmartsev, A. Luther, and A. Nersesyan, JETP Lett. **55**, 692 (1992); C. Castellani, C. di Castro, and W. Metzner, Phys. Rev. Lett. **69**, 1703 (1992); V. M. Yakovenko, JETP Lett. **56**, 5101 (1992); A. Finkelshtein and A. I. Larkin, Phys. Rev. B **47**, 10 461 (1993); A. A. Nersesyan, A. Luther, and F. V. Kusmartsev, Phys. Lett. A **176**, 363 (1993).
6. M. Fabrizio and A. Parola, Phys. Rev. Lett. **70**, 226 (1993); A. M. Tsvelik (unpublished).
7. S. Chakravarty, Phys. Rev. Lett. **49**, 681 (1982); S. Chakravarty and A. J. Leggett, Phys. Rev. Lett. **52**, 5 (1984); A. J. Leggett et al., Rev. Mod. Phys. **59**, 1 (1987), and references therein.
8. P. W. Anderson, Phys. Rev. **164**, 352 (1967).
9. Strictly speaking, the precise expression for N_{inc} is modified when $\Delta k \neq 0$, or if one attempts to deal with Z_1 in a self-consistent manner. However, such modifications do not change the conclusions which follow.
10. In determining $t_\perp^c$ here we have neglected the effect of coherent hops upon themselves, and also upon the incoherent hops, both of which will only serve to enhance

incoherence. The precise details are nontrivial: We are more concerned here with presenting a general *existence* argument for $t_\perp^c$, than a particular expression for it.

11. D. G. Clarke, S. P. Strong, and P. W. Anderson (to be published).

12. S. L. Cooper et al., Phys. Rev. Lett. **70**, 1553 (1993).

13. S. Chakravarty and P. W. Anderson (to be published).

L

Conductivity between Luttinger Liquids in the Confinement Regime and c-Axis Conductivity in the Cuprate Superconductors

David G. Clarke,* S. P. Strong,† and P. W. Anderson

(Received 9 September 1994)

ABSTRACT

We calculate the interliquid conductivity for Luttinger liquids, within the so-called "confinement" regime where interliquid hopping is completely incoherent. We argue that the interliquid conductivity behaves as $\sigma_\perp(\omega) \sim \omega^{4\alpha}$, where α is the Luttinger liquid exponent. We discuss the effect of finite temperature, which is found to introduce a coherent weight into the conductivity. These results are in good agreement with experimental measurements of the frequency dependent normal state c-axis conductivity in the high-temperature superconductors $YBa_2Cu_3O_{7-\delta}$ and $La_{2-x}Sr_xCuO_4$.

PACS numbers: 71.27.+a, 72.10.−d, 74.25.Fy, 74.72−h

One of the more important yet less frequently discussed anomalies observed in the normal state of the cuprate-based high-temperature superconductors (HTSC's) is the c-axis conductivity $\sigma_c(\omega, T)$. By comparison, a great deal more attention has been given to understanding the anomalous behavior of the in-plane conductivity $\sigma_{ab}(\omega, T) \sim (\omega^{-1}, T^{-1})$. It is a remarkable fact, given that the structures and chemistry of HTSC's can vary a great deal outside of the common CuO_2 planes, that $\sigma_{ab}(\omega, T)$ is an almost universal function over the HTSC family. In striking contrast to this, there is no single expression for $\sigma_c(\omega, T)$ with wide applicability. Various empirical fits for the interplane dc resistivity, $\rho_c(T)$, involving combinations of functions proportional to 1, T, T^{-1}, T^{-p}, and $e^{\Delta/T}$ have been made, the fit differing from material to material. It might be argued that this is not to be entirely unexpected since ρ_c probes the motion of electrons in the interplane direction, the nature of which, as stated above, varies considerably over the HTSC family. The challenge to such an attitude, however, is to demonstrate how the interplanar chemistry and/or structure can be such as to lead to c-axis conductivities well below the minimum metallic conductivity given by the Mott-Ioffe-Regel limit [1] and, more alarmingly, a *positive* derivative in temperature, $d\sigma_c/dT > 0$, observed in many HTSC's over a wide temperature range above T_c. Were it not for the intervention of superconductivity, one would have a situation in which $\sigma_{ab}(T)/\sigma_c(T)$ would increase, apparently without bound, as $T \to 0$. Within a Fermi-liquid picture, anisotropy alone cannot account for such a situation. However, the in-plane properties of HTSC's are not those of a Fermi liquid. One is thus led to the almost inescapable conclusion that the unusual behavior of σ_c is intimately connected to the non-Fermi-liquid properties of the 2D electron fluid in the cuprate planes. This is the starting point of the "confinement" hypothesis advanced several years ago [2–4].

Considerably more information about c-axis transport can be obtained by going beyond the measurement of dc conductivities. Recently, experimental determinations of the frequency dependence of the c-axis conductivity on the eV scale have been made in $YBa_2Cu_3O_{7-\delta}$ [5] and $La_{2-x}Sr_xCuO_4$ [6]. These experiments demonstrate unambiguously the absence of any Drude-like term in the electronic contribution to $\sigma_c(\omega)$ [7]. There is an enormous loss of spectral weight at low frequencies. In the samples studied, $\sigma_c(\omega)$ is a very slowly *increasing* function of ω over a wide frequency range. Again, most samples exhibit a dc conductivity below the Mott limit, when the data are extrapolated back to $\omega = 0$.

The absence of a Drude-like term in $\sigma_c(\omega)$ vindicates the confinement hypothesis of the normal state of HTSC's: single particle hopping in the interplane direction is completely incoherent. There is no band formation in the c direction, and as a consequence there is the potential for interplanar pair tunneling to give rise to a large kinetic energy gain and drive a mechanism for a high T_c [2, 4, 8].

The challenge to any candidate theory of the normal state of HTSC's is to explain how confinement can come about. In a recent paper [9], we have presented a strong argument for the existence of a coherent-incoherent (or "confinement-deconfinement") transition for single particle hopping between spin-charge separated Luttinger liquids. In this Letter we shall further this work to calculate the interliquid conductivity in the confinement regime. The connection to the HTSC's is made via the hypothesis that the electronic low energy physics of the cuprate planes is described by a 2D spin-charge separated tomographic Luttinger liquid [4]. In order to be able to perform definitive calculations we shall here restrict attention to the analogous problem in one dimension where the electron Green's functions are rigorously known.

In this Letter then we shall consider the problem of determining the interchain conductivity for a system of weakly coupled Hubbard chains. The low energy physics of a 1D Hubbard model, with parameters $(t_\parallel, U)$ as usually defined, as that of a spin-charge separated Luttinger liquid. Denoting the Hamiltonian for the Luttinger liquid on chain i by $H_{LL}^{(i)}$ we have

$$H = \sum_i H_{LL}^{(i)} + t_\perp \sum_{i,x} \{c_{i\sigma}^\dagger(x)c_{i+1\sigma}(x) + \text{H.c.}\}. \tag{1}$$

We have previously argued [9] that in a system governed by (1), and for sufficiently small $t_\perp$, electron motion in the interchain direction is completely *incoherent*. As such, the Fermi surface retains a one-dimensional form and there is no coherent interchain electron velocity. The underlying reason for this is that the interchain hopping term in (1) hops *real* electrons, however, the exact eigenstates of H_{LL} are *not* electronlike. Instead of exhibiting a sharp delta-function peak, the electron spectral function $\rho(k, \omega)$ of (1) has power law singularities at $\omega = \pm v_c k, v_s k$, a result of spin-charge separation and the infrared orthogonality catastrophe induced upon insertion of a single electron into the liquid [10]. As a result, in contrast to a noninteracting electron gas, in which all interchain hops are elastic, electron hopping between Luttinger liquids is a mixture of elastic and inelastic processes, the latter involving virtual transitions over an energy range $\Delta E \sim 1/t$. Of course, even for the case of single electron hopping between Fermi liquids there will be inelastic processes due to the incoherent part of the electron spectral function. However, in a Fermi liquid the coherent "quasi-particle" part of the electron Green's function $G(k, t)$ has a lifetime $\sim (k - k_F)^2$. As shown, e.g., in [11], in this case the quasiparticle part long outlives the incoherent contribution to $G(k, t)$ as $k - k_F \to 0$. Coherent interchain hopping results. In marked contrast to this, for coupled Luttinger liquids, and sufficiently small $t_\perp$, spin-charge separation implies a width of the electron spectral function $\sim (k - k_F) \gg (k - k_F)^2$, and this, in combination with interference between inelastic and elastic processes due to the orthogonality catastrophe, leads to the destruction of any signal of coherent interchain hopping [9]. Coupled with the Luttinger liquid hypothesis for HTSC's, this offers a natural explanation for the anisotropic transport observed in those materials, and significantly elucidates the proposal of "confinement" originally made on the basis of such transport data. We believe that such a regime will be ubiquitous to *any* sufficiently anisotropic, strongly correlated system, not just the cuprate superconductors. Indeed, we have argued elsewhere [12] that the anomalously rapid 3D to 2D crossover observed in the magnetoresistance of the (highly anisotropic) organic superconductor (TMTSF)$_2$PF$_6$, as a function of applied magnetic field, is a manifestation of the coherence-incoherence transition discussed in [9].

In the light of this result, we proceed to calculate the interliquid conductivity in the confinement regime. As in [9] we introduce the probability $P(t)$ that at time $t > 0$ the system is in the $t < 0$ ground state,

$$P(t) = \left| \left\langle \prod_i O_i | e^{-iHt} | \prod_i O_i \right\rangle \right|^2, \tag{2}$$

where $|O_i\rangle$ is the Luttinger liquid ground state of $H_{LL}^{(i)}$, $t_\perp = 0$ for $t < 0$; hence the $t < 0$ ground state is $|\prod_i O_i\rangle$. It is simple to show that, to $O(t_\perp^2)$, the effective rate of hopping out of a given chain is given by

$$-\frac{dP(t)}{dt} \equiv \Gamma(t) = 4t_\perp^2 L \mathrm{Re} \int_0^t dt' \left\{ \int dx \, G_e(x, t - t') G_h(x, t - t') \right\}, \tag{3}$$

where $G_e(x, t) \equiv \langle c(x, t) c^\dagger(0, 0) \rangle$ and $G_h(x, t) \equiv \langle c^\dagger(x, t) c(0, 0) \rangle$. Equivalently,

$$\Gamma(t) = t_\perp^2 L \int_{-\infty}^\infty \frac{d\omega}{2\pi} \frac{\sin \omega t}{\omega} \int_{-\infty}^\infty \frac{d\omega'}{2\pi} \int_{-\infty}^\infty \frac{dk}{2\pi} \rho(k, \omega') \rho(k, \omega' + \omega)$$
$$\times \Theta(-\omega') \{ 1 - \Theta(-(\omega' + \omega)) \}, \tag{4}$$

where $\rho(k, \omega)$ is the Luttinger liquid spectral function.

$P(t)$ and consequently $\Gamma(t)$ may readily be generalized to finite temperature via the usual Boltzmann weighting, the Θ functions in (4) being replaced by Fermi functions. The advantage of (4) over the space-time formalism of (3) is that it gives a more physical expression in terms of the electron spectral function and is simpler to use at finite temperature.

Let the chains be stacked a distance d apart along the z axis. We wish to calculate the interchain current induced by an electric field E parallel to $\hat{z}$. For $E = 0$ there is of course no net current, Eq. (4) describing incoherent motion of equal magnitudes into and out of a given chain. To determine the current in nonzero field we first observe that in the confinement regime interchain motion will be completely *diffuse*. We may write a constitutive equation for the interchain diffusion current J_D generated by a particle density gradient

$$J_D(z, t) = -\int_0^t dt' \, D(t - t') \frac{\partial}{\partial z} n(z, t'). \tag{5}$$

Here $n(z, t)$ is the particle density on the chain at z, and $D(t - t')$ is the diffusion kernel to be determined.

To zeroth order in $t_\perp$, if two adjacent chains differ in particle number by ΔN, then their Fermi momenta differ by $\Delta k = \pi \Delta N / L$. However, as was shown in [9] even for $\Delta N \sim O(L/a)$, where a is an in-chain length scale proportional to the lattice spacing, the additional terms introduced into $\Gamma(t)$ are rendered incoherent. Now, any *physical* particle number gradient will have ΔN only $O(1)$. We therefore believe that it is correct to argue that, to $O(t^2)$, *all* hopping occurs at the incoherent rate given by (3) and the diffusion kernel is therefore given by

$$ D(t - t') = 4t_\perp^2 d\mathrm{Re} \int_0^t dt' \left\{ \int dx\ G_e(x, t - t')G_h(x, t - t') \right\}, \quad (6) $$

which has the long time asymptotic behavior $D(t) \sim 1/t^{1+4\alpha}$.

Thus for a density gradient fluctuating at frequency ω, $J_D(\omega) = D(\omega)\nabla n(\omega)$ with $D(\omega) \sim \omega^{4\alpha}$. Utilizing the appropriate Einstein relation to connect $\sigma_c(\omega)$ to $D(\omega)$ we finally arrive at the following expression for the interchain conductivity:

$$ \sigma_\perp(\omega) \sim \alpha \frac{e^2}{\hbar} ad \left(\frac{t_\perp^2}{v_c v_F} \right) \left(\frac{\omega}{\Lambda v_c} \right)^{4\alpha}. \quad (7) $$

In writing (7) we have taken $v_c \gg v_s$ for simplicity, where $v_{c,s}$ refers to the charge and spin velocity, respectively. This corresponds to the limit $U/t \gg 1$ of the Hubbard model. Generically, we may replace the factor $t_\perp^2/v_c v_F$ by $\sim a^{-2} t_\perp^2 / t_\parallel^2$.

For the 1D Hubbard model, $0 < 4\alpha < \frac{1}{4}$ so that the power law behavior of $\sigma_\perp(\omega)$ is very weak—almost frequency independent. Within the Luttinger liquid hypothesis we expect this weak power law behavior to be applicable to the normal state interplane conductivity of the HTSC's. Indeed a very flat frequency dependence of the c-axis conductivity has been observed in $YBa_2Cu_3O_{7-\delta}$ [5] and $La_{2-x}Sr_xCuO_4$ [6]. In fact, a power law fit to the intermediate frequency data for $YBa_2Cu_3O_7$ yields a 2D Luttinger liquid exponent $\frac{1}{4} \lesssim 4\alpha \lesssim \frac{1}{3}$ which, while perhaps being a little larger than what one might expect, is not at all unreasonable. A similar fit to the electronic component [13] of $\sigma_c(\omega)$ in $La_{1.9}Sr_{0.1}CuO_4$ from [6] yields a similar exponent. Moreover, substitution of typical parameters $t_\perp, t_\parallel$, etc. for these HTSC's into the 2D generalization of (7) yields conductivities in reasonable *quantitative* agreement with those observed experimentally [14].

The expression (7) should be approximately valid over a range of intermediate frequencies $t_\perp \lesssim \omega \lesssim t_\parallel, U$. Our perturbative calculation cannot say anything precise about the low frequency region, $\omega \ll t_\perp$, but we expect on self-consistency grounds that $\sigma_\perp(\omega)$ would approach a small, but nonzero, value as $\omega \to 0$. At very high frequencies $\omega \gtrsim t_\parallel, U$ we expect on very general grounds that $\sigma_\perp(\omega)$ must eventually cross over to an ω^{-2} behavior.

We now turn our attention to thermal effects. Intuitively, one would expect that temperature will enhance the possibility of coherent hopping since it will cut off long time orthogonalities. This is indeed the case: evaluation of (4) at nonzero temperature yields the result

$$\frac{\Gamma(t)}{L} \sim \alpha \Lambda \left(\frac{t_\perp^2}{v_c \Lambda}\right)\left(\frac{T}{v_c \Lambda}\right)^{4\alpha} Tt + \text{incoherent.} \tag{8}$$

Note that this type of term is *not* affected by spin-charge separation, and can therefore be expected to persist to long times (i.e., $t \gtrsim t_\perp$). However, the coherent peak introduced into the conductivity $\sigma_\perp(\omega, T)$ by this linear-in-t term will generally still be broadened, so that it is a nontrivial matter to deduce even the dc conductivity $\sigma_\perp(\omega = 0, T)$ from (8). We believe that the correct interpretation of (8) is that it approximately prescribes the *total coherent weight* in the sense that

$$\int \sigma_\perp^{\text{coh}}(\omega, T)\, d\omega \sim \alpha \frac{e^2}{\hbar} ad \left(\frac{t_\perp^2}{v_c v_F}\right) T \left(\frac{T}{\Lambda v_c}\right)^{4\alpha}. \tag{9}$$

Thus, $\sigma_\perp^{\text{coh}}(\omega = 0, T)$ will be determined by the way this weight is broadened in frequency.

Finally, we remark that Eq. (8) is strictly valid only for times satisfying $Tt \lesssim 1$. Within perturbation theory we therefore require $T \lesssim t_\perp$. However, the complementary regime $T \gtrsim t_\perp$ can be dealt with within standard incoherence theory, since the in-liquid scattering rate $1/\tau_\parallel \sim T$. One finds in this case [4, 15] that $\rho_\perp(T) \sim (t_\perp/t_\parallel)^2 \rho_\parallel(T)$ so that $\rho_\perp(T)$ is *linear* in T. Such a high-temperature behavior is a common property of HTSC's. We emphasize, though, that it is completely misleading to use the term "metallic" to describe such a temperature dependence of the c-axis resistivity: it is the result of incoherent interplane transport, of the more standard type than that argued to take place within the confinement regime.

Our calculations should be most applicable to single-layer HTSC's such as $La_{2-x}Sr_xCuO_4$. In the multilayer materials the situation is complicated by the existence of two types of $t_\perp$: an intracell one, $t_\perp^{\text{intra}}$, coupling the "cells" of n closely spaced CuO_2 planes, and an intercell one, $t_\perp^{\text{inter}}$, which hops electrons between the cells. In the bilayers particularly, it is not inconceivable that intracell, interlayer superexchange terms are generated and may be responsible for the "spin-gap" signature in $\sigma_c(\omega)$ [16]. The point is that since the simple interliquid hopping term leads to only a very small conductivity at low frequencies (in particular, the zero temperature dc conductivity is very small, if not zero) the *observed* low frequency conductivity will be largely determined by whatever *other* interplane physics is involved in the real materials. Thus, it is not surprising that a universal behavior is not observed in the dc conductivity of the HTSC's. This is one of the few places where interplane chemistry or structure is important, but only insofar as it determines the *details* of σ_c, not the general fact of confinement.

From a pedagogical point of view it is worth remarking that there is some superficial similarity between the confinement phenomenon and the old Schrieffer theory of tunneling through an insulating barrier [17]. Indeed, Eq. (4) is strongly reminiscent of the Schrieffer tunneling formula. The crucial difference, however, is that in insulator barrier tunneling incoherence is introduced by the barrier (transverse momentum is not conserved in the tunneling process), while in Luttinger liquid tunneling the tunneling process itself generates the incoherence. We note, too, that one corollary of our theory is that the very theory of tunneling must be rethought with respect to c-axis tunneling into a HTSC above T_c. This is because the derivation [17, 18] of the Schrieffer tunneling formula assumes the existence of a good momentum quantum number in the direction normal to the interface, a property which does not exist in the confinement regime.

In summary, we have addressed the issue of determining the conductivity $\sigma_\perp(\omega, T)$ between Luttinger liquids in the confinement regime. Within a model of coupled 1D liquids we have argued that, for a large range of intermediate frequencies, $\sigma_\perp(\omega, T)$ follows a weak power law behavior as given in (7). We expect such a result to generalize to higher dimensions. Indeed, this weak power law behavior is both qualitatively and quantitatively consistent with experimental observations on $YBa_2Cu_3O_{7-\delta}$ [5] and $La_{2-x}Sr_xCuO_4$ [6]. The effect of nonzero temperature is to introduce a coherent component into $\sigma_\perp(\omega, T)$ with total weight $\propto (t_\perp/t_\parallel)^2 T^{1+4\alpha}$. For temperatures $T \gtrsim t_\perp$ the system enters the conventional incoherence regime in which the in-liquid scattering rate exceeds the interliquid hopping rate, and one finds $\rho_\perp(T)\rho_\parallel(T) \sim (t_\perp/t_\parallel)^2$.

Finally, we remark that in the light of these arguments and those presented in [9, 12] it would be of great interest to examine the conductivity along the weakest hopping direction in the quasi-1D and -2D organic conductors.

This work was supported by SERC Grant No. GR/H 73028 (D.G.C.) and NSF Grant No. DMR-9104873 (S.P.S. and P.W.A.).

REFERENCES

* Present address: Interdisciplinary Research Centre in Superconductivity, University of Cambridge, Cambridge CB3 0HE, United Kingdom.

† Present address: NEC Research Laboratories, Princeton, NJ 08540.

1. N. F. Mott and E. H. Davis, *Electronic Properties of Non-Crystalline Materials* (Taylor and Francis, London, 1975).

2. J. M. Wheatley, T. C. Hsu, and P. W. Anderson, Phys. Rev. B **37**, 5897 (1988); Nature (London) **333**, 121 (1988); P. W. Anderson et al., Physica (Amsterdam) **153–155C**, 527 (1988); T. C. Hsu and P. W. Anderson, Physica (Amsterdam) **162–164C**, 1445 (1989).

3. P. W. Anderson, Phys. Rev. Lett. **67**, 3844 (1991).

4. P. W. Anderson, "Princeton RVB Book" (unpublished).

5. S. L. Cooper, P. Nyhus, D. Reznik, M. V. Klein, W. C. Lee, D. M. Ginsberg, B. W. Veal, A. P. Paulikas, and B. Dabrovski, Phys. Rev. Lett. **70**, 1533 (1993).

6. K. Tamasaku, T. Ito, H. Takagi, and S. Uchida, Phys. Rev. Lett. **72**, 3088 (1994).

7. A Drude-like peak *is* observed in the nonsuperconducting *overdoped* samples of LaSrCuO [6]. This is not entirely unexpected, for these materials exhibit Fermi-liquid-like behavior in their in-plane properties. As pointed out in [6] there is a progression from incoherent to coherent c-axis transport as the Sr concentration is increased in LaSrCuO. We do not attempt to give a detailed account of the intermediate type of behavior seen, e.g., in $La_{1.8}Sr_{0.2}CuO_4$ [6] which, within our picture, would reflect the nontrivial crossover from incoherent to coherent c-axis transport due to proximity to a coherence-incoherence transition. It has also been argued [see, e.g., Schützmann et al., Phys. Rev. Lett. **73**, 176 (1994) and [5]] that fully oxygenated YBCO has a generalized Drude peak in $\sigma_c(\omega)$. However, apart from the fact that the width $\Gamma(\omega, T)$ is not Fermi-liquid-like, it is anomalously large: $\Gamma(0, 100 \text{ K}) \sim 1500 \text{ cm}^{-1}$. It is possible that $YBa_2Cu_3O_7$ is showing a hint of a crossover analogous to that observed in LaSrCuO but which, as YBCO does not have an overdoped regime, is not fully accessible experimentally. Nonetheless, the behavior of $\sigma_c(\omega)$ in $YBa_2Cu_3O_7$ is much closer to that of optimally doped than overdoped LaSrCuO, and it is not at all correct to consider the c-axis transport to be of conventional metallic type.

8. S. Chakravarty and P. W. Anderson, Phys. Rev. Lett. **72**, 3859 (1994).

9. D. G. Clarke, S. P. Strong, and P. W. Anderson, Phys. Rev. Lett. **72**, 3218 (1994).

10. V. Meden and K. Schönhammer, Phys. Rev. B **46**, 15753 (1992); J. Voit, Phys. Rev. B **47**, 6740 (1993); S. P. Strong (to be published).

11. A. A. Abrikosov, L. P. Gorkov, and I. E. Dzyaloshinski, *Methods of Quantum Field Theory in Statistical Physics* (Dover, New York, 1975).

12. S. P. Strong, D. G. Clarke, and P. W. Anderson, Phys. Rev. Lett. **73**, 1007 (1994).

13. S. Uchida (private communication).

14. For example, for $La_{1.9}Sr_{0.1}CuO_4$, taking $t_\perp \sim 0.05$ eV. $t_\parallel \sim 0.5$ eV, $U \sim 5$ eV gives $\sigma_c(0.1 \text{ eV}) \sim 5\text{--}10\Omega^{-1} \text{ cm}^{-1}$ which is consistent with [6] and well below the Mott-Ioffe-Regel limit.

15. N. Kumar and A. M. Jayannavar, Phys. Rev. B **45**, 5001 (1992).

16. C. C. Homes, T. Timusk, R. Liang, D. A. Bonn, and W. N. Hardy, Phys. Rev. Lett. **71**, 1645 (1993).

17. J. R. Schrieffer, Rev. Mod. Phys. **36**, 200 (1964), and references therein.

18. See especially J. Bardeen, Phys. Rev. Lett. **6**, 57 (1961); W. A. Harrison, Phys. Rev. **123**, 85 (1961); M. H. Cohen, L. M. Falicov, and J. C. Phillips, Phys. Rev. Lett. **8**, 316 (1962); R. E. Prange, Phys. Rev. **131**, 1083 (1963).

M

Infrared Conductivity of Cuprate Metals: Detailed Fit Using Luttinger Liquid Theory*

P. W. Anderson

(Received 21 October 1994; accepted 10 March 1995)

The infrared conductivity of the high-T_c cuprates in the normal state has a characteristic deviation from the normal "Drude" behavior of metals, which has sometimes been described as an additional, distinct "mid-infrared absorption" and sometimes as an extended tail of the low-frequency peak. Schlesinger,[1] some years ago, analyzed his data on the reflectivity of single crystals of $YBCO_7$ in terms of the conventional expression

$$\sigma = \frac{ne^2}{m(i\omega + \frac{1}{\tau})} \tag{1}$$

with frequency-dependent parameters $m(\omega)$ and $1/\tau(\omega)$, which showed remarkably simple behavior (see fig. 1): $1/\tau$ is proportional to ω, and m has a slow, approximately logarithmic variation. There is in fact little difference in the data among actual experiments, as opposed to interpretations, on good materials, so we may take fig. 1 as typical of optimally doped cuprates, since it is in essence a heuristic description of the data.

N. Bontemps and collaborators[2] have used a similar plot to describe data over a wide range of frequencies, up to around 8000 cm^{-1}, using transmission data on films of a number of cuprates, most but not all closely related to YBCO (see figs. 2 and 3). With this wide frequency range the family resemblance of all of the data becomes striking, particularly plotted using Schlesinger's parameters. I believe that there would be little disagreement as to the general characteristics of the actual data among these and other experimentalists, except that less highly

*This work was supported by the NSF, Grant # DMR-9104873.

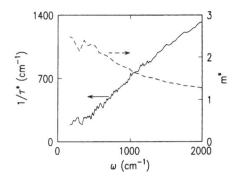

Figure M.1. The scattering rate $1/\tau^*$ (solid curve) and the effective mass m^* (dashed curve) extracted from the measured â-axis conductivity at $T = 100$ K [Fig. 3a] are shown. The broadly increasing scattering rate reflects the fact that the conductivity drops unusually slowly with increasing ω in the normal state. Both the linearity of the scattering rate vs ω and the large magnitude of the inelastic rate $(1/\tau^* \sim \omega)$ are highly unconventional.

doped YBCO samples show "spin gap" deviations at the lower end of the range $(\leq 500\text{cm}^{-1})$.[3]

We describe a detailed fit to the data of figs. 2a, b using the "Luttinger liquid" hypothesis for the electronic state of the 2D normal metal.[4] This result depends only on rather general properties of the theory but is totally dependent on the non-Fermi liquid nature of the spectrum.

The basis of the fit is the remark that the cuprates are in the "holon non-drag regime" of Luttinger liquid transport theory.[5] This is the regime where charge excitations ("holons") are scattered sufficiently rapidly that they do not recohere with the spinons after the accelerated electron decays into charge and spin excitations. (The condition for this regime is $\omega > (1/\tau)_{\text{holon}} >\sim \frac{\omega^2}{E_F}$; the source of $(\frac{1}{\tau_{\text{holon}}})$ is probably impurity scattering at small ω and phonons at large ω.)

Under these circumstances vertex corrections are damped out by holon scattering [4] and the conductivity is given by the simple one-loop diagram (fig. 3)

$$\sigma(\omega) \propto \frac{1}{\omega} \int dx \int dt\, G^e(x, t) G_h(x, t). \qquad (2)$$

That is, physical process which controls the rate of entropy production is the decay of the electron and hole into spin and charge excitations, but it is enabled by the fact that the charge is then scattered by the lattice. The process is analogous to phonon scattering in the phonon non-drag regime, where the momentum decay occurs by the scattering of the phonons by the lattice which prevents

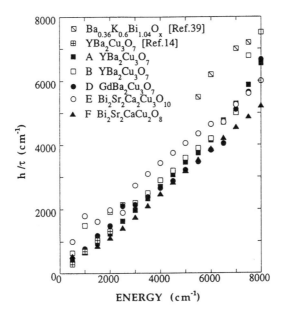

Figure M.2a. Relaxation rate $1/\tau$ (or inverse lifetime) of the A, B, D, E, and F films computed from the dielectric functions shown in Fig. 4. The data referring to the a axis in an untwinned single crystal and to $Ba_{1-x}K_xBiO_3$ film (Ref. 10) are shown for comparison. For recent data see Ref. (11).

phonon drag, while the entropy production is caused by phonon emission and controls the observed resistivity.

We can evaluate (2) very simply using the fact that $G_1(x, t)$ is a homogeneous function of (x, t) considered as a single variable. This is the consequence of the fact that all excitations have a finite Fermi velocity. For the Fermi liquid,

$$G_{FL} \propto \frac{e^{ik_F x}}{x - v_F t} \tag{3}$$

homogeneous of order (-1), while for the 1D Luttinger liquid,

$$G_{LL} \propto \frac{e^{ik_F x}}{\sqrt{(x - v_s t)(x - v_c t)}(x^2 - v_c^2 t^2)^{\frac{\alpha}{2}}}, \tag{4}$$

which is homogeneous of order $(-1 - \alpha)$. For the 2D liquid G is an average of an expression like (3) or (4) over the Fermi surface. For the Fermi liquid, the relevant G in momentum and frequency space may be approximated by

$$G(p, \omega) \simeq \frac{1}{\hbar\omega - (p - p_F)\vec{v}_F}$$

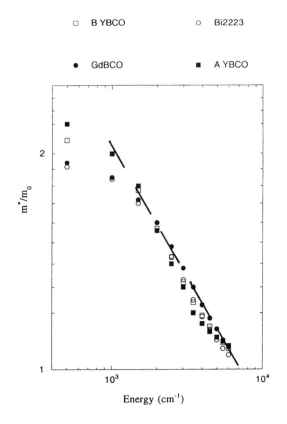

Figure M.2b

where p_F, v_F are at the projection of $\vec{p}$ on the Fermi surface along $\vec{v}_F$, p assumed close to the Fermi surface. A similar construction for the Luttinger liquid will give a pair of variables $\Delta p = p - p_F$, ω, in which the Green's function will again be homogeneous of order $-(1 - \alpha)$, but this function has no simple formal expression. By a simple scaling argument, we find

$$\sigma(\omega) = \frac{\text{const}}{(i\omega)^{1-2\alpha}}. \tag{5}$$

(5) holds up to an upper frequency cutoff $\Omega = \wedge/\hbar$ of the order of the electron band width $\wedge$. The sum rule on conductivity will be satisfied if the coefficient in (5) is set so that

$$\sigma(\omega) = \frac{ne^2}{i\omega m_0} \left(\frac{i\omega}{\Omega} \right)^{2\alpha} \frac{2\alpha}{\sin \pi \alpha}. \tag{6}$$

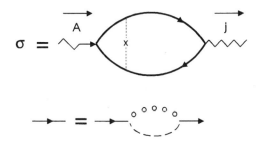

$$\circ \circ \circ = \text{HOLON}$$
$$- - - = \text{SPINON}$$

Figure M.3

Here m_0 is the sum rule mass,

$$\int \sigma(\omega) d\omega = \frac{ne^2}{m_0}$$

which should not be far from the band mass: (6) contains all intraband mass renormalization effects.

(6) contains only two free parameters, n/m_0 and α (the upper cutoff Ω merely scales m_0 and is not independent.) Neither can vary much: m_0 must not be much bigger than the band mass, and c-axis Hall data[6] among others tell us that n is the conventional band filling $\propto 1 - \delta$. α for the 1D Hubbard model is $\leq 1/8$, but models with $\alpha > 1/8$ exist. There is no fundamental theory of α in 2D. Vague indications from gauge theory[7] suggest $1/6(2\alpha = 1/3)$, while the tomographic picture[8] might suggest agreement with 1D.

The data give two independent measures of 2α, one from the slope of $1/\tau$ vs. ω and one from the dependence of m on ω. These two numbers are unrelated in the "marginal Fermi liquid" theory,[9] and their agreement argues against that theory, as well as does the relatively large value of 2α we find. The slope of $1/\tau$ (dashed line in fig. 2a) is $\sim .65 \pm .1$ which gives $\alpha \simeq .18 \pm .2$. The median slope is used for the dashed line in fig. 2b, which as you can see is an adequate fit, although the power-law form is not much constrained by the data. On the other hand, the analytic properties of σ require that if it has constant phase angle (as fig. 2a shows) it must be a power of $(i\omega)$.

Let us summarize the achievements of the Luttinger liquid hypothesis, coupled with the concept of the holon-non-drag regime. (A second way to think of this regime is as one in which the transport is by spinons[5] and is relaxed by holon emission and reabsorption.) The original motivation which was satisfied by this idea was to explain the absence of phonon scattering effects or, in most cases, of residual impurity scattering, both of which should be large in most of these materials. Let it be explicit that the separation of charge and spin, though it fails to appear in the formal expression (5) or (6), which depends only on the "Fermi surface" exponent, α, is essential to the entire theory.

Now we see that the theory leads to a unique scaling form for the conductivity which holds over almost two decades of frequency and for a number of cuprates. Particularly important, in my view, is the fact that the expression scales from $> 5000 \mathrm{cm}^{-1}$ to $< 500 \mathrm{cm}^{-1}$, a property which no alternative theory motivates in any natural way.

It is interesting that other groups (especially Bozovic[10]) see indications of similar behavior in the mid-infrared conductivity of a number of other materials, mostly those with other symptoms of strong correlation phenomena. With considerable caution because of the existence of other transport regimes, we would consider a Luttinger liquid explanation for some of these cases.

I would like to acknowledge especially extensive discussions with Nicole Bontemps, as well as the use of her data. I also was stimulated by discussions with R. Laughlin and E. Abrahams.

REFERENCES

1. Z. Schlesinger and R. Collins, Phys. Rev. Letters **65**, 801 (1990).
2. A. E. Azrak, R. Nahoon, A. C. Boccara, N. Bontemps, M. Guilloux-Viry, C. Thiret, A. Perrin, Z. Z. Li, and H. Raffy, Journal of Alloys and Compounds, **195**, 663 (1993).
3. G. A. Thomas et al. Phys. Rev. **B42**, 6342 (1990).
4. P. W. Anderson, chapter 6 of this book; P. W. Anderson, Proceedings of the "Materials and Mechanisms of Superconductivity, High Temperature Superconductors III," Kanazawa, July 22–26, 1991, Part I (eds. M. Tachiki, Y. Muto, and Y. Syono), (North-Holland, Amsterdam, 1991) p. 11, reprinted from Physica C **185–89**, 11–16 (1991).
5. M. Ogata and P. W. Anderson, Phys. Rev. Letters **70**, 3087 (1993).
6. J. M. Harris, Y. F. Yang, and N. Y. Ong, Phys. Rev. **B46**, 14293 (1992).
7. Z. Zou, private communication.
8. P. W. Anderson, chapter 6 of this book; P. W. Anderson and Y. Ren, The Normal State of High-T_c Superconductors: A New Quantum Liquid, Proceedings of the International Conference on The Physics of Highly Correlated Electron Systems, Los Alamos, **High Temperature Superconductivity Proceedings**, K. Bedell et al. eds. (Addison-Wesley 1990), pp. 3–33.
9. P. Littlewood and C. M. Varma. Journal of Applied Physics **69**, 1979 (1991).
10. J. H. Kim, I. Bozovic, J. S. Harris Jr., W. Y. Lee, C. B. Eoun, T. H. Geballe, E. S. Bellman, Physica **C185-189**, 1019 (1992); Bozovic, Harris et al., Phys. Rev. Letters **73**, 1436 (1993).
11. C. Baraducet et al., J. of Superconductivity **8**, 1 (1995).

N

Interlayer Tunneling Mechanism for High-T_c Superconductivity: Comparison with c-Axis Infrared Experiments

P. W. Anderson

ABSTRACT

Recent c-axis-polarized infrared measurements in the high-transition temperature (high-T_c) cuprate superconductor (La,Sr)$_2$CuO$_4$ can be interpreted on the basis that the entire condensation energy comes from the interlayer Josephson coupling. This gives a parameter-free determination of penetration depth λ and coherence length ξ for this superconductor that are in agreement with experiments.

Accurate measurements have been made of the far-infrared spectrum of $(La_{1.85}\text{-}Sr_{0.15})CuO_4$ for light polarized along the c axis (perpendicular to the planes (1). These data have recently been confirmed and even more carefully quantified (2, 3), and the frequency range has been extended (4). The original measurements (fig. 1) contain two anomalies relative to a usual (Fermi liquid) metal with a Bardeen-Cooper-Schriffer (BCS)–type attractive interaction.

1) The first anomaly is that in a conventional metal, the plasma edge observed near 50 cm^{-1}, below T_c, is present both in the normal and in the superconducting metal, reflecting the sum rule on conductivity σ over frequencies ω

$$\int \sigma(\omega)d\omega = \text{constant} \tag{1}$$

which can be satisfied in detail in conventional BCS superconductors if the conductivity in the gap region is moved into a δ function at zero frequency, leaving the dielectric constant not much changed for frequencies near and above the energy gap because of the Kramers-Kronig relation between the two. The relaxation rate of the carriers in the normal metal is known from the resistivity in the ab plane (perpendicular to the c-axis) to be

$$\frac{\hbar}{\tau} \simeq \max(kT, \hbar\omega) \tag{2}$$

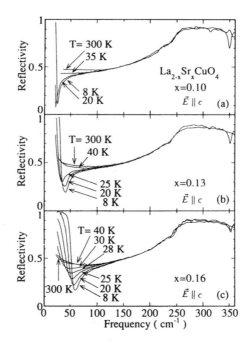

Figure N.1. Infrared reflectivity spectra of $La_{2-x}Sr_xCuO_4$ with polarization perpendicular to the CuO_2 planes (E c) above and below T_c for (a) $x = 0.10$ ($T_c = 27$ K), (b) $x = 0.13$ ($T_c = 32$ K), and (c) $x = 0.16$ ($T_c = 34$ K). [Adapted from (1), copyright 1992, American Institute of Physics.]

(where k is the Boltzmann constant, T is temperature, $\hbar$ is Planck's constant h divided by 2π, and τ is the relaxation time) which should smear but not remove this plasma edge in the normal state. In short, the reflectivity in the normal state should be 1 below the plasma frequency ω_p, not $\sim 1/2$. What is observed above T_c looks like a lossy insulator with a dielectric constant ϵ of about 20. In the original paper (1), Tanasaki et al. stated that the sum rule can be satisfied by the insertion of a "Drude" peak of conductivity of width ~ 170 cm^{-1}, of which there is no explicit evidence: It is simply the maximum size the experiments would allow. Later, it became clear (2–4) that the conductivity in the c direction $\sigma_c(\omega)$ extends over a much wider range and can hardly be called a "Drude" peak because its width is orders of magnitude larger than the observed relaxation rate. Timusk (2), in particular, showed that not enough σ_c exists in the low-frequency range to satisfy the sum rule. But in a sense, it is the absence of a Drude metallic conductivity that is the first anomaly, not primarily the details of spectral weight transfer, which are quite complex theoretically as well as experimentally.

2) The second anomaly is that the observed ω_p is much lower than expected, given the estimates of c-axis masses (or hopping matrix elements) obtained from band calculations, either local-density approximation (LDA) or tight-binding, and other experimental phenomena. The c direction bandwidth should be of order 500 to 1000 cm^{-1}, which would make ω_p an order of magnitude too high, at least in the range from 500 to 1000 cm^{-1}. The value of ω_p in the ab plane is around 10^4 cm^{-1} and ω_p as a function of direction should be simply proportional to the relative bandwidth (which is the hopping integral for an anisotropic Fermi liquid). The rough formula is

$$(\hbar\omega_p)^2 \simeq E_F \frac{e^2}{\epsilon_0 r_s} \tag{3}$$

where E_F is the Fermi energy, e is the electron charge, ϵ_0 is the dielectric constant, and r_s is the screening radius.

Both of these anomalies are caused by the "confinement"–interlayer tunneling (5) mechanism, and they allow a quantitative, parameter-free check on the accuracy of that theory. The principle of the theory is that, when the normal metal is a Luttinger liquid, coherent single-particle tunneling between layers is blocked (6), and the sum rule for conductivity in the c axis direction is spread out over a very wide frequency range, rather than having a Drude peak below $\hbar\omega \simeq kT$. At $T = 0$ and $\omega = 0$, $\sigma(\omega)$ would be literally zero in the normal state. On the other hand, Josephson-like pair tunneling is allowable between two Luttinger liquid layers if each layer has self-consistently developed an anomalous self-energy or "gap" Δ. Therefore, the Josephson energy

$$E_J = -\frac{\hbar J_J}{2e} \cos\varphi \tag{4}$$

(where J_J is the Josephson critical current and φ is the relative phase between layers) is the fundamental interaction that causes high-T_c superconductivity. (The sum-rule argument shows that for normal BCS superconductors, this interaction is not effective in causing condensation because it exactly cancels between normal and superconducting states.) The attractive energy for this interaction comes from a net reduction in kinetic energy of localization along the c axis.

On the other hand, one cannot expect the relatively small energy gap Δ to restore the whole of the missing coherent electron motion, or equivalently to restore all of the kinetic energy lost by confinement. The ac conductivity is spread over the entire bandwidth, and the high-energy states involved are not much affected by Δ. To estimate how much Josephson conductivity is restored is, in fact, trivial. The free energy of condensation will be roughly proportional to the cosine of the relative phase between layers and will vanish when the phase angle φ is 90°. At this point, we assume an interlayer mechanism: For any mechanism involving coupling pairs within the planes, the entire condensation energy does not depend crucially on φ. The current is $2e$ times the derivative of this energy with phase; hence, the superconducting binding energy is given by

$$F_{\text{cond}} = \frac{\hbar J_J}{e} \cos \varphi \tag{5}$$

(accounting for interactions with two neighboring layers). The "Josephson" plasma frequency (7), which in this case is the plasma frequency itself, is given by

$$(\hbar \omega_J)^2 = \frac{\hbar J_J}{e} \frac{2e^2}{\epsilon c} = \frac{2e\hbar}{\epsilon d} J_J \tag{6}$$

(ϵ is ~ 20, and d is the c-axis lattice constant). From the value 50 cm^{-1} for ω_J, $\hbar J_J / 2e \simeq 4/3$ K. This is then equated to the value of F_{cond} per unit cell. In any reasonable theory, and in fact in that of Chakravarty and Anderson (8)

$$F_{\text{cond}} \simeq \frac{1}{2} N(0) \Delta^2 \tag{7}$$

simply by energy-entropy balance. With a density of states at $\omega = 0$ of $N(0) \simeq 1/t_\| \simeq (1 \text{ eV})^{-1}$ (where $t_\|$ is the in-plane element of the kinetic energy matrix **t**), this gives

$$\Delta \simeq 100 \text{ K} \tag{8}$$

which is correct to within a factor 2. That is, the Josephson and condensation energies are the same, at least within the factor 2 accuracy of these estimates.

The anisotropy of the penetration depth will, in these materials, be greater than expected according to band theory and will give too large a value of the band mass relative to band calculations. The constant corresponding to n/m in the expression

$$\frac{1}{\lambda^2} = \frac{4\pi n e^2}{m c^2} \tag{9}$$

which is proportional to $t_\perp^2/t_\|$ in conventional theory ($t_\perp$ is the out-of-plane element), will be more like $t_\perp^4/t_\|^3$. Here I have observed that both the coherence length ξ and the penetration depth λ, the two parameters of Ginsberg-Landau theory, are uniquely determined in the c direction by the assumption of interlayer coupling as the mechanism for T_c if the material has only one type of layer. The coherence length (from the assumption in Eq. 5) is simply d (actually $d/\sqrt{2}$ or ~ 4 Å), and the penetration depth c/ω_J is determined from Eq. 6 [$\sim 2 \times 10^{-2}$ cm, in (LaSr)$_2$CuO$_4$]. The coherence length is in fair agreement with direct experiments (9).

I can see no alternative explanation of these very striking experimental data. Other single-layer materials, especially the Tl and Hg compounds, should be studied in the same way. Multilayers such as YBCO (yttrium barium copper oxide) will have much more complex infrared spectra, which nonetheless would be of interest to probe carefully. In these cases, there should be two Josephson-type frequencies, corresponding to the two different interlayer matrix elements $t_{\perp>}$ and $t_{\perp<}$, with the smaller one corresponding to the penetration depth and coherence length (because phase stiffness adds by inverses) and the larger one to T_c or Δ, because T_c adds directly [evidence of such a second frequency is suggested in some experiments (2)]. By this reasoning, the coherence length should satisfy

$$\xi \simeq \frac{d\,t_{\perp<}^2}{(t_{\perp>}^2 + t_{\perp<}^2)} \tag{10}$$

where $t_{\perp<}^2$ and $t_{\perp>}^2$ are, respectively, the larger and the smaller of the interlayer tunneling coupling constants. This result is in accord with the existing observations. Multilayers will have a corresponding increase in λ_c (penetration depth for current parallel to the c direction) as well.

REFERENCES AND NOTES

1. K. Tanasaki, Y. Nakamura, S. Uchida, *Phys. Rev. Lett.* **69**, 1455 (1992).
2. T. Timusk, personal communication.
3. J. H. Kim et al., in preparation.
4. K. Tamasaku, T. Ito, H. Takagi, S. Uchida, *Phys. Rev. Lett.* **72**, 3088 (1994).
5. J. M. Wheatley, T. C. Hsu, P. W. Anderson, *Nature* **333**, 121 (1988); P. W. Anderson et al., *Physica C* **153–155**, 527 (1988); P. W. Anderson, in *Strong Correlation and Superconductivity*, H. Fukuyama, S. Maekawa, A. P. Malozemoff, eds. (Springer-Verlag, Berlin, 1989), p. 2; P. W. Anderson and Y. Ren, in *High Temperature Superconductivity Proceedings*, K. Bedell et al., eds. (Addison-Wesley, Reading, MA, 1990), pp. 3–33.
6. D. G. Clarke, S. P. Strong, P. W. Anderson, *Phys. Rev. Lett.* **72**, 3218 (1994).
7. P. W. Anderson, *Weak Superconductivity: The Josephson Tunneling Effect* (5th International School of Physics, Revallo, Italy, 1963), E. R. Caianello, ed. (Academic Press, New York, 1964), vol. 2, p. 113. The term "Josephson plasma frequency" was first mentioned in this paper.
8. S. Chakravarty and P. W. Anderson, *Phys. Rev. Lett.* **72**, 3859 (1994).
9. M. Suzuki and M. Hikita, *Phys. Rev. B* **44**, 249 (1991).
10. I acknowledge communication of results by T. Timusk, as well as a discussion, and particularly extensive discussions with D. Van der Marel and T. M. Rice. I also was stimulated by queries from R. Laughlin. This work was supported by NSF grants DMR-9104873 and DMR-9224077.

O

Coexistence of Single Particle Confinement and Singlet Pair Off-Diagonal Long Range Order in Spin Chains

David G. Clarke

Interdisciplinary Research Centre in Superconductivity and Cavendish Laboratory,
Madingley Road, Cambridge, CB3 0HE, United Kingdom.
(4 November 1994)

ABSTRACT

On the basis of a model of impurity spins coupling to a Heisenberg chain, we present an argument for the existence of an off-diagonal long range order for singlet pair insertion into a Heisenberg chain. In contrast, single spin insertion leads to an orthogonality catastrophe. A simple model is presented in which incoherent, or at least strongly damped, tunneling of single spins coexists with coherent singlet pair tunneling. The implications for coupled Hubbard chains is briefly discussed.

The interplanar tunneling theory of high-temperature superconductivity has as its central mechanism the coexistence of two interplanar tunneling properties: (a) incoherent, or at least strongly damped, *single particle* tunneling, and (b) coherent *pair* tunneling. These properties were originally proposed on the basis of experimental observations, perhaps the clearest being the anomalously large anisotropy in normal state transport properties, and the large variance in T_c between materials with an essentially identical cuprate planar structure.

The heuristics of a scenario leading to (a) and (b) was set out some time ago by Anderson, and has been the subject of renewed investigations recently [1,2]. The basic idea involved spin-charge separation: only real electrons may tunnel between planes, but if the fundamental excitations of the 2D electron fluid in the cuprate planes are those of spinons and holons, which separately carry spin and charge, respectively, then single electron tunneling will require a "reconstitution" of the electron which will be difficult if the spinons and holons propagate at different velocities. In contrast, due to the existence of an off-diagonal long range order associated with the insertion of a pair of spinons into the 2D liquid, it was argued that pair tunneling would not be suppressed. Loosely, one can say that in this scenario the spinons are "pre-paired" and as a result pair tunneling, which requires a pair of spinons and a pair of holons, will not suffer from the effects of spin-charge separation in the way that single particle tunneling does.

In the interests of pedagogy, it is worthwhile searching for simple models which exhibit (a) and (b) above. We present here such a model involving impurity spins coupled to a Heisenberg chain. The complete model will be presented later. Let us begin by considering a spin-ζ chain with lattice spacing a. The sites of the spins in the chain are labelled $(k + 1/2)a$, for k an integer. Impurity spins are coupled into the chain by insertion at sites ma, m an integer. This model with a single impurity spin was studied in [3]. For a *pair* of impurity spins $\mathbf{S}(0)$, $\mathbf{S}(l)$ separated by l, the Hamiltonian is

$$H_2 = J \sum_{n=k+1/2} \mathbf{s}_n \cdot \mathbf{s}_{n+1} + g\{\mathbf{S}(0) \cdot [\mathbf{s}_{-1/2} + \mathbf{s}_{1/2}] + \mathbf{S}(l) \cdot [\mathbf{s}_{l-1/2} + \mathbf{s}_{l+1/2}]\}. \quad (1)$$

Each impurity spin couples only to the local magnetization. For simplicity we shall consider $S = \zeta = 1/2$, though most of our conclusions would be valid for any half-integer spins provided $S < 2\zeta$. We shall retain ζ as a parameter, to enable us to consistently invoke a spin-wave expansion. A mapping to a non-linear sigma model (NLσM) can be made in the usual way [4] and we find

$$H_2 = \frac{1}{2} \int dx \{J a \mathbf{m}^2(x) + 4a\zeta^2 J (\partial \Omega(x))^2 + 2Ja\zeta \mathbf{m}(x) \cdot \partial \Omega(x)\}$$
$$+ ga\mathbf{S}(0) \cdot \mathbf{m}(0) + ga\mathbf{S}(l) \cdot [\mathbf{m}(l) + (1 + (-1)^{(l+1)/a})\zeta \partial \Omega(l)]. \quad (2)$$

The third term under the integral is responsible for the so-called topological term in half-integer spin chains. This term ensures that the excitation spectrum of the spin chain alone is gapless. Of great importance in (2) is the fact that the coupling of the chain to the impurity spins depends upon whether l/a is even or odd, i.e., whether there is an even or odd number of chain spins between the impurities. We shall see later how this affects the physics and why.

We now proceed to make a spin wave (SW) approximation [5]: $\Omega_z \rightarrow 1$, $m_z \rightarrow -(\mathbf{m}_\perp \cdot \Omega_\perp)$, where $\mathbf{O}_\perp$ denotes the component of $\mathbf{O}$ transverse to $\langle \Omega \rangle \equiv \hat{z}$. Discarding the couplings of the impurity spins to m_z, which are formally $O(\zeta^{-1/2})$ smaller than those to $\mathbf{m}_\perp$ and to $\zeta \partial \Omega_\perp$, and making a canonical transformation to $\tilde{\mathbf{m}}_\perp(x) = \mathbf{m}_\perp(x) + \zeta \partial \Omega_\perp(x)$ we obtain

$$H_2^{SW} = \frac{1}{2} \int dx \left\{ \frac{1}{\chi} \mathbf{m}_\perp^2(x) + \rho(\partial \Omega_\perp(x))^2 \right\}$$
$$+ ga\mathbf{S}_\perp(0) \cdot [\mathbf{m}_\perp(0) - \zeta \partial \Omega_\perp(0)] + ga\mathbf{S}_\perp(l)$$
$$\cdot [\mathbf{m}_\perp(l) + (-1)^{(l+1)/a} \zeta \partial \Omega_\perp(l)] \quad (3)$$

where (Ω_x, m_y), $(-\Omega_y, m_x)$ are canonically conjugate pairs,

$$[\tilde{m}_x(x), \Omega_y(x')] = -[\tilde{m}_y(x), \Omega_x(x')] = i\delta(x - x')$$

all other commutators vanishing.

To set the stage, let us first discuss the case of a *single* impurity spin. Consider insertion of a single impurity spin up at $x - 0$ at time $t = 0$. The Hamiltonian governing this problem is then

$$H_1 = J \sum_{n=k+1/2} \mathbf{s}_n \cdot \mathbf{s}_{n+1} + g\mathbf{S}(0) \cdot [\mathbf{s}_{-1/2} + \mathbf{s}_{1/2}] \tag{4}$$

which has SW approximation

$$H_1^{SW} = \frac{1}{2} \int dx \left\{ \frac{1}{\chi} \mathbf{m}_\perp^2(x) + \rho(\partial\,\Omega_\perp(x))^2 \right\}$$

$$+ ga\mathbf{S}_\perp(0) \cdot [\mathbf{m}_\perp(0) - \zeta\partial\Omega_\perp(0)] \tag{5}$$

and the Green's function appropriate to describe the process is just

$$\mathcal{G}_1(t) = \langle \uparrow | e^{-iHt} | \uparrow \rangle$$

where $|\uparrow\rangle$ denotes the $g = 0$ ground state of the system with $2S_z(0)|\uparrow\rangle = |\uparrow\rangle$. We shall evaluate $\mathcal{G}_1(t)$ to $O(g^2)$. Alternatively, as was done in [3], one could calculate the impurity spin susceptibility perturbatively in g. The central question is the same: when coupled to the chain does the impurity spin continue to behave like a free spin or not?

A direct calculation of $\mathcal{G}_1(t)$ using (5) gives

$$\mathcal{G}_1(t) = 1 - (ga)^2 \left(\chi + \frac{\zeta^2}{\rho} \right) \int_0^\Lambda \frac{dk}{2\pi} \left\{ -it + \frac{1}{ck}(1 - e^{-ickt}) \right\}$$

$$\sim 1 - i\Delta Et - \frac{(ga)^2}{2\pi} \frac{1}{c} \left(\chi + \frac{\zeta^2}{\rho} \right) \log(it/\tau_\Lambda) \tag{6}$$

where $\Delta E = -g^2\frac{a}{2\pi}(\chi + \zeta^2/\rho)$ represents an energy shift, $\Lambda \sim a^{-1}$ is a momentum cutoff, and (6) is valid for $t \gg \tau_\Lambda \sim (\Lambda c)^{-1}$. The long time behavior of $\mathcal{G}_1(t)$ clearly betrays a breakdown of perturbation theory—it is not possible to adiabatically continue to the two ground states of H_1^{SW} from the direct product of the spin chain ground state and the $S_z(0) = \pm 1/2$ free spin states.

However, we can go beyond the SW approximation and perturbation theory in g by bosonization of H_1. It is then possible to show [3] that the behavior of the impurity spin is equivalent to that of the Kondo spin in the two channel Kondo model. In particular, the zero temperature impurity spin susceptibility is $\chi_{imp}(\omega) \sim \frac{1}{T_*} \log(T_*/\omega)$, where $T_* \sim J e^{-\text{const } J/g}$ in weak coupling. Such a non-Curie susceptibility implies an orthogonality of the two ground states $|O_1\rangle$ and $|O_2\rangle$ of H_1 in the sense that they are not connected by simply flipping the impurity spin, i.e., $\langle O_1|S^\pm(0)|O_2\rangle = 0$. It then follows that the $g = 0$ and $g \neq 0$ ground states are also orthogonal

$$\langle O_{1,2}| \uparrow \rangle = \langle O_{1,2}| \downarrow \rangle = 0. \tag{7}$$

It is this orthogonality which is being signalled by the long time divergence of $\mathcal{G}_1(t)$ at $O(g^2)$ in the SW approximation. This is analogous to the X-ray edge problem where the orthogonality of the Fermi seas in the presence/absence of a deep hole is signalled at lowest order in perturbation theory by a logarithmic divergence in time of the deep hole propagator.

What is the nature of the ground states $|O_1\rangle$ and $|O_2\rangle$? This was discussed in [3]. On the basis of the behavior of $\langle \Omega_z(x)\Omega_z(-x)\rangle$ for $g = 0$ and $g \neq 0$, the fact that the susceptibility of a spin in a spin-1/2 chain is indeed logarithmic, and the study of a similar model in [7], we argued that the ground state is that of a spin chain with one additional spin. The impurity spin is "pulled into" the chain. This is illustrated schematically in fig. 1—a kink is introduced and all spins to one side of the impurity are flipped. It is this long-range effect of the impurity spin on the chain which is the reason for the various orthogonalities. Were the impurity to disturb only the spins in its local vicinity there would not be any such orthogonalities.

With our understanding of the single impurity model we now consider the two impurity model. Via bosonization, one can show that the two impurity model is equivalent to the *two-impurity* two-channel Kondo model [8], restricted to one dimension. Rather than attack the latter model directly, we shall simply extend the above pedestrian method. Consider a *singlet* pair of impurity spins inserted at $x = 0, l$ at time $t = 0$ and coupling to the spin chain according to (1). We calculate

$$\mathcal{G}_2(l/a, t) = \frac{1}{2}\langle \uparrow_0\downarrow_l - \downarrow_0\uparrow_l \,|e^{-iHt}|\, \uparrow_0\downarrow_l - \downarrow_0\uparrow_l \rangle$$

in obvious notation. To $O(g^2)$ and within the SW approximation (i.e., using (3)) we obtain

$$\mathcal{G}_2(l/a, t) = 1 - (ga)^2\left(\chi + \frac{\zeta^2}{\rho}\right)\int_0^\Lambda \frac{dk}{2\pi}(1 - \cos kl)\left\{-it + \frac{1}{ck}(1 - e^{-ickt})\right\}$$

$$+ (ga)^2\frac{\zeta^2}{\rho}\int_0^\Lambda \frac{dk}{2\pi}[1 + (-1)^{(l+1)/a}]\cos kl\left\{-it + \frac{1}{ck}(1 - e^{-ickt})\right\}. \tag{8}$$

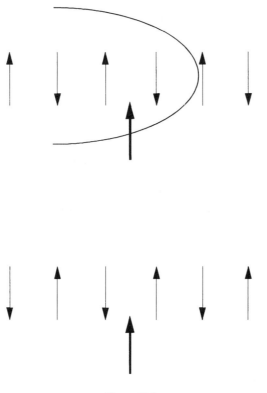

Figure O.1

At this order $\mathcal{G}_2(l/a, t) - 1$ is a sum of two types of terms. The first type are those where an impurity spin flips, emitting a magnon which is later reabsorbed by the *same* impurity. These are the same processes which give rise to the $\log(t)$ divergence in $\mathcal{G}_1(t)$. The second type of contribution is from processes where a magnon is exchanged between the impurity spins, i.e., one impurity spin flips, emitting a magnon which is later absorbed by the *other* impurity spin, which therefore also undergoes a spin flip. There is a relative minus sign between these two types of processes, from the nature of the singlet wavefunction, and some spatial factors $e^{\pm ikl}$. Otherwise the two terms are identical, and as a result the first term above in $\mathcal{G}_2(l/a, t) - 1$ is the same as that in $\mathcal{G}_1(t) - 1$ *except* for the extra factor $(1 - \cos kl)$. This factor renders the long time behavior harmless and one finds, for $l \ll ct$ and $l/a = 2n$ *even*,

$$\mathcal{G}_2(2n, t) = 1 - \frac{(ga)^2}{2\pi} \frac{1}{c} \left(\chi + \frac{\zeta^2}{\rho} \right) \left\{ 2\log(l\Lambda) + \frac{1}{4} \left(\frac{l}{ct} \right)^2 \right\}. \qquad (9)$$

We therefore find that, to $O(g^2)$ (and in contrast to the behavior of $\mathcal{G}_1(t)$) $\mathcal{G}_2(2n, t)$ is well behaved at long times. If this were to remain true to all orders in g the result would be a non-zero overlap between the $g = 0$ and $g \neq 0$ ground states of H_2^{SW}. At $O(g/J)^{2n}$ the leading log contribution to $\mathcal{G}_1(t)$ is $\sim (g/J)^{2n} \log^n(t/\tau_\Lambda)$. However, it is simple to show that, in the expansion of $\mathcal{G}_2(2n, t)$ in powers of $(g/J)^2$ the coefficients of all such leading log terms are zero. This is strongly supportive of the conjecture that there is a non-zero overlap between the $g = 0$ and $g \neq 0$ groundstates of H_2^{SW} and, by extension, of H_2. In fact, this conjecture has recently been verified numerically by Talstra, Strong, and Anderson (TSA) for Heisenberg, Inverse-Square-Exchange and XY spin chains [9]. While the study of TSA involves insertion of a single pair of spins into a chain of $N - 2$ spins in a slightly different manner to that given by (1), we believe, for reasons outlined below, that the physics is the same.

In stark contrast to the case of l/a even, the expansion for $\mathcal{G}_2(l/a, t)$ given by (8) has a logarithmic divergence at $O(g^2)$ when l/a is *odd*. Thus we expect the behavior of $\mathcal{G}_2(2n + 1, t)$ to be similar to $\mathcal{G}_1(t)$ and in particular to exhibit ground state orthogonalities.

The various cases can be summarized simply in spectral function language. If $G_1(\omega)$, $G_2(l/a, \omega)$ denote the Fourier transforms of $\mathcal{G}_1(t)$ and $\mathcal{G}_2(l/a, t)$, respectively, and we write

$$\frac{1}{\pi} Im G_1(\omega) = Z_1 \delta(\omega) + incoherent$$

$$\frac{1}{\pi} Im G_2(l/a, \omega) = Z_2(l/a)\delta(\omega) + incoherent$$

then $Z_1 = 0$, $Z_2(2n + 1) = 0$, but $Z_2(2n) \neq 0$. This entails the existence of an off-diagonal long range order for singlet pair insertion when l/a is even.

Based upon our understanding of the single impurity problem, we can actually build up some simple heuristics to understand these results. Recall that the ground state of H_1 is such that the impurity spin is pulled into the chain. This necessitates a sublattice interchange on a half-infinite line, which is the root of all the orthogonalities. Now, for insertion of a singlet pair separated by l, where l/a is even, we see schematically in fig. 2 that it is possible to absorb both impurity spins into the chain with a sublattice interchange necessary over only a *finite* $O(l)$ length of the chain. That is, the two impurities can be naturally absorbed onto opposite sublattices of a new spin chain with two additional sites. On the other hand, there is no analogue of fig. 2 for insertion of a singlet when l/a is odd: it is not possible to absorb the singlet without a finite energy cost, since there is no way to accommodate them on opposite sublattices of a chain with two additional sites. We believe that this simple picture provides the correct heuristics to understand not only the results obtained above in a pedestrian manner but also the more rigorous results obtained by TSA.

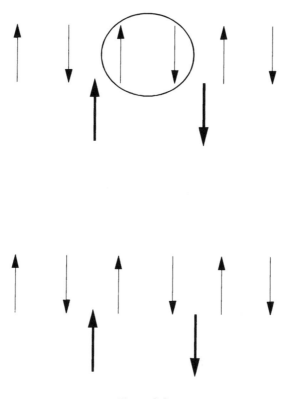

Figure O.2

Finally, let us turn our attention to a model in which the impurity spins hop into and out of the impurity sites at a rate Δ. Consider the Hamiltonian

$$H = J \sum_{n=k+1/2} \mathbf{s}_n \cdot \mathbf{s}_{n+1} + g \sum_m \mathbf{S}(m) \cdot [\mathbf{s}_{m-1/2} + \mathbf{s}_{m+1/2}] + \Delta \sum_m (c_{m\sigma}^\dagger d_{m\sigma} + h.c.)$$

(10)

where $\mathbf{S}(m) = \frac{1}{2} c_{m\sigma}^\dagger \boldsymbol{\sigma} c_{m\sigma}$. The d-fermions provide a reservoir of fermions which may hop into and out of the impurity sites. It is to be understood that there are projection operators restricting impurity sites to single occupancy. The chemical potential is such that for $\Delta = 0$ all d-sites are doubly occupied. We shall argue that this model will exhibit both of the properties (a) and (b) discussed earlier.

Suppose that for $t < 0$ we have $\Delta = 0$ and all impurity sites empty. The question we wish to address is whether hopping of fermions between the reservoir and the impurity sites is a coherent process. We can answer this perturbatively for $\Delta \ll g, J$, by examining

$$P(t) = |\langle O_-|e^{-iHt}|O_-\rangle|^2$$

where $|O_-\rangle$ is the $t < 0$ ground state, i.e., the ground state of the spin chain alone multiplied by the doubly occupied d-sites. To $O(\Delta^2)$ we have

$$P(t) = 1 - \Delta^2 \int_0^t dt_1 \int_0^{t_1} dt_2 \mathcal{G}_1(t_1 - t_2). \tag{11}$$

The orthogonality (7) is not, by itself, sufficient to guarantee incoherence. This issue will be determined by the *strength* of the orthogonality. Unfortunately, the knowledge of the impurity spin susceptibility in H_1 does not provide $\mathcal{G}_1(t)$, but only correlation functions of the type $\mathcal{F}(t) \equiv \langle 0|S^+(0, t)S^-(0, 0)|0\rangle \sim 1/t$ where $|0\rangle$ is the ground state of H_1. Analogues of $\mathcal{G}_1(t)$ and $\mathcal{F}(t)$ in the context of the two-level system model are discussed in [10], where it is argued that in many cases of interest the long time behaviors are the same. We do not know whether this is the case for $\mathcal{G}(t)$ and $\mathcal{F}(t)$. However, we expect on physical grounds that the long time behavior of $\mathcal{G}_1(t)$ and $\mathcal{F}(t)$ cannot be markedly different. A behavior $\mathcal{G}_1(t) \sim 1/t$ would imply that single particle tunneling is on the border between being completely incoherent and being strongly damped, so we believe that the true behavior of single particle tunneling is incoherent or strongly damped. Which of these alternatives it is requires further study: nonetheless the tunneling is certainly not in the (weakly damped) coherent regime.

At $O(\Delta^2)$ the interactions between the hopping of two or more fermions into impurity sites is not probed. The first effect of inter-impurity interactions induced by the chain will occur at $O(\Delta^4)$. Without making a detailed analysis it is clear that, since $Im G_2(2, \omega)$ has a non-vanishing $\delta(\omega)$ part, there will be a *coherent* contribution to $P(t)$ from correlated singlet pair hopping. Thus, this toy model exhibits the properties (a) and (b): incoherent, or at least strongly damped, single particle tunneling, but coherent (singlet) pair tunneling into impurity sites.

While we have argued [1] that single particle hopping between Luttinger liquids can be rendered incoherent by strong in-liquid correlations, this confinement is more subtle than that discussed above, due to the in-liquid itinerancy of the fermions. Incoherence results from a combination of orthogonality catastrophes and spin-charge separation. In the $U \to \infty$ limit of the 1D Hubbard model the ground state factorizes into a product of a spinless fermion determinant describing the charge degrees of freedom, and a "squeezed" Heisenberg model ground state [11]. There is therefore a real possibility for coherent pair hopping between Hubbard chains since there will effectively be no spin-charge separation, the singlet pair of spins of the two electrons hopping having "condensed" via the ODLRO of the Heisenberg model. This issue clearly warrants more detailed study.

The author is grateful to P. W. Anderson and S. P. Strong for stimulating discussions. This work was supported by SERC grant GR/H 73028 (while the author was at Oxford University), St. Catherine's College, Cambridge, and the IRC.

REFERENCES

1. D. G. Clarke, S. P. Strong, and P. W. Anderson, Phys. Rev. Lett. **72**, 3218 (1994).
2. S. Chakravarty and P. W. Anderson, Phys. Rev. Lett. **72**, 3858 (1994).
3. D. G. Clarke, T. Giamarchi, and B. I. Shraiman, Phys. Rev. B **48**, 7070 (1993); D. G. Clarke, Ph.D. thesis, Princeton University (1993), unpublished.
4. See, e.g., I. Affleck, in *Fields, Strings and Critical Phenomena*, Les Houches Session XLIX, 1988 (North-Holland, Amsterdam, 1990).
5. Of course, the absence of a true long range order of Ω, even at zero temperature, means that a SW approximation is, strictly speaking, invalid. The reader concerned with rigor may insert a small staggered field into the Hamiltonian and follow the arguments through. The essential conclusions are unchanged.
6. To the extent that we employ SW methods, one could have begun with a Holstein-Primakoff (HP) approach. However, we have found that the non-linear sigma model language leads to the conclusions we wish to draw in a much less opaque way than the HP language.
7. S. Eggert and I. Affleck, Phys. Rev. B **46**, 10866 (1992).
8. K. Ingersent, B. A. Jones, and J. W. Wilkins, Phys. Rev. Lett. **69**, 2594 (1992).
9. J. Talstra, S. P. Strong, and P. W. Anderson, to be published.
10. A. J. Leggett et al., Rev. Mod. Phys. **59**, 1 (1987).
11. M. Ogata and H. Shiba, Phys. Rev. B **41**, 2326 (1990).

P

New Types of Off-Diagonal Long Range Order in Quantum Spin Chains

J. C. Talstra,* S. P. Strong,† and P. W. Anderson

(Received 17 November 1994)

ABSTRACT

We discuss new possibilities for off-diagonal long range order (ODLRO) in spin chains involving operators which add or delete sites from the chain. For the one-dimensional, periodic Heisenberg, inverse square exchange, and XY models we give numerical evidence for the hidden ODLRO conjectured by Anderson. A connection to the singlet pair correlations in one-dimensional models of interacting electrons is made and briefly discussed.

PACS numbers: 71.27.+a, 74.20.Mn, 75.10.Jm

In 1991 one of us (P.W.A.) conjectured, based on resonating valence bond (RVB) ideas for the one-dimensional Hubbard model, that there should be a nonzero overlap between the ground state of the one-dimensional Heisenberg model ($H = \sum_i \vec{S}_i \cdot \vec{S}_{i+1}$) on a chain of N sites and the state obtained by inserting a pair of nearest-neighbor spins in a singlet configuration into the ground state of the $N - 2$ site chains [1]. Both this model and the inverse squared exchange (ISE) model [2] $[H = \sum_{i,j} [(N/\pi) \sin \pi(i - j)/N]^{-2} \vec{S}_i \cdot \vec{S}_j]$, which is in the same universality class, were therefore expected to have a hidden form of true off-diagonal long-range order (ODLRO) [3] for an operator which not only involved sampling the state of the system (in this case checking that a given pair of spins was in a singlet) but also changing the Hilbert space of the system in an essential way (adding two sites). The overlap has not been previously investigated (but see [4] where evidence for the resulting ODLRO was found); therefore we numerically tested the original conjecture that the overlap for the ISE and Heisenberg models between the N site ground state and the $N - 2$ site ground state with a local singlet inserted should be nonzero.

*Joseph Henry Laboratories, Princeton University
†NEC Research Institute, Princeton, N.J.

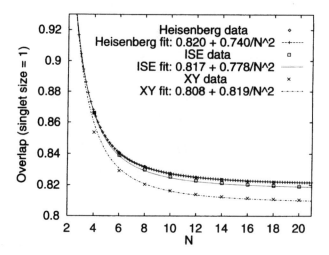

Figure P.1. Calculated overlaps as functions of system size. Shown are the calculated overlaps between the N spin ground states of the $1/r^2$, nearest-neighbor Heisenberg and *XY* models and the states obtained by inserting a nearest-neighbor singlet pair of spins into the $N - 2$ spin ground states of those models.

These overlaps are shown as functions of N in fig. 1 together with fits to the results of the form $0.817 + 0.778N^{-2}$ for the ISE model and $0.820 + 0.740N^{-2}$ for the Heisenberg model. The results strongly suggest that both overlaps remain finite in the limit as the system size goes to infinity. Note that since the phases of the N and $N - 2$ site wave functions may be chosen independently the phase of the overlaps is meaningless and further, since the ground state momenta of the N and $N - 2$ site ground state wave functions differ by π, the overlap is multiplied by -1 if the location of the singlet pair is shifted by one site. We have chosen the overlap real and positive for convenience. As a consequence of this alternation and the finiteness of the overlap, the singlet insertion–singlet deletion correlation function has fixed magnitude and a sign determined by the number of sites between the insertion and the deletion modulo 2 in the limit of infinite separation of the insertion and deletion: the models therefore have a hidden ODLRO which breaks a Z_2 symmetry.

We now present an analytical calculation in the ISE model that gives some understanding of the origin of the finite overlap. The ground state wave function of the ISE model in a basis of local spins $\{|\sigma_1 \cdots \sigma_M\rangle\}$, $\sigma_i = \pm\frac{1}{2}$, is given by [5]

$$\Psi_M^0\{\sigma_1 \cdots \sigma_M\} = \prod_{i<j}^{M}(z_i - z_j)^{\delta_{\sigma_i,\sigma_j}} e^{(\pi i/2)\mathrm{sgn}(\sigma_i - \sigma_j)}. \tag{1}$$

For the N site ISE ground state $M = N$ and $\{z_i\} \equiv C_N = \{e^{2\pi i n/N}\}_{n=1}^N$, while for the $N - 2$ site ground state $M = N - 2$ and $\{z_i\} \equiv C_{N-2} = \{e^{2\pi i n/N-2}\}_{n=1}^{N-2}$. To compute the ISE overlap we add σ_{N-1}, σ_N, sitting in a singlet, to the $N - 2$ site ground state. This overlap is not calculable since the z_i's from both sets are not commensurate with each other. However, if we slightly deform the set C_{N-2} to be $\{e^{2\pi i n/N}\}_{n=1}^{N-2}$ and leave σ_{N-1}, σ_N in a singlet, then we obtain a new state Ψ_N^2 that can be recognized as a *localized two-spinon state* [6, 7]. This is an uncontrolled approximation whose justification lies in agreement with our numerical results. This state is an admixture of eigenstates that contain only 0 spinons or 2 spinons. Here the localized spinons sit at sites $N - 1$ and N in a singlet. In general we could have put them at sites α, β by deforming C_{N-2} into $\{e^{2\pi i n/N} \mid i = 1, \ldots, N\}/\{e^{2\pi i \alpha/N}, e^{2\pi i \beta/N}\}$.

In the basis of states with M overturned spins with respect to the ferromagnetic state labeled by their positions along the chain, $n_1, \ldots, n_M$, Ψ_N^0 and Ψ_N^2 are given by

$$\Psi_N^0(n_l, \ldots, n_{N/2}) = \prod_{i=1}^{N/2} (-)^{n_i} \prod_{i<j}^{N/2} \sin^2\left(\frac{n_i - n_j}{N}\pi\right), \tag{2}$$

$$\Psi_N^2(n_1, \ldots, n_{(N/2)-1}) = \prod_{i=1}^{N/2-1} (-)^{n_i} \prod_{i<j} \sin^2\left(\frac{n_{2i} n_j}{N}\pi\right)$$

$$\times \prod_{i=1}^{N/2-1} \sin\left(\frac{n_i - \alpha}{N}\pi\right)$$

$$\times \sin\left(\frac{n_i - \beta}{N}\pi\right). \tag{3}$$

We now calculate the overlap between Ψ_N^0 and Ψ_N^2 for arbitrary separation α between the spinons (because of translational invariance we can fix β to be 0). We have for the overlap

$$\frac{\langle\Psi^0|\Psi^2(\alpha)\rangle}{\sqrt{\langle\Psi^2|\Psi^0\rangle\langle\Psi^2(\alpha)|\Psi^2(\alpha)\rangle}} = \frac{\langle\Psi^0|\Psi^2(\alpha)\rangle}{\langle\Psi^0|\Psi^0\rangle}\left(\frac{\langle\Psi^2(\alpha)|\Psi^2(\alpha)\rangle}{\langle\Psi^0|\Psi^0\rangle}\right)^{-1/2}. \tag{4}$$

First we will determine $\langle\Psi^0|\Psi^2(\alpha)\rangle$. Setting $M = N/2 - 1$,

$$\langle\Psi^0|\Psi^2(\alpha)\rangle = \left(\frac{N}{2}\right)! \sum_{n_1,\ldots,n_M} \Psi^2(n_1, \ldots, n_M|0, \alpha)$$

$$\times \{\Psi^0(n_1, \ldots, n_M, 0) - \Psi^0(n_1, \ldots, n_M, \alpha)\}^*$$

$$= \left(\frac{N}{2}\right)! \sigma(\alpha)[1 - (-)^\alpha]. \tag{5}$$

Here

$$\sigma(\alpha) = \sum_{n_1,\ldots,n_M} \left[\prod_{i<j} \sin^4\left(\frac{n_i - n_j}{N}\pi\right) \right]$$
$$\times \prod_{i=1}^{M} \sin^3\left(\frac{\pi n_i}{N}\right) \prod_{i=1}^{M} \sin\left(\frac{n_i - \alpha}{N}\pi\right). \tag{6}$$

Since Ψ^0 is a singlet we know that if the two spinons in Ψ^2 were in the $S_z = 0$ *triplet* state the overlap should be zero, therefore $\sigma(\alpha)$ vanishes for all *even* α. Since $\sigma(\alpha)$ is a polynomial in $\cos(\pi\alpha/N)$:

$$\sigma(\alpha) \propto \prod_{j=1}^{N/2} \left[\cos\left(\frac{\pi\alpha}{N}\right) - \cos\left(\frac{2\pi j}{N}\right) \right] \propto \frac{\sin(\pi\alpha/2)}{\sin(\pi\alpha/N)}. \tag{7}$$

Now we turn our attention to the second piece. Only its numerator is α dependent:

$$\langle \Psi^2(\alpha) | \Psi^2(\alpha) \rangle = 2\left(\frac{N}{2} - 1\right)! \sum_{\{n_1,\ldots,n_M\}} [\Psi^2(n_1,\ldots,n_M|0,\alpha)]^2$$
$$= 2\left(\frac{N}{2} - 1\right)! \sum_{n_1,\ldots,n_M} \prod_{i<j} \sin^4\left(\frac{n_i - n_j}{N}\pi\right)$$
$$\times \prod_{i=1}^{M} \sin^2\left(\frac{\pi n_i}{N}\right) \sin^2\left(\frac{n_i - \alpha}{N}\pi\right). \tag{8}$$

But this sum can be recognized as the $\langle S_z(\alpha) S_z(0) \rangle$ static correlation function in the ISE model [2], or—after an (exact) conversion of the sums to integrals—as the one-particle density matrix in the Calogero-Sutherland model at half-filling [8]. The result for large N is proportional to $[\mathrm{Si}(\pi\alpha)/\pi\alpha]$, where $\mathrm{Si}(x)$ is the sine-integral function. The entire expression for the overlap, with normalization computed by considering $\alpha = 0$, then becomes

$$\frac{2}{N} \frac{\sin(\pi\alpha/2)}{\pi\alpha/N} \sqrt{\frac{\pi\alpha}{\mathrm{Si}(\pi\alpha)}}. \tag{9}$$

Thus for a nearest neighbor the overlap is $(2/\pi) \times \sqrt{\pi/\mathrm{Si}(\pi)} \simeq 0.82917$, which is within 1.5% of the Heisenberg and ISE ground state singlet insertion overlaps.

Many of the properties of the two-spinon matrix element might be expected from those of the Girvin-MacDonald-Read [9] order parameter for the *bosonic*

$v = \frac{1}{2}$ Laughlin [10] state. In the spinon language, the insertion of up-spin spinons acts on the down-spin wave function in the same way as the quasihole operator of the $v = \frac{1}{2}$ state acts on the Laughlin state, while the insertion of a down-spin spinon analogously to the insertion of a quasihole and a hard-core boson. Consequently, the two-spinon insertion for $\alpha = 0$ takes the bosonic $v = \frac{1}{2}$ state with $N/2 - 1$ particles, periodic on a ring of size $N - 2$, to the $v = \frac{1}{2}$ state with $N/2$ particles periodic on a ring of size N, and is essentially the Read order parameter for the $v = \frac{1}{2}$ state. This analogy suggests that a Ginzburg-Landau theory for spin chains might be constructable based on this connection [11]. The Goldstone bosons should be the underlying bosons of the Luttinger liquid description [12], while spinons might appear as the analog of the semionic quasiparticles of the $v = \frac{1}{2}$ state.

For $\alpha = 0$, the two-spinon insertion is analogous to two spatially separated operators: a quasihole operator and, a finite distance away, a hard-core boson creation operator and another quasihole operator. Such an object should still have an expectation value given essentially by the quasihole propagator. This would decay exponentially in the bulk of a $v = \frac{1}{2}$ state, but near the edge would decay only as $\alpha^{-1/2}$, exactly as the two spinon matrix element does [see Eq. (9) and fig. 2]. The vanishing of the matrix element for any even α and its alternation for odd α then makes sense given the lattice structure of our model, the phase of the quasihole propagator, and the fact that the spinon operator is a superposition of operators for two edges so that the result is real, rather than complex.

The reason the two-spinon overlap agrees so well with the actual singlet insertion overlap is unclear; however, our numerical results for the singlet insertion in the Heisenberg and ISE models display all of the same qualitative features as the two-spinon calculation, so it appears that for these models the spin chain ODLRO is essentially that of the $v = \frac{1}{2}$ Laughlin state.

Since the two-spinon insertion operator generates states with either zero or two spinons, the zero-spinon state being the ground state, its action is analogous to the action of a pair of fermionic creation and annihilation operators on the fermionic ground state. The matrix element to the ground state as a function of the space-time separation of the fermionic creation and annihilation operators is by definition the fermion propagator. The overlap we calculate is, by analogy, proportional to the equal time spinon propagator.

In an effort to better understand our results for the Heisenberg and ISE models, we examined the overlap for the same insertion operator acting on the $N - 2$ site XY model ($H = \sum_i S_i^x S_{i+1}^x + S_i^y S_{i+1}^y$) ground state with the N site XY model ground state. The strikingly similar result is shown in fig. 1, together with a fit to $0.808 + 0.819 N^{-2}$. We have also found numerical evidence for the same ODLRO for general XXZ models. Since the XY model can be mapped onto spinless fermions, the "order parameter" can be recast in that language and a more detailed study made. In that language the nonzero overlap is between

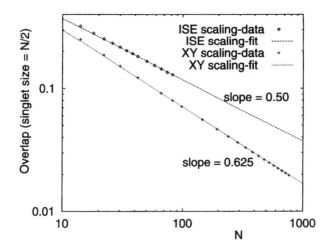

Figure P.2. Decay of the overlap with separation. The overlaps between the N spin ground states and the states obtained from the $N - 2$ spin ground states by inserting two spins in a singlet configuration separated from each other by $(N - 2)/2$. The XY results are from exact numerical results, while those of the ISE model are Monte Carlo.

the ground state for $N/2$ spinless fermions on N sites and the state obtained by adding two sites and one fermion (in a superposition of being on the two added sites with a relative minus sign for the two different sites) to the ground state for the $N/2 - 1$ spinless fermions on $N - 2$ sites—with the opposite boundary conditions from the N site case.

Since the wave functions for the two fermion states to be overlapped are those of free particles, considerable progress can be made in its computation and it can be bounded rigorously from below. The calculation is straightforward, using simple properties of polynomials whose roots lie on the unit circle, and the fact that the corresponding momenta for the $N - 2$ and N site models differ only by $O(1/N^2)$ near the Fermi surface. The lower bound we obtained was $e^{-5/2} + O(N^{-1})$.

The nonzero overlap for the XY model is particularly striking since we know that the overlap between the N site, $N/2$ particle ground state and the state created by acting on the N site, $N/2 - 1$ particle ground state with a single, localized fermion creation operator vanishes like $N^{-1/2}$, while the overlap between the N site, $N/2$ particle ground states between models with periodic and antiperiodic boundary conditions vanishes like N^{-1}.

Order of the type we find can have important consequences for more general models than spin chains. We now show that the ODLRO of the Heisenberg model is responsible for the leading contribution to the singlet pair susceptibility of the one-dimensional Hubbard model. First, we make the connection between the ODLRO found for the Heisenberg model and the equal time, singlet pairing correlation function of the one-dimensional Hubbard model in the limit as $U \to \infty$. In that limit the ground state wave function of the Hubbard model is given by a product of spin and charge wave functions, the latter being given by a spinless fermion determinant and the former by the ground state, Bethe ansatz wave function for the "squeezed" Heisenberg model, i.e., a Heisenberg model defined only on those sites occupied by the spinless fermions [13]. The equal time singlet pair correlation function is given by the overlap of the ground state with the state obtained from the ground state by removing a nearest-neighbor, singlet pair of electrons at sites which we can take to be 0 and 1 and then inserting a nearest-neighbor, singlet pair of electrons at sites j and $j + 1$. Because of the hidden ODLRO of the Heisenberg model, the spin wave function overlap for a fixed charge configuration has a piece which for large separation j is given by a constant time $(-1)^{\sum_{j>l>1} n(l)}$. This leads to a contribution to the correlation function given by $\langle \Psi^{\dagger}(j+1)\Psi^{\dagger}(j)\Psi(1)\Psi(0)(-1)^{\sum_{j>l>1} n(l)} \rangle$. The leading asymptotic behavior of this expectation value can be computed straightforwardly from Abelian bosonization and is proportional to $\cos(k_F x)x^{-5.2}$, where k_F is the k_F of the spinless fermions. This agrees with the predictions of the Luttinger liquid description of the Hubbard model [12, 14, 15], provided that we take the charge sector rediagonalization parameter K_ρ to be $1/2$.

In the Hubbard model, this contribution to the singlet pair correlation functions arises because the bosonized form of $\psi_\uparrow(j)\psi_\downarrow(j+1)$ contains an operator proportional to $\exp(i\Theta_{R,\rho})$, involving only charge degrees of freedom [16]. This operator is present because of the operator product expansion for $\exp[\frac{i}{2}\Theta_{R,\sigma}(x)]\exp[-\frac{i}{2}\Theta_{R,\sigma}(x')]$ contains the identity times a coefficient asymptotically proportional to $(x - x')^{-1/2}$. The decay of this coefficient with separation implies that, if the electrons are inserted m sites apart where $m \gg 1$ but $j \gg m$, the singlet pair correlations decay with j in the same way, but there is a multiplication of the prefactor A by $m^{-1/2}$ arising from spin degrees of freedom. This is in agreement with the decay we find for the singlet and two-spinon insertions in the ISE and Heisenberg models, and suggests the identification of the insertion of an up-spin spinon into the ISE model with the action of the operator $O_{\text{insert}} = i^j \exp[\frac{i}{2}\Theta_{R,\sigma}(ja)] + (-i)^j \exp[\frac{i}{2}\Theta_{L,\sigma}(ja)]$, where a is the lattice spacing and j is the number of sites to the left of the insertion site *in the original chain*. Since this operator is a semion, it is natural to identity it as the spinon creation operator; an identification compatible with the observation of [17] that the generalized commutation relations of the Fourier modes of this operator provide a natural realization of the Yangian.

The identification is also compatible with the alternation with odd separation and vanishing for even separation that we find for the singlet and two-spinon insertions. Both are in agreement with the singlet pair correlations of the Hubbard model as computed with Abelian bosonization.

The connection to Abelian bosonization should be extendable to generalizations of the Hubbard model which include spin-dependent interactions so that the Luttinger liquid rediagonalization parameter of the spin sector of the Hubbard model is renormalized from 1. At half-filling the low energy sector of the model then becomes a general XXZ model changing the exponent for the decay of the singlet insertion overlap from $\frac{1}{2}$. For XY symmetry the new exponent is equal to $\frac{5}{8}$, exactly as we observe (fig. 2). Based on the connection between the two-spinon overlap and the singlet insertion for the ISE model, where spinons are noninteracting, we argued that the decay exponent was just the exponent of the spinon-spinon propagator. The continuously varying exponent of the overlap decay in general XXZ chains is just the exponent of the spinon-spinon propagator in models of interacting spinons.

We note that the alternating piece in the singlet-pair correlation function is the slowest decaying piece of that correlation function for the Hubbard model and may have important consequences for superconducting correlations in models based on the one-dimensional Hubbard model. In particular, if one considers an array of Hubbard chains coupled with operators which properly correlate the neighboring chains, then the scaling dimension of pair hopping between the chains will be renormalized and pair hopping will become relevant and lead to an instability similar to that envisioned in the interlayer tunneling mechanism for superconductivity. Relevant operators having this effect occur naturally when the spin-spin superexchange interaction between chains is considered in the bosonization language [18].

We gratefully acknowledge helpful correspondence with M. Ogata and discussions with F.D.M. Haldane, C. Nayak, and D. G. Clarke, as well as financial support from NSF Grants No. DMR-9104873 (P.W.A.) and No. DMR-922407 (J.C.T.) and the NEC corporation (S.P.S.).

REFERENCES

1. P. W. Anderson, this book.
2. F.D.M. Haldane, Phys. Rev. Lett. **60**, 635 (1988); B. S. Shastry, Phys. Rev. Lett. **60**, 639 (1988).
3. P. W. Anderson, Rev. Mod. Phys. **37**, 298 (1966).
4. T. Pruschke and H. Shiba, Phys. Rev. B **46**, 356 (1992).
5. F.D.M. Haldane, in *Proceedings of the 16th Tanaguchi Symposium on Condensed Matter, Kashikojima, Japan, 1993*, edited by A. Okiji and N. Kawakami (Springer-Verlag, Berlin, 1994).
6. F.D.M. Haldane, Phys. Rev. Lett. **66**, 1529 (1991).
7. J. C. Talstra and F.D.M. Haldane, Phys. Rev. B **50**, 6889 (1994).
8. B. Sutherland, Phys. Rev. A **4**, 2019 (1971).

9. S. Girvin and A. MacDonald, Phys. Rev. Lett. **58**, 1252 (1987); N. Read, Phys. Rev. Lett. **62**, 86 (1989).

10. R. B. Laughlin, Phys. Rev. Lett. **50**, 1395 (1983).

11. We thank C. Nayak for calling our attention to the possibility of a Ginzburg-Landau theory of our order.

12. F.D.M. Haldane, Phys. Rev. Lett. **47**, 1840 (1981); J. Phys. C **14**, 2585 (1981).

13. M. Ogata and H. Shiba, Phys. Rev. B **41**, 2326 (1990).

14. H. J. Schulz, Phys. Rev. Lett. **64**, 2831 (1990); S. Sorella, A. Parolla, M. Parrinello, and E. Tosatti, Europhys. Lett. **12**, 729 (1990); H. Shiba and M. Ogata, Prog. Theor. Phys. Suppl. **108**, 265 (1992); M. Gulasci and K. S. Bedell, Phys. Rev. Lett. **72**, 2765 (1994).

15. Holger Frahm and V. E. Korepin, Phys. Rev. B **42**, 10533 (1990).

16. Notation is based on that of Ref. [12].

17. P. Bouwknegt, A.W.W. Ludwig, and K. Schoutens, Phys. Lett. B (to be published); Report No. hep-th 9406020.

18. S. P. Strong, J. C. Talstra, and P. W. Anderson (unpublished).

Q

Coherence and Localization in 2D Luttinger Liquids

P. W. Anderson, T. V. Ramakrishnan,* Steve Strong,† D. G. Clarke‡

ABSTRACT

Recent measurements on the resistivity of $(La - Sr)_2CuO_4$ are shown to fit well within the general framework of Luttinger liquid transport theory. They exhibit a cross over from the spin-charge separated "holon non-drag regime" usually observed, with $\rho_{ab} \sim T$, to a "localizing" regime dominated by impurity scattering at low temperature. The proportionality of ρ_c and ρ_{ab} and the giant anisotropy follow directly from the theory.

Recently Ando et al.[1] have measured the resistivity of two $(La - Sr)_2CuO_4$ samples, one of them in the fully metallic optimally doped range, in magnetic fields up to $60\ T$, which completely destroys superconductivity. The result confirms other measurements which showed that at sufficiently low temperature the resistivity of the cuprate ab plane in the normal state crosses over to a negative temperature derivative with an approximately logarithmic T-dependence, as opposed to the conventional linear T behavior. (At least where the spin gap does not intervene as it does in bilayer materials such as BISCO 2212.) What is new about the Boebinger experiments is that the c-axis and ab-plane conductivities become proportional to each other at low temperature although with an extraordinarily large anisotropy of the order 1000.

The c-axis conductivity in these measurements follows a smooth extrapolation from the normal state values above T_c. It is reasonable to suppose that the c-axis conductivity is therefore not crossing over to a new behavior; we must then suppose that the ab plane does so.

*Dept. of Physics, Indian Inst. of Science, Bangalore, 560 012 India
†NEC Research, 4 Independence Way, Princeton, N.J. 08540
‡TCM, Cavendish Laboratory, Madingley Road, Cambridge CB3 0HE, England
Work at Princeton was supported by the NSF, Grant # DMR-9104873 and DMR94 00362.

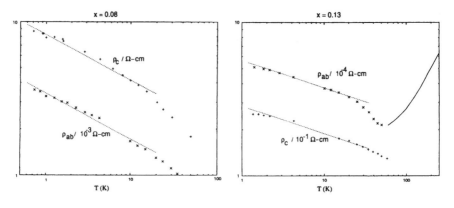

Figure Q.1. $x = 0.08$, $2\alpha = 0.30$ Figure Q.2. $x = 0.13$, $2\alpha = 0.18$

We[2] have derived the c-axis conductivity as a function of frequency and at $T = 0$, for Luttinger liquid planes or chains weakly coupled by a tunneling matrix element $t_\perp$, presumed diagonal in the momentum parallel to the chain or plane. We get

$$\sigma = \frac{e^2}{h} \frac{t^2}{t_\parallel^2} \left(\frac{\omega}{\wedge}\right)^{2\alpha} \tag{1}$$

where this is measured per electron and between one pair of planes: i.e., it is a conductance in $(\Omega)^{-1}$. α is the Fermi surface exponent giving the power law for $n(k)$ at k_F, and $\wedge$ is a cutoff of order $t_\parallel$. This ω-dependence will be reflected in a T-dependence since the derivation of (1) depends only on the scaling properties of the one-particle Green's function, which at finite temperatures $kT > \hbar\omega$ will have an infrared cutoff $\sim kT$ rather than $\hbar\omega$ (no extraneous cutoff such as $(\hbar/\tau)$ enters the expressions, since they converge at the low-frequency end). Thus our interpretation of the c-axis conductivity is that it follows, over most of the range, a power law

$$\sigma_c \propto \omega^{2\alpha}$$

and in figs. (1) and (2) we plot the c-axis results of Ando et al.[1] as $\ln \sigma$ $vs.$ $\ln T$ to show an approximate fit, with $2\alpha = .35$ ($\delta = 0.8$) and $.22$ ($\delta = .13$). At higher temperatures $T > t_\perp$, σ_c eventually crosses over to a positive temperature coefficient. $t_\perp$ is several hundred degrees K according to band calculations.[3]

The remarkable observation of Ando et al. is that as $T \to 0$ $\sigma_{ab} \propto \sigma_c$. From the Luttinger liquid theory we predict a crossover phenomenon in σ_{ab}. The essence of Luttinger liquid transport t theory is summarized in Ref. (4). (See also (5).) There are three regimes (see fig. 3) for in-plane transport, crossing over from one regime to another with increasing elastic scattering rate $\hbar/\tau_{el}$ or decreasing temperature kT.

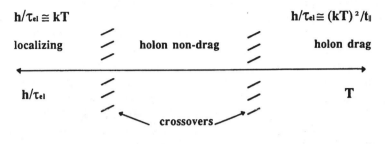

$$\text{Figure Q.3}$$

The regime characteristic of the usual normal state of the cuprates is the "holon non-drag" regime, the middle one of the three. In this regime the fastest dissipative process is the decay of the electron into spinon + holon, at a rate $\propto (kT, \hbar\omega)$. The holons are then incoherently scattered at a rate $1/\tau_{el}$ with

$$kT > \hbar/\tau_{el} > \frac{(kT)^2}{E_F} \tag{2}$$

the latter being the rate of the inverse process holon + spinon $\rightarrow$ electron (by detailed balance). Under these circumstances the holons lose their momentum before they are able to recombine, and the rate of current decay is simply kT. As shown in Ref. (5)[6] σ has a small power-law correction: $\sigma \propto \omega^{-1+2\alpha}$; the expression is

$$\sigma = \frac{e^2}{\hbar} \sin 2\alpha (\wedge/\omega)^{1-2\alpha}. \tag{3}$$

Here as elsewhere we use $\hbar\omega$ and kT interchangeably in the power law dependencies. The ω-dependence of (3) is evident in the infrared data quoted in Ref. (5). Note that under the condition (2) this conductivity is *less* than that it would have been under impurity scattering alone, i.e., spinon-holon decay is an extra dissipative mechanism. It is anomalous in that Matthiesson's rule fails, and the two mechanisms are not additive.

The other regime of interest here is the "localizing" regime, where $kT < \hbar/\tau_{el}$. Here the electron has no time to decay before scattering, so charge-spin separation is no longer relevant: the transport properties are those of electrons. However, it is still true that the Green's function in the absence of scattering would exhibit anomalous scaling with x, t, or conversely with k, ω:

$$G(r, t) = t^{-1-\alpha} \qquad F\left(\frac{x}{t}\right)$$
$$G(k, \omega) = \omega^{-1+\alpha} \qquad F'\left(\frac{k}{\omega}\right). \tag{4}$$

The scattering processes are sufficiently frequent that they can be treated as matrix elements connecting whole electron states, i.e., simply as one-electron operators between different channels of electron propagation at different points on the Fermi surface.

We argue that in two dimensions, as opposed to one, there is no severe renormalization of the actual scattering potential such as was described by Kane and Fisher.[7] In one dimension, the singular $2k_F$ response enhances or screens the $2k_F$ scattering matrix elements and leads to an infinite renormalization, but this effect is not serious for scattering around a 2D Fermi surface, where at the general Q-vector there is no singular response. The Kane-Fisher effect is linear in η, the phase shift caused by the interaction, hence is of opposite sign for attractive and repulsive potentials and overwhelms in 1D the effect we shall discuss.

In 2D the effect of singular interactions is to modify the propagation of electrons between scatterings. In a sense, the Fermi surface exponent α can be thought of as modifying the dimensionality of the scattering problem. For small η it is always positive and proportional to η^2.

Localization in a Luttinger liquid is a complex problem, and the following is only a first attempt at a discussion, expected to be valid only in the "classical" limit $g = \sigma\hbar/e^2 \gg 1$ of weak scattering (fortunately this condition is satisfied in our case).

In this limit we can draw a close analogy between the calculation by Clarke and Strong[2] of the c-axis conductivity, and the impurity scattering problem of ab plane conductivity, using the generalized Landauer formula[8]

$$\sigma = \frac{e^2}{\hbar} \sum_{ij} |t_{ij}|^2 \tag{5}$$

where $\sum_{ij}$ is a sum over outgoing and incoming channels. In any system approaching macroscopic dimensions, any one matrix element t_{ij} is small, since every incoming channel connects randomly with every outgoing one: the electrons are certain to have been scattered. In the weak scattering case we can think of the channels as labelled by directions k_F on the Fermi surface.

The question then is: after scattering, what is the amplitude for the electron to arrive in the final channel as a coherent entity? The final, outgoing channel will normally be at a different k-vector on the Fermi surface from the initial one, so the strong forward-scattering interaction which causes spin-charge separation and anomalous dimensionality does not act between initial and final states. The relevant diagram is as shown in Fig. (4), with no vertex corrections to t_{ij}. But this is the identical diagram to that calculated by Clarke and Strong. In essence, we think of the impurity scattering as a barrier between ingoing and outgoing channels, in exact correspondence to the barrier between layers, with a weak transmission coefficient t_{ij} which can be treated in lowest-order perturbation

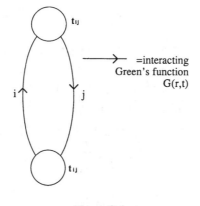

Figure Q.4

theory. The result of that treatment is

$$\sigma \propto (\hbar\omega, kT)^{2\alpha}$$

as for the c-axis conductivity. This can be normalized to the value of conductivity at the crossover temperature to give us a quantitative formula:

$$\sigma_{ab} \simeq \frac{e^2}{\hbar} \sin 2\alpha \left(\frac{\wedge\tau_{el}}{\hbar}\right) \left(\frac{kT}{\wedge}\right)^{2\alpha}. \tag{6}$$

In figs. (1) and (2) we give a log-log plot of the ab plane conductivity observed by Ando et al., showing the crossover (measurements at high T are inferred from other work).[9] Fig. (5) shows data, also agreeing roughly with (6), on the one-layer BISCO compounds, taken by N.-P. Ong's group.[10] This also seems to fit Eq. (6) but with quite small values of (2α).

Comparing (6) and (1), we obtain an expression for the anisotropy (factors of order 1 are possibly missing, since we have no reason to assume $\wedge$ is the same in the two formulas).

$$\frac{\sigma_{ab}}{\sigma_c} = 2\pi \sin 2\alpha \left(\frac{\wedge\tau_{el}}{\hbar}\right) \frac{t_{\parallel}^2}{t_{\perp}^2}.$$

The extra factor $\frac{\wedge\tau_{el}}{\hbar} = \frac{\wedge}{(kT)_{\text{crossover}}} \simeq 50$ accounts for the enormous numerical value of the anisotropy, which is far too great to be a ratio of masses or Fermi surface areas.

A final note on the crossover to localization. At this crossover the mechanism we have proposed for the temperature-dependent Hall effect, which, when analyzed, requires that the spinon and holon currents be non-colinear, ceases to hold, and in the localizing regime we expect the Hall resistivity to become

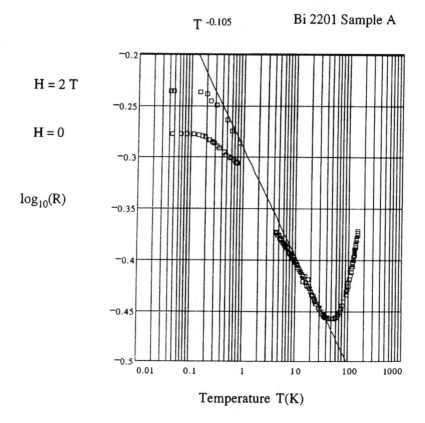

Figure Q.5

temperature-independent following conventional theory. The Hall angle $\omega_c \tau$, them, ceases to follow the T^{-2} behavior characteristic of the holon non-drag regime, and instead crosses over to

$$\Theta_H = \frac{R_{xy}}{P_{xx}} \propto T^{2\alpha}.$$

This would be a spectacular effect and should be investigated. Some evidence that $R_{xy} \to$ const does exist.

We acknowledge valuable discussions with S. Chakravarty, A. Balatsky, and E. Abrahams, as well as generous sharing of their data by the authors of Ref. (1). This work was supported by grants from the NSF under DMR 9104873 and MRSEC 9400362.

REFERENCES

1. Y. Ando, G. S. Boebinger, A. Passner, T. Kimura, and K. Kishio, Phys. Rev. Lett., **75**, 4662 (1995).

2. D. G. Clarke, S. Strong, P. W. Anderson, Phys. Rev. Lett., **74**, 4499 (1994); Clarke and Strong, to appear in Ferroelectrics (1995).

3. O. K. Anderson, Phys. Rev. **B49**, 4145 (1995). Neutron measurements of $J_\perp$ confirm this estimate.

4. P. W. Anderson, chap. 7 of forthcoming book; see also chap. 4 of forthcoming book.

5. M. Ogata and P. W. Anderson, Phys. Rev. Lett., **70**, 3087 (1993).

6. P. W. Anderson, Infrared Conductivity of Cuprate Metals: Detailed Fit Using Luttinger Liquid Theory, submitted to Phys. Rev. B.

7. C. Kane and M.W.P. Fisher, Phys. Rev. Lett. **68**, 1220 (1992).

8. P. W. Anderson, Phys. Rev. B **23**, 4828 (1981).

9. We have simply extrapolated the data of Ref. (1) using the well-known "linear T" law found in all metallic samples.

10. N. P. Ong, Phys. Rev. B, to be submitted.

R

The "Spin Gap" in Cuprate Superconductors

P. W. Anderson

ABSTRACT

We discuss some generalities about the spin gap in cuprate superconductors and, in detail, how it arises from the interlayer picture. It can be thought of as spinon (uncharged) pairing, which occurs indpendently at each point of the 2D Fermi surface because of the momentum selection rule on interlayer superexchange and pair tunneling interactions. Some predictions can be made.

The problem with the Spin Gap [1] is that there are too many right ways to understand it, not too few: when one realizes what is going on it seems all too obvious in several ways that one should have known all along.

(1) The most obvious: spinon pairing. We have realized all along that the normal state has charge-spin separation, so why didn't we expect two pairings, one for spin and the second for charge?

(2) Also obvious: there is no phase transition, hardly even a crossover. So the gap opens without change of symmetry or condensation. It must be not a self-consistent mean field but a property of the separate Fermi surface excitations.

(3) Finally, when one looks at the interlayer theory, and it takes seriously, one realizes that the phenomenon jumps out at you and is a trivial consequence of the interlayer interaction. The Strong-Anderson [2] model is not a complete theory, but can be used to calculate with: $\chi(T)$, for instance.

Let me start, then, in the inverse of chronological order and try to make the synthetic argument first. We start from the fact that every experimental, computational, and theoretical bit of evidence we have supports the dogma that the 2D interacting electron gas in the cuprates is a liquid of Fermions with a Fermi surface, and with little or no tendency to superconductivity or to exhibit antiferromagnetism, once it is metallic—i.e., there is no clear indication of "antiferromagnetic spin fluctuations," as relatively soft bosonic modes, in the isolated plane. Rather, in the plane the magnetic interaction modifies the elementary excitation spectrum as it does in the ferromagnetic case. The symmetry of this state is the Haldane-Houghton Fermi liquid symmetry $(U(2))^Z \neq (U(1) \times SU(2))^Z$, one of "Z" for each point on the Fermi surface. Every point on the Fermi surface is independent, and charge and spin are separately conserved.

Our theory postulates that in fact the $U(2)$ *is* broken into $U(1) \times SU(2)$ with the charge and spin excitations having different Fermi velocities and the charge also having anomalous dimension, i.e., the charge bosons are a Luttinger liquid; but this does not change the symmetry argument. What is little realized is that the spin excitations are *always* describable as spinons, even for free electrons,

$$\psi_{\hat{k}}^*(r) \simeq s_{\hat{k}}^+(r) e^{i\theta_{\hat{k}}(r)}.$$

The spin part is always a spinon, the charge is a bosonized Luttinger liquid. This, then, is our high-temperature, high-energy state above temperatures and energies where the interplane interactions come into play.

Spinons in 2D are paired but gapless. What the nonexistence of a phase transition when we lower T to the interplanar scale tells us is that the spin gap state has the same symmetry. It must leave the crucial fact of Fermi or Luttinger liquids intact: the independence of different Fermi surface points. Then all that can happen is that the spectrum at each point changes, and the simplest way for that to happen is for the spinon to acquire a "mass," i.e., the spinons which used to have a free electron like linear spectrum

$$v_s(k - k_F) \quad \text{or} \quad v_s \sin \frac{\pi}{2} \frac{(k - k_F)}{k_F}$$

open a gap and have energies

$$E^2 = \Delta^2(\hat{k}) + v_s^2(k - k_F)^2. \tag{1}$$

This is possible because of the peculiar nature of spinons, that they are BCS quasi-particle like even in the normal state (as shown long ago by Rokshar [3]). That is, they are semions, or Majorana Fermions, which have no true antiparticles (we use the convention $-k = -k, -\sigma k = k, \sigma$)

$$s_k^+ = s_{-k} \qquad s_k = s_{-k}^+$$

so that the Hamiltonian for free spinons may be written

$$v_s(k - k_F)(s_k^+ s_{-k}^+ + s_{-k}s_k) \tag{2}$$

just as well as in terms of $s_k^+ s_k$ and it is *not a symmetry change* to add a term

$$\Delta_k s_k^+ s_{-k}^+.$$

Spinons are always effectively paired (Strong and Talstra [4]). It is natural that spinons are more easily paired in the underdoped regime, because the spinon velocity becomes progressively lower (J smaller) as we go toward the Mott insulator; therefore the density of states is higher, χ_{pair} larger, on the underdoped side.

Finally, let me make one last remark of a synthetic, rather than analytic, nature. As I have already said the basic description either of a Fermi or a Luttinger liquid is the independence of different Fermi surface points. If we are to go smoothly from a two-dimensional electron liquid to a gapped state *without change of symmetry*—without introducing any new correlations—we must do so without coupling the different Fermi surface points, that is, we need interactions which conserve two-dimensional momenta k_x k_y. There is only one source of such interactions, namely, the interlayer tunneling.

$$\mathcal{H}_{IL} = \sum_{k,\sigma,i,j} t_\perp(k) c_{ki\sigma}^+ \, c_{kj\sigma}^+ \tag{3}$$

which, in second order, leads to two types of interlayer coupling:
Pair tunneling

$$\mathcal{H}_{PT} = \lambda_J(k) \sum_{(ij),k,k'} c_{k\uparrow i}^+ c_{-k'\downarrow i}^+ c_{-k'\downarrow j} c_{k\uparrow j} \tag{4}$$

and superexchange

$$\mathcal{H}_{SE} = \lambda_S(k) \sum_{(ij)k,k'} c_{k\uparrow i}^+ c_{-k'\downarrow j}^+ c_{-k'\downarrow i} c_{k\uparrow j}. \tag{5}$$

(In both, $k' \simeq k$) which represent exchange of charge and spin, respectively, between two layers. The empirical (and theoretical) fact that coherent single-particle hopping does not take place in the cuprates leaves these as the two second-order terms which can lead to coherent interactions—such as we are looking for—between two layers.

It is important to recognize that (4) and (5) have one extra conservation relative to conventional interactions. This seems to be very difficult for many theorists to grasp.

This superexchange interaction does not much resemble that used by Millis and Monien,[1] and it does not have anything to do with the "J" of the $t - J$ model. Superexchange occurs as a result of frustrated kinetic energy, and the kinetic energy which is frustrated in the cuprate layer compounds is only the c-axis kinetic energy $t_\perp$. They are very like Mott insulators in one of three spatial dimensions: and they exhibit superexchange in that dimension. But they retain no Mott character in the two dimensions of the planes.

It is an unpublished conjecture of Baskaran that λ_S / λ_J increases as we approach the insulating phase, i.e., as "α", the Fermi surface exponent, increases. This may be one other reason why underdoped materials show the spin gap.

Now, finally, let us do the calculational problem. At this point we have to stop talking in generalities and make some rather severe assumptions in order to make progress. They seem innocuous, and are quite standard in conventional BCS theory, but here we have no particular reason to believe that they will serve as better than a rough guide. These assumptions are: (1) the Schrieffer pairing condition, i.e., we use only the BCS reduced interaction $-k' = -k$. This is justified at high enough T by the fact that a given state k can only pair with one other $-k'$ to give a quasicoherent matrix element; our picture of the kind of process involved is that a transition into a high-energy state intervenes between two low-energy states which are connected by two—and only two—single-particle tunneling processes, $k_a \rightarrow k_b; -k_b \rightarrow -k_a$. It is perhaps best to think of the pairing as always $k, -k$ but with center of mass momentum thermally fluctuating. (2) More orthodox but more serious: We neglect $|v_c - v_s|$ and treat c_k^+ as though it were an eigenoperation, i.e.,

$$\mathcal{H}_K = \sum_k \epsilon_k n_k. \tag{6}$$

Actually, we use the Nambu-PWA form

$$\mathcal{H}_K(k) = \epsilon_k(n_k + n_{-k} - 1) = \epsilon_k \tau_{3k}.$$

Now we have a straightforward Hamiltonian which is trivially diagonalized, because it separates into separate Hamiltonians for every k.

$$\mathcal{H} = \sum_k \mathcal{H}_k$$

$$\mathcal{H}_k = \mathcal{H}_K + \lambda_j c_{k1}^+ c_{-k1}^+ c_{-k2} c_{k2} + 1 \leftrightarrow 2 + \lambda_s c_{k1}^+ c_{-k2}^+ c_{-k1}^+ c_{k2}$$

(Here we use the convention $k = k \uparrow -k = -k \downarrow$). The first attempt was made by Strong and Anderson neglecting λ_s and this leads to a beautiful spin gap. The KE spectrum of the 4 fermions $1, 2, k, -k$ has $16 = 2^4$ states which are grouped into 5 sets, $n_{tot} = 0, 1, 2, 3, 4$ (see fig. 1). Of these only the $n = 2$ states are affected by the interactions, and of these 2 will be split off by H_J and 2 by H_S. In either case, these gaps are completely T-independent and are simply manifested as the individual states drop out:

$$Z = 16 \cosh^4 \frac{\beta \epsilon_k}{2} + 2(\cosh \beta \lambda_J - 1)$$

(because with the added "-1" $n = 2$ states are at 0 energy.

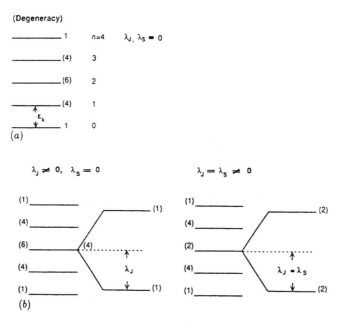

Figure R.1

χ for this case is

$$\chi = \int_{-\infty}^{\infty} d\epsilon \frac{\cosh^2 \beta\epsilon/2}{\cosh^4 \beta\frac{\epsilon}{2} + \frac{1}{8}(\cosh \beta J - 1)}.$$

A second calculation may be carried out with both terms, $\lambda_J \simeq \lambda_S$ and the result is to split out two levels rather than one and to replace $1/8$ with $1/4$. This is the curve for susceptibility I show in fig. 2 and it is not a bad fit to susceptibility data.

But actually I am not totally convinced that this is the right formalism, although it may be the right arithmetic. The reason it works seems clearly to me to be that we have picked a form for the pairing Hamiltonian that connects states which are "neutral"—i.e., only the $n = 2$ states are connected to each other within the k manifold. But in some real sense these are states with the spinons paired but with no holon pairing—no charge pairing—at all, even though nominally different layers are connected. I think it is more nearly valid to describe the correct state by rewriting $\mathcal{H}_j + \mathcal{H}_s$ as

$$(\mathcal{H}_J + \mathcal{H}_S)_k \simeq c_{ke}^+ c_{-ke}^+ c_{-ke} c_{ke}$$

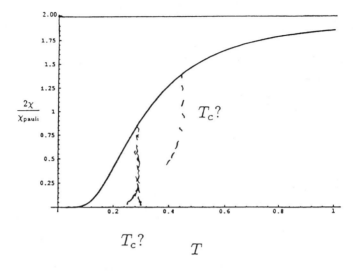

Figure R.2

where $c_{ke}^{+} = \frac{c_{k1}+c_{k2}}{\sqrt{2}}$. That is, the spin-gap state is a state in which spinons belonging to the *even* linear combination are paired, the *odd* unpaired. This has a strong relationship to the Keimer neutron selection rule observed for the superconducting state. [5] Keimer has begun neutron investigations on spin-gap material, but his results are completely preliminary. I anticipate that he will see peaks at energies corresponding to the spin gap and that they will satisfy his even↔odd sum rule, which results from this pairing.

One consequence of the assumption of Fermi rather than Luttinger liquid is the T independence of the spin gap. Actually, the broadening of single-particle states $\propto kT$ will damp out the spin gap when $KT > \Delta_{SG}$, as seems to be observed. But at low T, Δ_{SG} will not vary with T.

This has been a very preliminary account of this work, which is emphatically in progress. I have benefitted from discussions with many people, especially Steve Strong, but also T. V. Ramakrishnan, S. Sarker, G. Baskaran, and D. Clarke; S.-D. Liang helped me with the integral.

REFERENCES

1. For a general description of the spin gap, see A. J. Millis and H. Monien, Phys. Rev. B **50**, 16606 (1994); Phys. Rev. Lett **70**, 2810 (1993); or chapter 4 of this book.

2. S. Strong and P. W. Anderson, Proceedings of the Taiwan International Conference, Chinese Journal of Physics, Vol. 34, No. 2-II, p. 159, April 1996.

3. S. A. Kivelson, D. S. Rokshar, and J. P. Sethna, Phys. Rev. B **35** 8867 (1967).

4. J. C. Talstra, S. P. Strong, and P. W. Anderson, Phys. Rev. Lett. **74**, 5256 (1995).

5. H. L. Fong, B. Keimer, et al., Phys. Rev. Lett. **75**, 316 (1995).

S

Remarks on Spin Gaps and Neutron Peak Selection Rules in YBCO—Do Interlayer Tunneling and Interlayer RVB as Mechanisms for Cuprate Superconductors Differ?

P. W. Anderson

(15 March 1996)

We point out that both superexchange between CuO_2 layers, and interlayer tunneling, derive from frustrated one-particle hopping between layers, and that they should be treated on an equal footing. Doing so, we arrive at a new view of the nature of pairing in the cuprate superconductors, which explains the striking even ↔ odd selection rule observed by Keimer in neutron scattering by YBCO.

Recently, two different hypotheses have been put forward for the "spin gap" phenomenon exhibited by the bilayer structures of cuprate superconductors. (As in YBCO and BISCO 2-layer materials.) Millis and Monien and coworkers [1] have proposed that what is occurring is pair formation between the two layers with one of the pair on each layer, motivated by the interlayer antiferromagnetic superexchange. Strong and myself [2] have proposed that the interlayer pair tunneling Hamiltonian produces a correlated state with pairing on each layer for each momentum k, but the different $\vec{k}$'s are independently phased. Both postulate "preformed BCS pair" states as the essential nature of the spin gap phenomenon.

Although it is not explicitly mentioned in the former paper, both theories have one vital element in common: the coherent motion of single electrons between the two layers must be blocked. As I have emphasized elsewhere, [3] antiferromagnetic "superexchange" interaction exists only if the single-particle kinetic energy is frustrated, and is a consequence of the second-order, virtual action of this kinetic energy, causing hopping of electrons between the relevant sites. If the kinetic energy were diagonalized, the Fermi surface would be split into separate surfaces for even and odd linear combinations of the two layers, which, being orthogonal, would exhibit only ferromagnetic exchange interactions. (The reader should note that although the rest of this article will be couched in language referring only to bilayer materials, only a slight modification allows one to present similar arguments for coupled layers, either single or multiple layers.)

In fact, the Millis-Monien et al. papers miss another feature of interlayer superexchange. The hopping matrix element $t_\perp$ between layers is of course diagonal in $k_\parallel$, the interlayer momentum, so that superexchange, $\sim t_\perp^2/U$, must also contain an extra momentum δ-function. That is,

$$\mathcal{H}_{SE} = \sum_k \sum_{k'} c_{k\uparrow}^{+(1)} c_{k\uparrow}^{(2)} c_{k'\downarrow}^{+(2)} c_{k'\downarrow}^{(1)} \times t_k \cdot t_{k'}^* / \text{energy denom} \qquad (1)$$

and the only way to satisfy the pairing condition is for $k' = -k$. Thus these two mechanisms share the characteristic k-diagonality, which leads to approximate independence of different parts of the Fermi surface; and the Strong-Anderson mechanism for producing a spin gap is common to both.

A third relevant theory paper is the weak-coupling solution of the two-chain Hubbard model by Balents and Fisher [4], in which much of the relevant region of parameter space seems to exhibit the *same* fixed point with a spin gap and one gapless charged excitation, which seems to interpolate between the pairing schemes of Ref. (1) and (2).

We argue here that the two mechanisms for spin gaps are not incompatible but complementary. They result from the same phenomenon of virtual hopping between layers, which is of course only virtual because the direct coherent hopping at zero frequency is blocked by interaction effects. Both the spin gap phenomenon and superconductivity itself are explicable on the same basis.

We may understand the superexchange phenomenon in Mott insulators by realizing that the orbitals of opposite-spin electrons need not be orthogonal. Thus a possible path for exchange of two electrons of opposite spins is for the second electron (of down spin, say) to hop back from site 2 to site 1, during the period while the electron of up spin, while nominally on site 1, has made a virtual transition to site 2. Thus if $a_{1\sigma}^+$ and $a_{2\sigma}^+$ are the properly orthogonalized one-electron operators for the two sites, the actual one-electron operators are

$$c_1^+ = (a_1^+ + \epsilon a_2^+)/\sqrt{1 + \epsilon^2}$$
$$c_2^+ = (a_2^+ + \epsilon a_1^+)/\sqrt{1 + \epsilon^2}$$

where

$$\epsilon = t/U.$$

These are not orthogonal because they need not be if they belong to opposite-spin electrons. The superexchange energy is then

$$J = \epsilon t$$

This picture can be renormalized, as discussed in Herring [5], by taking into account interactions between the two electrons along the exchange path, but the physics remains the same.

J is the amplitude for interchange of spins keeping the same charge state. We may also ask for the pair tunneling amplitude, i.e., the amplitude for an up-spin down-spin pair to tunnel from one site to another. This is of course charge disfavored for atomic sites, but for tunneling between metallic layers there is no charge rigidity. In this case the down-spin electron hops in the opposite direction, and the amplitude is also

$$E_J = t_\perp \epsilon,$$

with one hop taking place by virtue of non-orthogonality of wave functions in the two layers, the second by the actual frustrated one-electron matrix element. This two-particle tunneling process is the basis of the interlayer theory. It, like superexchange, can be described by an effective Hamiltonian for low energy states.

It becomes clear that there is no sharp way to distinguish between antiferromagnetic superexchange and pair tunneling mechanisms for spin gaps and superconductivity if we look at the order parameters which can result.

First let us set up some notation. Let k be a momentum near the Fermi momentum of the two chains or planes, 1 and 2. Orthonormal electron operators are defined as

$$\begin{aligned} a_{k\uparrow}^{(1)+} &= a_1^+ & a_{k\uparrow}^{(2)+} &= a_2^+ \\ a_{-k\downarrow}^{+(1)} &= a_{-1}^+ & a_{-k\downarrow}^{(2)+} &= a_{-2}^+ \end{aligned} \tag{2}$$

and these can combine into the even and odd eigen-operators of the kinetic energy,

$$a_e^+ = \frac{a_1^+ + a_2^+}{\sqrt{2}}, \text{ etc.} \tag{3}$$

$$a_o^+ = \frac{a_1^+ - a_2^+}{\sqrt{2}}, \text{ etc.}$$

We know that the kinetic energy operator may be written

$$\mathcal{H}_k = \epsilon_k(n_{k_1} + n_{k_2}) \tag{4}$$
$$+ t_\perp(a_2^+ a_1 + a_{-1}^+ a_{-2}) + \text{H.C.}$$
$$= (\epsilon_k + t_\perp)(a_o^+ a_o + a_{-o}^+ a_{-o})$$
$$+ (\epsilon_k - t_\perp)(a_e^+ a_e + a_{-e}^+ a_{-e})$$

but in the case where interactions within the layers or chains are sufficiently strong, the splitting given by the last expression into even and odd eigenstates is not expressed in the actual state: $t_\perp$ causes primarily virtual transitions [6], as in the Mott insulator. (More precisely: $t_\perp$ causes no *coherent* transitions.) This is an experimental fact in several of the cuprates, according to photoemission and infrared spectroscopy. [7,8]. Instead of splitting, the electrons hop virtually, so that the electron operators which describe the actual eigenexcitations are roughly

$$c_1^+ = (a_1^+ + \epsilon a_2^+)/\sqrt{1 + \epsilon^2} \tag{5}$$

$$c_2^+ = (a_2^+ + \epsilon a_1^+)/\sqrt{1 + \epsilon^2}.$$

Now let us imagine that by the pair tunneling argument we have derived an order parameter in which the electrons are paired in their separate layers:

$$(\text{O.P.})_{\text{TL}} = \langle c_1^+ c_{-1}^+ + c_2^+ c_{-2}^+ \rangle \tag{6}$$

$$\frac{\langle (a_1^+ a_{-1}^+ + a_2^+ a_{-2}^+) \rangle + \epsilon (a_1^+ a_{-2}^+ + a_2^+ a_{-1}^+) \rangle}{\sqrt{1 + \epsilon^2}}.$$

This order parameter is, in terms of orthonormal states, a mixture of the Strong and the Millis Pairings; it may also be written

$$(\text{OP})_{\text{IL}} = \frac{\langle a_e^+ a_{-e}^+ + a_o^+ a_{-o}^+ \rangle}{\sqrt{1 + \epsilon^2}}$$

$$+ \langle (a_e^+ a_{-e}^+ - a_o^+ a_{-o}^+) \frac{\epsilon}{\sqrt{1 + \epsilon^2}}$$

$$= \langle a_e^+ a_{-e}^+ \rangle \left(\frac{1 + \epsilon}{\sqrt{1 + \epsilon^2}} \right) + \langle a_o^+ a_{-o}^+ \rangle \left(\frac{1 - \epsilon}{\sqrt{1 + \epsilon^2}} \right).$$

Equally, we may imagine a Millis-Monien pairing

$$\langle c_1^+ c_{-2}^+ + c_2^+ c_{-1}^+ \rangle = \frac{\langle (a_1^+ a_{-2}^+ + a_2^+ a_{-1}^+) \rangle + \epsilon \langle (a_1^+ a_{-1}^+ + a_1^+ a_{-2}^+) \rangle}{\sqrt{1 + \epsilon^2}}. \tag{7}$$

Again, in terms of orthogonal orbitals this is a mixture of the two pairings. In terms of even and odd

$$(\text{O.P.})_{\text{MM}} = \langle a_e^+ a_{-e}^+ \rangle \left(\frac{1 + \epsilon}{\sqrt{1 + \epsilon^2}} \right) - \langle a_o^+ a_{-o}^+ \rangle \left(\frac{1 - \epsilon}{\sqrt{1 + \epsilon}} \right). \tag{8}$$

(Incidentally, neither pairing is of the "$s, -s$" type suggested by Scalapino and Yakovenko.) The M-M pairing will also lead to superconductivity because of the ϵ term which represents interlayer pair tunneling, if there is coupling to other layers as well as to other momenta, through conventional short-range interactions.

These two pairings have the interesting feature that they reinforce each other for the even-even pairing, but have opposite sign for the odd-odd pairing. We suggest here that the effective pairing Hamiltonians coming from the two sources have approximately the same coefficient so that in effect we have only even-even pairing: the order parameter is

$$(\text{O.P.}) \simeq \langle a_e^+ a_{-e}^+ \rangle,$$

since the odd-odd pairing is favored with $-$ sign by the superexchange term and favored with $+$ sign by the interlayer term approximately equally.

This pairing is that appropriate to explain the selection rules observed by Keimer et al. [9] in the scattering of neutrons by the bilayer material $YBa_2Cu_3O_7$. The observation is that a pronounced peak appears, rather sharp in energy at ~ 42 mev but broader than instrumental resolution in momentum space, near π, π in the ab plane Brillouin zone. The peak appears only below T_c (magnetic scattering reported above T_c seems to have been an artifact) and is either simply quasiparticle pair production, or the same somewhat enhanced by excitonic interaction effects. More exotic explanations seem incompatible with the experimental facts, in particular, exotic collective excitations seem to have no reason to appear sharply below T_c, and at an energy so close to the supposed value of 2Δ. The $k_\parallel$-dependence is very compatible with the idea that coherence factors forbid magnetic scattering between k's with energy gaps which have the same sign, and enhance strongly those at energy gaps of opposite sign, which presumably (using BISCO as a model, and relying on Josephson interference measurements) occur near points X and Y, separated by π, π in the zone. The observed BISCO gap would fit the $k_\parallel$-dependence well.

The dependence on $k_\perp$ in the c-direction is remarkable. This is a sinusoidal curve with a period given by the inverse of the interplanar spacing, showing that the scattering changes sign between the two planes. An equivalent statement is that scattering satisfies the selection rule even $\leftrightarrow$ odd as far as symmetry in the pair of planes is concerned.

A little thought convinces one that this cannot be a coherence factor selection rule. That is, if the coherence factor is large between k-points $(0, \pi, 0$ and $\pi, 0, \pi)$, the gaps $\Delta_{0\pi0} = -\Delta_{\pi0\pi}$. But then the sign of $\Delta_{\pi00}$ must be opposite to either one or the other, leading to a second peak, which is not observed, either at $\pi, \pi, 0$ or $0, 0, \pi$. We propose that this rule holds because *at* the energy gap the even state is preferentially occupied, the odd state empty. Then the dominant scattering process for a neutron is even $\rightarrow$ odd, when the pair of particles created is primarily at the energy gap.

The understanding of how this comes about requires us to go rather deeply into the BCS mechanism and the slight generalization that is necessary in this problem.

In conventional BCS, the effective Hamiltonian for a single pair of states is

$$\mathcal{H}_{\text{one pair}} = (\epsilon_k - \mu)(n_{k\uparrow} + n_{-k\downarrow} - 1) \tag{9}$$
$$+ \Delta_k c^+_{k\uparrow} c^+_{-k\downarrow} + \Delta^+_k c_{-k\downarrow} c_{k\uparrow}.$$

An irrelevant constant term has been added to make it clear that $\mathcal{H}$ may be taken such that it has no effect on the subspace $n_{k\uparrow} + n_{-k\downarrow} = 1$ of states not satisfying the Schrieffer pairing condition.

In the pseudospin representation, [10] one may write (1) in terms of the Nambu spinors τ as

$$\mathcal{H} = (\epsilon_k - \mu)\tau^z_k + \frac{\Delta}{2}\tau^+_k + \frac{\Delta^*}{2}\tau^-_k \tag{10}$$

and it can be diagonalized by

$$\Psi_o = u_k \Psi(\tau^z_k = +1) + v_k \Psi(\tau^z_k = -1) \tag{11}$$

with

$$u_k = \cos\frac{\theta}{2}, \, v_k = \frac{\sin\theta_k}{2}$$

$$\tan\theta_k = \frac{\Delta_k}{\epsilon_k - \mu}.$$

In the present instance, for a pair of planes we have four Fermions belonging to a given k-vector, $c^{+(\alpha)}_{k\uparrow}$, $c^{+(\alpha)}_{-k\downarrow}$ with $\alpha = 1, 2$ a plane index. Alternatively, we may use $c^{e+}_{k\uparrow}$, $c^o_{k\uparrow}$ with

$$c^{e,o}_{k\uparrow} = \frac{1}{\sqrt{2}}(c^{(1)}_{k\uparrow} \pm c^{(2)}_{k\uparrow}). \tag{12}$$

There are two calculations which may be carried out. One is in the spin-gap situation, where we neglect all coupling to other momenta. Then the states of momentum k, as a decoupled subspace, couple through the pair tunneling and superexchange interactions. These are, first, from (1)

$$\mathcal{H}_{\text{SE}} = \sum_k \frac{\lambda_k}{2}(c^{+(1)}_{k\uparrow} c^{+(2)}_{-k\downarrow} c^{(1)}_{-k\downarrow} c^{(2)}_{k\uparrow}$$

$$+ (1 \leftrightarrow 2) + \text{H.C.}$$

and, second, as previously proposed,

$$\mathcal{H}_{PT} = \sum_k \frac{\lambda_k}{2} (c_{k\uparrow}^{+(1)} c_{-k\downarrow}^{+(1)} c_{-k\downarrow}^{(2)} c_{k\uparrow}^{(2)}).$$

In a similar way to the BCS case, states not satisfying $n = 2$ are annihilated by the interaction. In this case the Hamiltonian is number-conserving overall and the states which diagonalizes the sum of the two interactions are simply

$$\Psi_{sg}^1 = c_{k\uparrow}^{e+} c_{-k\downarrow}^{e+} \Psi_{vac}$$
$$\Psi_{sg}^2 = c_{kt}^{o+} c_{-kt}^{o+} \Psi_{vac}$$

and all other states are unaffected by the pairing Hamiltonian. These are the spin-gap states, identical to that proposed by Strong and Anderson except that the pairing is in the even *or* odd state, while the two other $N = 2$ states are either repulsive or not lowered in energy. Otherwise the story is unchanged. In the superconducting state we presume the gap function is correlated by couplings to other nearby momentum states and to other planes. We can model this by assuming that the momentum-conserving δ-function is not exact but has a finite width $\delta k \sim \frac{1}{L}$ (an additional length scale which enters the interlayer theory, whose physical consequences we do not explore here). But $L \to \infty$ leads to physical nonsense when explored in too great detail.

Note that (for instance) any electron excited from a spin-gap state must leave behind a hole in an even state $c_k^{(e)}$ and any k in the spin-gap state can only accept an electron in an odd state $c_k^{+(o)}$. Thus the spin-gap states can exactly satisfy the Keimer selection rule $e \to o$ for scattering of electrons by neutrons. This reflects the source of the pairing energy in the kinetic energy term.

$$\text{K.E.} = t_\perp (n_e - n_o).$$

Also note that although the odd states are nominally unpaired, nonetheless adding or removing an electron in an odd state from a spin-gapped k-value destroys the pairing criterion $n = 2$ and costs one unit of pairing energy. Thus there is a gap for all one-electron excitations, even though only the even state is paired.

Now we consider the more conventional BCS-like theory. Here we must require phase-coherence of the pairing among all k-states, so that the pairing Hamiltonian is no longer number-conserving, and may be treated by the usual mean field theory. But when we transcribe our second-order pairing Hamiltonian into "even" and "odd" language, it turns out to read

$$-\mathcal{H} = \sum_k \lambda_k \sum_{k-k'<L} (c_{k\uparrow}^{e+} c_{-k\downarrow}^{e+} - c_{k\uparrow}^{o+} c_{-k\downarrow}^{o+})$$
$$\times (c_{-k'\downarrow}^e c_{k'\uparrow}^e - c_{-k'\downarrow}^o c_{-k'\uparrow}^o).$$

Thus it is favorable for pairing to occur in even *or* odd states but not in both together. The resulting problem is more complicated than BCS; in fact, unlike BCS mean field is not quite adequate to the solution. Basically, the system goes from 4 states empty

(a) $(n_k^e = n_k^o = o)$ to all full **(b)** $(n_k^e = n_k^o = 2)$

via *even pairing only*

(c) $n_k^e = 2, n_k^o = o$

but continuously, with coherent occupation of (a), (b), (c). A self-consistent BCS calculation give us

$$\Delta_{eff}^{+(k)} = \lambda_k \sum (\langle c_k^{+e} c_{-k}^{+e} \rangle - \langle c_k^{o+} c_{-k}^{o+} \rangle)$$

changing sign and therefore vanishing at $k = k_F$. [11] What we expect will happen, however, is that at $k \simeq k_F$ there will be phase-correlated fluctuations of b_k^+ and $b_{k'}^+$ as in the spin gap state. Thus the actual energy gap does not actually vanish at any k. How to deal in any exact way with the phase fluctuations of the gap is as yet an unsolved problem. I speculate that the mathematics of gap formation may resemble more nearly that of Bose condensation of preformed pairs than the simple pair condensation of BCS.

If we had only the same-layer pairing Hamiltonian which we have used in the past, i.e.,

$$\sum_k \lambda_k \sum_{k'-k<L} (c_k^{+(1)} c_{-k}^{+(1)} c_{-k'}^{(2)} c_{k'}^{(2)} + cc),$$

we obtain $\lambda_k \chi_k^{pr} \simeq 1$ as the effective gap equation, with $\chi_k = \frac{1}{E_k}$; this can only be satisfied with $\epsilon_k^2 \langle \lambda_k^2$. The pairing amplitude $b_k = \Delta_k = \sin \theta_k$.

$$\cos \theta_k = \frac{\epsilon_k}{\lambda_k} = n_k.$$

A crude approximation is to assume that this same mathematics occurs twice, once for even pairs and once for odd ones. Basically, n_k^o goes linearly from 2 to 0 from $\epsilon_k = -\lambda_k$ to 0; and n_k^e goes linearly from 2 to 0 from $\epsilon_n = 0$ to $\epsilon_k = +\lambda_k$.

The ratio of amplitudes $e \to e$ and $o \to o$ to $e \leftrightarrow o$ can be calculated using this assumption naively. The ration of occupation factors is equal to

$$\frac{\int_{-\lambda}^{\lambda} n_k^o d\epsilon \int_{-\lambda}^{\lambda} d\epsilon (2 - n_k^e) + \int_{-\lambda}^{\lambda} n_k^e d\epsilon \int_{-\lambda}^{\lambda} n_k^o) d\epsilon}{\int_{-\lambda}^{\lambda} n_k^e d\epsilon \int_{-\lambda}^{\lambda} (2 - n_k^e) d\epsilon + \int_{-\lambda}^{\lambda} n_k^o \int_{\lambda}^{\lambda} (2 - n_k^o)} = \frac{5}{3}.$$

This leaves out coherence factors which may considerably enhance the ratio (I estimate by a factor 2). This is still less skewed than the data. The state may resemble more the correlated "spin-gap" state than this uncorrelated mean field approximation. The "spin gap" gives a ratio of 1:0, and it is unreasonable that the superconducting state should be very much lower.

In a forthcoming paper, in collaboration with S. Chakravarty, we shall show how the amplitude, shape, and intensity of the neutron peak follows from the above ideas. Many new complexities are caused by the unique nature of the pairing Hamiltonian in this theory, and we look forward to exploring an unexpectedly rich field of physics.

I have benefitted greatly from discussions with N. P. Ong, D. C. Clarke, S. Sarker, S. P. Strong, and, especially, S. Chakravarty. The work grew out of discussions with B. Keimer. This work was supported by the NSF, Grant # DMR-9104873.

REFERENCES

1. A. Millis and H. Monien, Phys. Rev. Lett., **70**, 2810 (1993); also recent preprint.
2. P. W. Anderson and S. P. Strong, Chinese Journal of Physics, forthcoming.
3. P. W. Anderson, Advances in Physics, forthcoming.
4. L. Balents and M.P.A. Fisher, preprint.
5. C. Herring, in *Magnetism*, G. Rado and H. Suhl eds., Vol. II (64), Vol. IV (1966) Acad. Press, New York.
6. D. G. Clarke, S. P. Strong, and P. W. Anderson, Phys. Rev. Lett., **74**, 4499 (1995).
7. J. R. Campuzano et al., Phys. Rev. Lett., **76**, 1533 (1996).
8. D. Van der Marel, et al., Physica C **235–40**, Pt. II, p. 1145 (1994) and references therein.
9. H. F. Fong, B. Keimer, et al., Phys. Rev. Lett., **75**, 316 (1995).
10. P. W. Anderson, Phys. Rev. **112**, 1900 (1958); also see Y. Nambu.
11. Thus as far as the mean field gap function is concerned, the speculation which I proposed in 1991 (P. W. Anderson, Physica C **185–89** 11–16 [1991]) is roughly confirmed; this was followed up by A. Balatsky and E. A. Abrahams, Phys. Rev. **B45**, 13125 (1992). However, the physical picture of a conventional energy gap is more descriptive.

T

The Neutron Peak in the Interlayer Tunneling Model of High-Temperature Superconductors

Lan Yin,* Sudip Chakravarty,* and Philip W. Anderson

(18 June 1996)

ABSTRACT

Recent neutron scattering experiments in YBCO exhibit an unusual magnetic peak that appears only below the superconducting transition temperature. The experimental observations are explained within the context of the interlayer tunneling theory of high-temperature superconductors.

PACS: 74.72.-h, 74.20-z, 74.25Ha

Many experimental observations in high temperature superconductors are commonly fit with a phenomenological model that derives from the original theory of Bardeen, Cooper, and Schrieffer (BCS). The model has certain ingredients. It is characterized by a gap equation corresponding to a presumed symmetry of the order parameter, with an adjustable dimensionless coupling constant, or alternatively an adjustable ratio of the gap to the superconducting transition temperature, and a given Fermi surface. Inherent in this description are the coherence factors that determine a number of interference processes unique to the BCS theory. Although a microscopic derivation of this effective model for the high-temperature superconductors does not exist, the model is still used with a considerable degree of confidence. The main difficulties, to which we return below, are the unusual normal state properties of these materials and the enormously high transition temperatures, but there are many others. Such difficulties are extensively surveyed in the literature [1].

In the absence of a microscopic derivation, it is useful to ask if this phenomenological BCS model is unique and if an alternative phenomenological model exists that is capable of capturing features of these superconductors. One such physically motivated model was elaborated in a recent paper [2].

*Department of Physics and Astronomy, University of California, Los Angeles.

In the present paper we examine this model, called the interlayer tunneling model, to interpret the startling neutron scattering experiments in optimally doped YBCO [3].

These neutron scattering experiments are startling for a number of reasons. The experiments appear to have established that there are no sharp, or even broad, features in the magnetic excitation spectrum in the normal state. In contrast, the superconducting state exhibits a sharp magnetic feature that appears to be localized both in energy and momentum. It is located at an energy of 41 meV and near a wavevector $(\pi/a, \pi/a, \pi/c_b)$, where a is the lattice spacing of the square-planar CuO lattice, and c_b is the distance between the nearest-neighbor copper atoms within a bilayer. The intensity under the peak vanishes at the transition temperature, but its frequency softens very little.

In principle, there can be many explanations, and such explanations have been proposed [4]. None of these explanations are fully microscopic, nor are they fully consistent with some of the other prominent experimental observations in these materials. In particular, we draw attention to the face that explanations that rely on forming *coherent* linear superpositions of the bilayer bands would also predict large splittings of the bands in certain regions of the Brillouin zone. Such splittings, studied extensively in experiments, are missing to a high degree of accuracy [5]. The argument that the observed bands are highly renormalized has little force. If the renormalization effects are so strong as to reduce the splitting to nearly zero, it is not meaningful to speak of a coherent superposition of states at the experimentally relevant temperatures.

The explanation we have to offer within the interlayer tunneling model is exceedingly simple and purely kinematic in origin. We also present illustrative but quantitative results to compare with experiments. The comparison shows that many of the features of the experiments are captured well. While many improvements can possibly made, the kinematic aspects should survive future scrutiny.

The model Hamiltonian, motivated earlier [2], is that of a bilayer complex. Its generalization to many layers is straightforward and not essential for the present purpose. It is

$$
\begin{aligned}
H &= \sum_{k\sigma i} \varepsilon_k c^\dagger_{k\sigma i} c_{k\sigma i} \\
&= \sum_{q,k,k',\sigma,\sigma',i} V_{q,k,k'} c^\dagger_{k\sigma i} c^\dagger_{-k+q\sigma' i} c_{-k'+q\sigma' i} c_{k'\sigma i} \\
&= \sum_{q,k,\sigma,\sigma',i \neq j} T_J(q,k)[c^\dagger_{k\sigma i} c^\dagger_{-k+q\sigma' i} c_{-k+q\sigma' j} c_{k\sigma j} + \text{h.c.}].
\end{aligned}
\tag{1}
$$

Here $i = 1, 2$ is the layer index. The fermion operators are labeled by the spin σ and the wavevector k; V is the in-plane pairing interaction. The last term describes tunneling of pairs between the layers. Such a Hamiltonian should be understood from the point of view of an effective Hamiltonian, as emphasized recently [6]. It incorporates the unique feature of the interlayer mechanism, that tunneling occurs with conservation of transverse momentum $\mathbf{k}$. Therefore, the momentum sum in the T_J term is only over $\mathbf{k}$ and $\mathbf{q}$. Disorder between the layers is weak, and even disorder would not affect this crucial difference between the tunneling and interaction terms.

Only in the subspace in which both the states $(\mathbf{k} \uparrow)$ and $(-\mathbf{k} \downarrow)$ are both simultaneously occupied or unoccupied, is the following reduced Hamiltonian sufficient:

$$H_{\text{red}} = \sum_{\mathbf{k}\sigma i} \varepsilon_\mathbf{k} c^\dagger_{\mathbf{k}\sigma i} c_{\mathbf{k}\sigma i} - \sum_{\mathbf{k},\mathbf{k}',i} V_{\mathbf{k},\mathbf{k}'} c^\dagger_{\mathbf{k}\uparrow i} c^\dagger_{-\mathbf{k}\downarrow i} c_{-\mathbf{k}'\downarrow i} c_{\mathbf{k}'\uparrow i}$$

$$- \sum_{\mathbf{k},i \neq j} T_J(\mathbf{k}) [c^\dagger_{\mathbf{k}\uparrow i} c^\dagger_{-\mathbf{k}\downarrow i} c_{-\mathbf{k}\downarrow j} c_{\mathbf{k}\uparrow j} + \text{h.c.}]. \qquad (2)$$

This is because these are the only matrix elements that survive. It must be remembered that the actual Hamiltonian is H and that H_{red} is merely a convenient tool to calculate the matrix elements in the paired subspace. We do not anticipate that the Schrieffer zero-momentum pairing hypothesis will be necessarily as accurate in this case as in BCS, but it surely will be a guide.

An important feature is the absence of the single particle tunneling term from the Hamiltonian, which allows for a coherent superposition of the states of the layers that leads to a splitting, not observed in experiments. Therefore, such terms are not included in our model. Although the single particle tunneling term is absent, incoherence does not preclude a particle-hole two particle tunneling term [2,6], because the two particle term is generated by a second order virtual process and the question about coherence is irrelevant; the energy need not be conserved in a virtual process. However, this term does not appear to be crucially important for optimally doped materials (except, see later). For underdoped materials exhibiting "spin gap" phenomena [7], this neglected term can be quite important [8]. We also note that we shall treat the Fermion operators as ordinary anticommuting fermion operators. This can only be a crude approximation due to the non-Fermi liquid feature of the normal state. A possible improvement along the lines discussed earlier [6] is exceedingly complex.

The mean field treatment of the reduced Hamiltonian is straightforward [2] and leads to the gap equation:

$$\Delta_k = \frac{1}{1 - \chi_\mathbf{k} T_J(\mathbf{k})} \sum_{k'} V_{\mathbf{k},\mathbf{k}'} \chi_{k'} \Delta_{k'}, \qquad (3)$$

where $\chi_{\mathbf{k}} = (\Delta_{\mathbf{k}}/2E_{\mathbf{k}}) \tanh(E_{\mathbf{k}}/2T)$ is the pair susceptibility, with $E_{\mathbf{k}} = \sqrt{(\varepsilon_{\mathbf{k}} - \mu)^2 + \Delta_{\mathbf{k}}^2}$. Here μ is the chemical potential. Note that, until now, we have not specified the symmetry of the in-plane pairing kernel. We shall now assume it to be of the symmetry $d_{x^2-y^2}$, $V_{\mathbf{k},\mathbf{k}'} = V g_{\mathbf{k}} g_{\mathbf{k}'}$, where $g_{\mathbf{k}} = \frac{1}{2}[\cos(k_x a) - \cos(k_y a)]$.

Further specifications are necessary. The in-plane one electron dispersion is chosen to be $\varepsilon_{\mathbf{k}} = -2t[\cos(k_x a) + \cos(k_y a)] + 4t' \cos(k_x a) \cos(k_y a)$. The quantitative calculations are carried out with a representative set of parameters appropriate to YBCO. These are $t = 0.25$ eV, $t' = 0.45t$, and $\mu = -0.315$ eV, corresponding to an open Fermi surface, with a band filling of 0.86. The quantity $T_J(\mathbf{k})$ was argued [2] to be $T_J(\mathbf{k}) = (T_J/16)[\cos(k_x a) - \cos(k_y a)]^4$. The validity of this expression is now strengthened by detailed electronic structure calculations [9] for the single particle hopping matrix element $t_\perp(\mathbf{k})$.

The magnetic neutron scattering intensity is proportional to the imaginary part of the wavevector and the frequency dependent spin susceptibility $\chi(\mathbf{q}, \omega)$, which for the above model is simply the expression:

$$\chi(\mathbf{q}, \omega) = \sum_{\mathbf{k}} \left[\frac{A_{\mathbf{k},\mathbf{q}}^+ F_{\mathbf{k},\mathbf{q}}^-}{\Omega_{\mathbf{k},\mathbf{q}}^1(\omega)} + \frac{A_{\mathbf{k},\mathbf{q}}^-(1 - F_{\mathbf{k},\mathbf{q}}^+)}{2} \left(\frac{1}{\Omega_{\mathbf{k},\mathbf{q}}^{2+}(\omega)} - \frac{1}{\Omega_{\mathbf{k},\mathbf{q}}^{2-}(\omega)} \right) \right], \quad (4)$$

where

$$A_{\mathbf{k},\mathbf{q}}^\pm = \frac{1}{2} \left[1 \pm \frac{(\varepsilon_{\mathbf{k}} - \mu)(\varepsilon_{\mathbf{k+q}} - \mu) + \Delta_{\mathbf{k}} \Delta_{\mathbf{k+q}}}{E_{\mathbf{k}} E_{\mathbf{k+q}}} \right], \quad (5)$$

$\Omega_{\mathbf{k},\mathbf{q}}^1(\omega) = \omega - (E_{\mathbf{k+q}} - E_{\mathbf{k}}) + i\delta$, $\Omega_{\mathbf{k},\mathbf{q}}^{2\pm}(\omega) = \omega \pm (E_{\mathbf{k+q}} + E_{\mathbf{k}}) + i\delta$, $F_{\mathbf{k},\mathbf{q}}^\pm = f(E_{\mathbf{k+q}}) \pm f(E_{\mathbf{k}})$, and $f(x)$ is the Fermi function.

Note that near $T = 0$ only the A^- terms contribute, and these are negligible unless $\Delta_{\mathbf{k}}$ and $\Delta_{\mathbf{k+q}}$ are of opposite sign, as noted in Ref. [3]. In fact, nothing is observed for $q \approx 0$ or 2π, as expected, and the peak appears at $\mathbf{q} = (\pi/a, \pi/a)$, connecting two opposite-sign lobes of $x^2 - y^2$. At higher temperatures, the A^+ part is a temperature dependent, but almost frequency independent function for experimentally relevant frequencies. The χ with the A^+ term omitted will be denoted by $\bar{\chi}$. We shall first calculate the imaginary part of this susceptibility by restricting all wavevectors to be in the CuO-plane. Later, we shall discuss the dependence on the momentum transfer perpendicular to the plane.

At $T = 0$, Im $\bar{\chi}$ is calculated by solving the gap equation for a set of T_J, and with $\lambda \equiv N(0)V = 0.184$, where $N(0)$ is the density of states per spin at the Fermi energy ($V = 0.2$ eV, $N(0) = 0.92$). These are shown in fig. 1. While the calculations show a peak at the threshold for $T_J \neq 0$, in the pure BCS case, $T_J = 0$, the intensity is a step discontinuity at the threshold, and other excitonic enhancement mechanisms are necessary to produce a peak at the threshold.

For illustrative purposes the gap equation is solved at finite temperatures for $\lambda = 0.184$ and for a fixed $T_J = 0.075$ eV. No attempts were made to fit

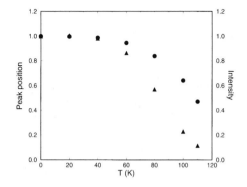

Figure T.1. Im $\bar{\chi}$ at $T = 0$. From left to right, $T_J = 0.025, 0.05, 0.075$ eV. The curve corresponding to a step discontinuity at the edge is for BCS, with the maximum d-wave gap equal to 0.025 eV.

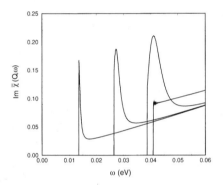

Figure T.2. Im $\bar{\chi}$ for $T_J = 0.075$ eV. From left to right, $T = 120, 110, 100, 80, 60, 40,$ 20, and 0 K. The results for $T = 0$ and 20 K are almost indistinguishable.

precisely the experimental data. The results are plotted in fig. 2. We calculate the intensity under the peak and plot it as a function of temperature. The intensity was calculated by fitting a linear background at high energies. The results are shown in fig. 3. The intensity falls off to zero as the temperature is raised to T_c and the results compare well with experiments. Also shown in fig. 3, is the position of the peak whose softening is weaker than the fall-off of the intensity. However, the experimental softening of the position of the peak is even weaker.

The question to address is why Im $\bar{\chi}$ produces a peak in the interlayer tunneling model, while the BCS model exhibits only a step discontinuity at the threshold. The reason is simple kinematics that holds for an open Fermi surface combined with an unusual feature of the interlayer gap equation. This

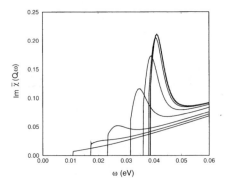

Figure T.3. The intensity (solid triangles) and the position of the peak (filled circles) normalized to the zero temperature values. The position of the peak is at 41.2 meV at $T = 0$.

feature was emphasized in early works on the gap equation [10], but has not been specifically noted in recent calculations: it is that when T_J dominates, as is necessarily so for highT_c, the density of states for a given value of $T_J/2$ is approximately a δ-function at $T_J/2$. Concomitant with this is the fact that electrons with $|\varepsilon_\mathbf{k} - \mu| > T_J(\mathbf{k})/2$ are unaffected by pairing, in contrast to BCS where effects of the gap extend out to high energies.

For simplicity, consider $T = 0$, set the coherence factor to unity and look at the wavevector $\mathbf{Q} = (\pi/a, \pi/a)$. In the BCS case, that is with $T_j(\mathbf{k})$ set equal to zero, $(E_\mathbf{k} + E_{\mathbf{Q}+\mathbf{k}})$ has either minima, or saddle points. The minima give rise to a step discontinuity at the threshold at approximately twice the maximum of the d-wave gap (see fig. 1). The saddle points give rise to van Hove singularities which are generally not at the threshold, but can be brought close to the threshold by adjusting parameters.

Now consider the case of a very small in-plane pairing kernel in the interlayer tunneling equation. The unusual feature is that the superconducting region is a very narrow region around the Fermi line as shown in fig. 4. The scattering surface is now radically different. The contribution to the peak must come from regions in which both $\Delta_\mathbf{k}$ and $\Delta_{\mathbf{k}+\mathbf{Q}}$ are finite. Also, note that $E_{\mathbf{k}+\mathbf{Q}} = E_{\mathbf{Q}-\mathbf{k}}$. The image of the superconducting region including the Fermi line under the mapping $\mathbf{k} \rightarrow \mathbf{Q} - \mathbf{k}$ is shown in fig. 4. The regions in which both $\Delta_\mathbf{k}$ and $\Delta_{\mathbf{Q}-\mathbf{k}}$ are finite are the diamond shaped overlap regions, with one of its diagonals given by $k_x + k_y = \pi$. Consider the lower diamond. In this region, we can write $E_\mathbf{k} \approx T_J \cos^4(d/\sqrt{2})$, where d is the distance from $(\pi/a, 0)$. The corresponding constant ω-contours are the arcs shown in fig. 4.

Note that the Im $\bar{\chi}$, as a function of ω, is proportional to the length of the arcs for which $(E_k + E_{\mathbf{Q}-k}) = \omega$. The lower and the upper cutoffs are defined by the arcs touching the diamond, and the peak is contained between these two

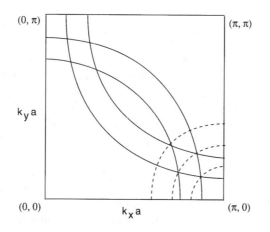

Figure T.4. The superconducting region and its mapping under $\mathbf{k} \rightarrow \mathbf{Q} - \mathbf{k}$. The diamond-shaped overlap regions contribute to the peak of the imaginary part of the spin susceptibility. The dashed arcs are the constant ω-contours centered at $(\pi, 0)$.

cutoffs. As ω increases, the arc length increases approximately linearly within the diamond, and then drops approximately linearly. Moreover, because the superconducting region around the Fermi line is very narrow, the width of this peak is also very narrow, as in experiments. The peak is at $T_J(\mathbf{k}_0)$, where $\mathbf{k}_0$ is located at the center of the diamond. This elementary analysis substantiates the numerical results in which none of the above simplifying approximations were made.

To understand the dependence of the scattering intensity on the momentum transfer perpendicular to the plane, it is necessary to consider the mixing of the electronic wavefunctions between the layers. This cannot be a coherent superposition of states, otherwise the bands will be split, which, as stated earlier, is not observed in experiments. However, a virtual mixing of the states due to second order processes, similar to superexchange, is possible. This virtual mixing can be described by approximately constructing the operators [8]: $\alpha_{\mathbf{k},\sigma,1} = (c_{\mathbf{k},\sigma,1} + \eta(\mathbf{k})c_{\mathbf{k},\sigma,2})/\sqrt{1 + \eta^2(\mathbf{k})}$, and $\alpha_{\mathbf{k},\sigma,2} = (c_{\mathbf{k},\sigma,2} + \eta(\mathbf{k})c_{\mathbf{k},\sigma,1}/\sqrt{1 + \eta^2(\mathbf{k})}$, where $\eta(\mathbf{k})$ is the mixing parameter. The order parameter that takes the mixing into account is $\Lambda_{\mathbf{k}} = \langle \alpha^\dagger_{\mathbf{k},\uparrow,1}\alpha^\dagger_{-\mathbf{k},\downarrow,1} + \alpha^\dagger_{\mathbf{k},\uparrow,2}\alpha^\dagger_{-\mathbf{k},\downarrow,2} \rangle$. It is also possible to rewrite this in terms of the operators $c^e_{\mathbf{k},\sigma} = \frac{1}{\sqrt{2}}(c_{\mathbf{k},\sigma,1} + c_{\mathbf{k},\sigma,2})$ and $c^o_{\mathbf{k},\sigma} = \frac{1}{\sqrt{2}}(c_{\mathbf{k},\sigma,1} - c_{\mathbf{k},\sigma,2})$. Thus,

$$\Lambda_{\mathbf{k}} = \frac{(1 + \eta(\mathbf{k}))^2}{1 + \eta^2(\mathbf{k})}\langle c^{e\dagger}c^{-e\dagger} \rangle + \frac{(1 - \eta(\mathbf{k}))^2}{1 + \eta^2(\mathbf{k})}\langle c^{o\dagger}c^{-o\dagger} \rangle, \tag{6}$$

where we have used the notation $c^{e\dagger} \equiv c^{e\dagger}_{\mathbf{k},\uparrow}$, $c^{-e\dagger} \equiv c^{e\dagger}_{-\mathbf{k},\downarrow}$, etc.

This virtual mixing of the wave functions on the layers is a necessary concomitant of the source of the pairing energy in the hopping terms in the original Hamiltonian $-\sum_{\mathbf{k},\sigma,i\neq j} t_\perp(\mathbf{k})(c^\dagger_{\mathbf{k},\sigma,i}c_{\mathbf{k},\sigma,j} + \text{h.c.}) = -\sum_{\mathbf{k},\sigma} t_\perp(\mathbf{k})(n^e_{\mathbf{k},\sigma} - n^o_{\mathbf{k},\sigma})$, where n are the number operators. The pairing energy must be numerically augmented by its expectation value. However, the two-particle tunneling term alone does not give quite enough difference in occupancy to explain the experimental modulation with q_x. In a forthcoming paper [8], we will show that this tendency is strongly enhanced by including the "superexchange" type terms caused by virtual particle-hole hopping. In fact, it may be that the pairing occurs primarily in the "even" linear combination alone [11].

Then, any electron excited from the condensate must leave behind a hole in the even state. Similarly, it can accept only an electron in the odd state, hence the striking even $\leftrightarrow$ odd selection rule observed in the neutron scattering experiments. Note that although the odd states are nominally unpaired, the addition or removal of an electron from the odd state destroys the pairing criterion and costs one unit of pairing energy. Thus, there is a gap for all one-electron excitations, even though only the even state is paired.

In conclusion, we find that these neutron results give us a surprisingly direct and complete picture of the nature of the pairing responsible for superconductivity, not only showing us the symmetry and approximate form of the energy gap, but even fixing the nature of the gap equation and the source of the pairing energy. In particular, the strong correlation between the layers excludes any purely intralayer mechanism for superconductivity, under our assumption that the quasiparticle pairs do not have strong final state interactions.

This work was supported by the National Science Foundation: Grant Nos. DMR-9531575 and DMR-9104873. We thank B. Keimer for many stimulating discussions.

REFERENCES

1. P. W. Anderson, Science **256**, 1526 (1992).
2. S. Chakravarty, A. Sudbø, P. W. Anderson, and S. Strong, Science **261**, 337 (1993).
3. H. F. Fong et al., Phys. Rev. Lett. **75**, 321 (1995) and H. F. Fong et al., to be published.
4. I. I. Mazin and V. Yakovenko, Phys. Rev. Lett. **75**, 4134 (1995); E. Delmer and S.-C. Zhang, Phys. Rev. Lett. **75**, 4126 (1995); D. Z. Liu, Y. Zha, and K. Levin, Phys. Rev. Lett. **75**, 4130 (1995); N. Bulut and D. J. Scalapino, Phys. Rev. B **53**, 5149 (1996); B. Normand, H. Kohno, and H. Fukuyama, J. Phys. Jpn. **64**, 3903 (1995).
5. H. Ding et al., Phys. Rev. Lett. **76**, 1533 (1996); D. Van der Marel et al. Physica C **235–240**, 1145 (1994) and references therein.
6. L. Yin and S. Chakravarty, Int. J. Mod. Phys. **10**, 805 (1996).

7. A. G. Loeser et al., Science, in press, H. Ding et al., Nature in press, and references therein.

8. P. W. Anderson, preceding preprint.

9. A. I. Liechtenstein, O. Gunnarsson, O. K. Andersen, and R. M. Martin, preprint (cond-mat/9509101).

10. P. W. Anderson, Physica C **185–189**, 11 (1991); J. Phys. Chem. Solids **54**, 1073 (1993); chap. 7 of a forthcoming book.

11. If we augment our Hamiltonian (Eq. 2) with an equal particle-hole pair hopping (superexchange) term and ignore the small in-plane interaction, it acquires an additional approximate symmetry with respect to the interchange of particles of a given spin species and of wave vector **k** between the layers. The odd order parameter does not respect this symmetry and must vanish unless the symmetry is spontaneously broken.

PART III
A Critical Postscript

A Critique of Alternative Theories

In the body of this book I have almost completely ignored alternative theories of high-T_c superconductors, feeling that the story I had to tell was most cleanly written as a single intellectual thread without confusing the reader by discussing the alternative lines that I as well as others have pursued. If the book is to have any permanence, it should retain its usefulness long after the controversies have evaporated, and they should not be featured.

Nonetheless, few readers will see clearly, at first, why these very popular theories which are, or were, pursued with such vigor must each be rejected on experimental grounds (and often on theoretical grounds as well).

Here in a postscript I try to assemble and organize the main contenders, and to give at least one crucial experiment which rejects each of them. Almost all of these experiments have been available since the first few years of the subject. One of the quotations in the press which I found most remarkable was in a Physics Today news story in which, in 1993, a prominent experimentalist was quoted as saying, "finally (referring to the d-wave interference measurements) we have some information from experiment!" Experiment had uniquely determined the form of the theory several years before that time.

I shall restrict myself to serious theories: theories promulgated by several investigators, making some pretense either to explain a variety of experimental phenomena or to be based on a realistic physical model of the materials, and at least to some extent referred to by others. The field has been beset by eminent outsiders, with highly publicized but not very useful suggestions. We, in the field, are much to blame for this because of our frequent public statements about "lack of consensus." A much larger number of obscure individuals, of course, have also published such papers but not in general been taken much notice of. The scientific literature is, probably for good reason, very tolerant of irrelevance, balderdash, and even chicanery, and deals with it, as I shall, by ignoring it.

A second type of theory which I will reject en masse is the type of theory variously called "d-p," "extended Hubbard," or "two-story house," in which the Cu_{3d} and O_{2p} electrons are treated as dynamically independent. Theory and experiment dispose of this possibility very simply and straightforwardly, by remarking that the hybridization bonding of these two types of electron is responsible for the square planar structure, which is very stable and rigid. Thus the bonding-antibonding splitting is one of the largest parameters in the problem ($\sim$4–6 ev) and only the antibonding linear combination is involved in mobile electron states, while the bonding combination is always occupied by a pair of electrons.

But direct experimental evidence also exists. Slichter et al. have shown that the NMR relaxation rate goes to the same value at high temperatures ($> \sim 500°K$) for all values of δ. In the insulator, this relaxation is undeniably due to the magnetic electrons, which are assumed in the above theory to be "d," since these are the only mobile degrees of freedom. For large values of δ, this same relaxation fits smoothly on to the Korringa relaxation from the free electrons: presumably "p" in these theories. The coincidence is too much to swallow: they must be the same electrons.

We may classify the serious theories in several ways, and I choose arbitrarily the following three categories:

I. Theories following the BCS scheme of pairing electrons via the exchange of X, where X is some boson not particularly involved in the peculiar electric and magnetic properties of the $Cu O_2$ planes, and where the "electrons" are assumed to be coming from a conventional normal metal. The main contenders for X are phonons, some kind of nonlinear excitation associated with the "apical oxygen," plasmons, or various more complex and much vaguer excitonic degrees of freedom, such as "d-mons."

All of these theories are, to my mind, excluded on the "tooth fairy" principle, at least. The remarkable and unique magnetic and anisotropic electrical properties of the planes, giving a universal "generalized phase diagram," accompany absolutely unique superconducting T_c's. Ockham's razor tells us these two sets of phenomena must be related, and the phonon mechanism does not explain that relationship. In general, there is no explanation, suggested or implied, for any of the systematics of T_c or other properties, for instance, why does overdoping invariably lower T_c, where in BCS T_c increases with carrier density?

More rigorously, in chapter 4 we showed that the mechanism responsible for resistance in the normal metal is—experimentally—*not phonons* but low-energy electrons (it gaps with the superconducting gap, leading to high quasiparticle conductivity below T_c). Conventional BCS cannot deal with that fact. The d-wave gap symmetry, if true, also excludes all of category I, since phonons are uniformly attractive for all pairs, as are all of the bosons suggested, in spite of claims to the contrary.

An interesting point needs to be made here. A bit of conventional wisdom which seems to crop up unchallenged in unlikely places (as for instance in A. J. Leggett's history of superconductivity in the Institute of Physics 3-volume history of physics in the 20th century) is that all interactions are carried by bosons of some kind, leading to some kind of inevitability of the BCS-Eliashberg structure.

This is valid only if the interacting entities are themselves simple Fermions or bosons. Spins, for instance, have superexchange and double exchange interactions which are not thus describable. Nuclear physics was confused for many years by the attempt to describe the complex effects of quantum chromodynamics by boson exchanges. If electrons in CuO_2 layers are not eigenexcitations, as we argue, there is no reason their interactions need fit into this kind of straightjacket, and they do not.

II. Theories which retain the framework of BCS, a "normal" metal with quasi particles which pair and which retain a finite "Z" over most at least of the Fermi surface, but where interactions or quasiparticle properties take into account the peculiar nature of the planes. There are two main contenders here: van Hove singularity theories, and spin fluctuations.

There is here another "tooth fairy" problem, less obvious but still quite strong: T_c varies widely depending on the structure *between* the planes, in a very systematic way (as seen in the fig. in chapter 7). The properties of the planes do not vary, as shown in Batlogg's plot (as updated recently by Ong) of T_c vs $\frac{d\rho_{ab}}{dT}$. (Fig. 1.) Any theory which relies for T_c on the same mechanism which causes transport within the plane is relying on a whole band of tooth fairies, since one varies widely with the reservoir layers, and the other does not. There is also a sharp experimental problem for each of the two theories separately.

Spin fluctuations can be persuaded to fit $\rho_{ab}(T)$, with some difficulty. They have no obvious (or nonobvious) way to fit the remainder of the transport properties, especially the characteristic anomalous Hall angle and the infrared reflectivity. And they give no plausible non-tooth-fairy explanation of the simple power laws for these phenomena nor of the striking resistivity anisotropy, which is ruled out in a finite Z theory. Computations purporting to "explain" these properties by some magic which has no obvious physical basis but is hidden in complex machine calculations become increasingly desperate and leave me saddened and, I am sorry to say, suspicious of their reproducibility by independent workers.

Two further comments need to be made. Modern neutron data demonstrate conclusively that in optimally doped $YBCO_7$ the normal state shows no peak in χ'' near $Q = \pi, \pi$; the "spin fluctuations" on which the theory is based simply do not exist. Second, as remarked in the text, the naive RPA formalism used is not equivalent to what has been done with ferromagnetic spin fluctuations, and is not internally consistent.

The van Hove singularity "theory" has evolved an electronic and a phonon version and is held in different versions by different authors. This is one of the most serious problems of the "van Hove scenario." The literature on it has a tendency to point proudly to the fact that there *is* a van Hove singularity

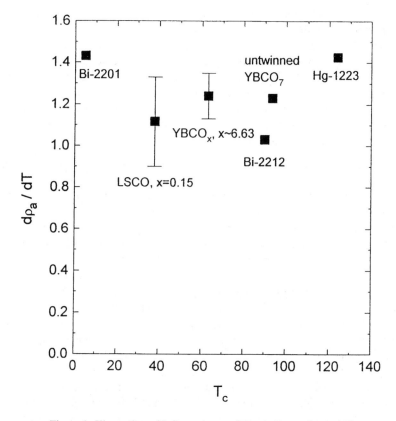

Figure 1. Illustration of Independence of Planar Properties and T_c

somewhere near E_F, but not to present a coherent or consistent program for actually relating that fact to the high T_c's and other anomalous properties. The spin fluctuation theories suffer from excessive detail; the van Hove, from vagueness.

The phonon interaction version is excluded by the same arguments as other phonon theories. For both, the problem is that van Hove singularities, i.e., sharp density of states peaks—have experimental consequences, which are clearly not visible in these materials. The other type of material which is thought (though probably not correctly) to exhibit van Hove singularity peaks is the strong coupling transition metal compounds such as V_3Si and Nb_3Sn. These exhibit large paramagnetic spin susceptibility which peaks at low temperature (a "Curie-Weiss" susceptibility), evinced both in NMR and χ data. This is presumed to be the consequence of thermal smearing of density of states peaks. The susceptibility (and Knight shift) of the high-T_c normal metal is a rising function of temperature quite opposite in appearance to that of these metals. Other DOS-peak related phenomena such as peaks in angle-averaged photoemission, tunneling conductivity, etc., are also not seen. There is clear experimental evidence, in fact, that the Fermi surface does run fairly close to a singularity at the M point. But with doping it seems to move casually through this point without any rapid change in T_c or any of the above phenomena appearing, which indicates that the phenomenology is dominated by many-body interactions and responds only mildly to detailed band structure effects.

Finally, the van Hove peak is particularly implausible because a peak in density of states is necessarily a relatively unstable position for the Fermi level thermodynamically, i.e., it is a point of maximum curvature

$$\frac{d^2 F}{dn^2}$$

of the free energy as a function of electron density. As has been well known since Hume-Rothery, the Fermi level seeks gaps and minima in the density of states, not maxima. Moreover, it is observed that the high-T_c materials are, in general, "self-doping": they naturally dope themselves to the highest T_c, which is a point of high stability. A state near a van Hove peak would have quite the opposite behavior.

Let me meliorate the above very negative-sounding attitude to these theories, both spin fluctuation and van Hove. Both theories have a definite relationship to reality, in that the phenomena that are actually taking place would be described, within the "standard model" of renormalized quasiparticles, by these two theories. It is not the physics of these theories but the underlying model which is the failure. We are learning that spin-charge separation within the fundamental excitations may be the general response to strong repulsive interactions, in new theories and experiments on the Mott transition; antiferromagnetic spin fluctuations à la RPA are the best approximation to this phenomenon within the standard model but turn out not to be good enough.

Equally, undoubtedly the one-dimensional character of the excitation spectrum is enhanced and made more obvious by the approximate nesting of the Fermi surface; Luther has made it clear that this has a tendency to be self-reinforcing. But any treatment of these phenomena must, as a starting point, take into account the non-Fermi liquid nature of the excitation spectrum. It is also clear empirically that the apparent band flattening is a many-body phenomenon which is not reflected in density of states. So again, a part of the story, treated from the point of view of the standard model, cannot carry the whole weight.

Finally, category III is the states based on quite new physics. Ours is one of these; four others are the "anyon" theory originated by Laughlin and Mele and picked up by many authors such as Zee, Wilczek, Wen, Wiegmann, etc.; the "bipolaron" theory of Mott and Ranninger; the "marginal Fermi liquid" theory of Varma, Littlewood, and others; and, in a category between II and III, the "spin bags" of Schrieffer and coworkers.

Anyon theory is a natural development from the early idea that the key to the phenomena was the antiferromagnetic interaction J. In the absence of Néel antiferromagnetism, a reasonable ground state of the Heisenberg 2D Hamiltonian is a chiral "RVB" state based on Gutzwiller projection of a filled Landau level, rather than the corresponding projection of the Fermi sea. This is an equally natural generalization of the 1D Bethe solution to two dimensions. Various mean field theories also suggested that the Néel antiferromagnet on doping might break down first toward a spirally ordered state; and some gauge theories led in this direction. But in general the weakness of the attempt to fit fractional statistics theory on to high-T_c superconductivity has been that these theories do not start with a clear relationship to the Hubbard model or to any good physical model of the CuO_2 planes. What instead was done was (as in the quantum Hall effect, at first) to postulate wave functions and to test them variationally (if even that degree of realism), and in terms of their experimental consequences (but it didn't work in this case). Therefore it is rather hard to find sharp experimental contradictions. However, one is enough, and this is a very general one: in any fractional statistics theory, one adds an effective flux line to each particle, compensating the resulting flux by an enormous net average fictitious field. The result is to remove the Fermi surface and replace it by flat energy-levels with gaps between. The theory depends on gaps and postulates gaps. But in the normal state of high-T_c materials, at least those with the highest T_c's, there is unquestionably a Fermi surface, and this Fermi surface has the correct volume and shape to be that based on the simplest possible band structure. In the presence of a magnetic field, Fermi surface even seems to be present in the superconducting mixed state. There are many experimental demonstrations of this, so many that one hardly need detail them, but the Korringa relaxation is the cleanest. The gapped state below T_c is not exotic: it has Josephson effects to Nb, so this too is not "anyon superconductivity," whatever that is. It also, in some cases, retains gapless behavior over some fraction of the surface. These gauge theories are fundamentally incompatible with the existence of a Fermi surface.

The bipolaron theory postulates preformed pairs (coupled presumably by strong electron-phonon coupling) which then Bose condense at T_c. These preformed pairs must of course be singlets. One's instincts are of course completely contrary to this in that there are no indications of the Frank-Condon effects which should accompany this, but the wily arguments of the advocates are not easy to refute. Most straightforward perhaps is NMR: all of the electrons seem to be essentially ungapped in the normal state at least of $YBCO_7$, and to show a reasonable approximation to a Yosida curve of Knight shift for the O sites, so that they all have electron-type masses and pair only at or near T_c, and they all pair, at least in some cases. This group is historically uninterested in magnetic phenomena, and has made no effort to relate their theory to the magnetic data, but there is no possibility of doing so. Neutron measurements give similar, if less accurate, conclusions about spin phenomena.

The "marginal Fermi liquid" theory has much in common with our own non-Fermi liquid theory but has two key differences:

(a) In Fermi liquid theory the Fermi velocity is corrected by a factor $\frac{\partial \Sigma}{\partial \omega} / \frac{\partial \Sigma}{\partial k}$. We postulate that the velocities remain finite, i.e., that spinons, i.e., Fermion-ized spin fluctuations, exist. Hence $\frac{\partial \Sigma}{\partial k}$ must also diverge and the Luttinger liquid is in the single-particle sense incompressible: it does not respond normally to scattering or other one-electron perturbations such as interchain coupling, in our theory, but does in MFL.

(b) An extraneous but unspecified strong interaction vertex apparently unrelated to the Hubbard interaction which causes magnetism, is responsible for the phenomena. With this vertex one can reproduce a reasonable fit to many of the normal state responses.

To my mind, prejudiced as it is against tooth fairies, the introduction of a key mechanism *for the normal state* independent of Mott-Hubbard is enough; and there is also no discussion of systematics of T_c, which introduces another band of tooth fairies. Although the resistivity ρ_{ab} is now natural and not custom-fitted, there is, in common with spin-fluctuations, no further discussion of transport. The chief advantage of non-Fermi liquid theories in blocking interlayer conductivity is discarded because the Fermi surface singularity is too weak, only logarithmic. Although the mysterious interaction vertex is sufficiently vague to allow almost any structure, it is not explicitly motivated to give d-wave symmetry, and the recent interference experiments may also be taken as conclusively negative for this theory, if true as they seem to be.

Finally, this theory has no mechanism which would explain the striking contrast in impurity scattering effects between Zn and other defects. This is a general difficulty with many theories, but particularly obviously so here.

With "spin bags" one is at a bit of a loss because their consequences have been little discussed—e.g., whether they are or are not quasiparticles. As with all theories focused on the planes, there is no explanation of the strong variation of T_c with reservoir layer properties. In fact, the lack of attempts at realistic data fitting relative to others in general basically disqualifies this idea at this stage. (It has serious problems of principle as well, which we do not go into here.)

We can also put these remarks in tabular form. Theories are listed across the top of Table I, crucial experimental data to the left, and an "X" implies exclusion, an "XXX" that this is the most damning experimental evidence in this case. d-wave symmetry is taken as still dubious so is marked as "X?" or "XXX?," but at this moment (3/95) these question marks can be ignored in view of the convergence of many types of data.

TABLE I

	BCS phonons	BCS x-ons	Spin Fluct	Van Hove	Bipolarons	Anyons	Marginal FL	Spin Bags
Occam's Razor: Relevance of A.F.	X	X			X		X	
Heuristics: Variation w. reservoir	X	X	XXX	X	X	?	X	X
Vanishing of $\hbar/\tau$, $T < T_c$	XXX	XXX		X(1/2)	X			
d-wave Symm.	XXX?	XXX?		X?	XXX?		XXX?	X?
Hall Eff. $+\rho_{ab}$	X	XXX	X	X	X	X	X	X
c-axis resistivity	X	X	XXX	X	X		XXX	X
$\chi(T)$				XXX	XXX	X		
Stability of optimal doping	X	X		X				?
Obs. of Fermi surface					X	XXX		
Impurity effects	X	X	X	X	X	?	X	?

It has been remarkable how little discussion has been devoted to this kind of critical comparison of theories. In particular, it seems acceptable to publish paper after paper without confronting the fact that falsifying experiments exist for all of the popular theories. To my mind the most striking example of this universal blindness has been unwillingness to confront the fact that drives most empirical materials people: that T_c varies most radically with layer number. This means that a theory of *superconductivity* which focuses on the single layer is not viable.

More striking still has been the "anyon superconductivity" work which continued apace in the full knowledge that Fermi surfaces exist. Perhaps equally irrational have been the group of experimentalists and band theorists who continue to ascribe a three-dimensional band structure to materials which have no measurable Fermi velocity along one dimension, to an accuracy of at least 10^{-2}.

It is unfair to comment on the response of my own theory to these tests; someone else should suggest the appropriate ones. My own feelings as to the weakest links are these:

(1) A d-wave theory using the repulsive interactions as the "residual interaction" needs to be worked out. It is not clear how to avoid double counting

of diagrams: T_c may come out too high, i.e., d-wave is far from being excluded in principle.

(2) The spin gap phenomena are puzzling, but a theory exists within the model.

Neither contains crucial experimental refutation. If such exists, I do not know about it. If I were asked what is the crucial evidence in *favor* of this theory, it would be the *absence* of crucial negative evidence; put more simply, it is the only theory that is devoid of "tooth fairy" assumptions, in that all parameters used are known or calculable, and are in completely reasonable ranges.

I have not discussed the gauge theories as elaborated particularly by Nagaosa and Lee, as well as, in one form or another, by Wilczek and Zou, Wiegmann and Joffe, and others. The high T results resemble, and are probably roughly equivalent to, ours. The gauge idea is to satisfy the projective constraint of strong interaction or, in many cases, RVB singlet coupling by a Lagrange multiplier which, in a quantum field theory, becomes a gauge field. As in our early work, it is natural to fermionize the magnetic fluctuations in the form of spinons. The problem with the gauge method is that one ends up either with a holon Bose field or a gauge field which tends to condense at too high a temperature, in any mean field treatment. I feel that a correct gauge theory will end up with bosons with a Fermi surface, which will therefore not condense, and hence will lead to a tomographic Luttinger liquid. That is, the gauge approach is correct but more difficult than our methods and in no sense do I condemn it here, out of hand. But I feel that none of the authors have made a conclusive case that they are working with the right gauge theory. In addition, because we know the properties of spinons we know that the gauge symmetries are still there at low temperatures, so any mean field or spontaneous mass generation is incorrect, which makes gauge theories very difficult to deal with mathematically. The gauge theories will, of course, lead naturally to the interlayer mechanism, which was, actually, first motivated by the idea of a non-Abelian gauge field which would confine by a string mechanism.

Yet another candidate theory which can be taken somewhat seriously has appeared over the past two or three years, originating with V. J. Emery but elaborated by others. This is the phase segregation idea, in which the system which shows unusual metallic properties and leads to high T_c is supposed to be an intimate mixture of the antiferromagnetic Mott insulator and the metallic state, the phase boundaries being the operative objects. There does seem to be good evidence for a "stripe" phase of this sort at $\delta = .125$ in $(La - Ba)_2 CuO_4$ and neighboring $La - Sr$ compositions. But this stripe phase causes a marked *decrease* in T_c. The optimally doped metallic phases, as I remarked in discussing the van Hove scenario, are extraordinarily stable thermodynamically and all transport and other evidence (for instance long quasiparticle mean free paths) shows they are homogeneous. Laser experiments which undoubtedly produce a mixture of phases seem invariably to give *macroscopic* segregation of droplets of superconductor in the magnetic insulator. Another crucial fact is that this is yet another boring plane-based theory, with no explanation of the striking dependence on reservoir layers. Like many of its sibling theories, this is a response to the perceived intellectual crisis rather than a reasoned response to experimental fact.

Summary

The book chapters themselves reflect the state of my knowledge when they were last revised, variously from early '93 to early '95. I have tried to supplement them with reprints rather than rewriting them completely to reflect recent developments unless the picture given is completely misleading. Chapters 3 and 4 are the most nearly up-to-date and represent to a great extent my most recent opinions or the most recent experimental data.

The overall scheme has not changed since 1991, but a bewildering variety of details have been cleared up, where my earlier thoughts were vague or preliminary. I should emphasize some of these here, to guide the reader to the appropriate reprint or preprint.

The most fundamental changes have come in the material of chapters 6 and 7.

In chapter 6, the major development is the cleaning up of the derivation of non-Fermi liquid properties in 2D. (I have yet to get my thoughts about higher dimensions in clear form; the reprints (d, g, i) express my near-certainty that the Luttinger liquid phenomenon will occur in that case, but an explicit formal theory does not exist.) This is given in reprint (g), which shows that the key is the effect of interactions on the commutation algebra of the bosons.

On the transport theory, there are several supplements. Recent ARPES data show that the Fermi surface, at least of BISCO, is hole-like, so the sign-change of the Hall effect shown in the text is unnecessary, at least in that case. YBCO as yet has no reliable ARPES Fermi surface, so the question remains open; if Ong's c-axis Hall effect stands, it suggests that the F.S. closes the other way for that YBCO. The Hall effect sign is strongly dependent on the ratio of current carried by spinons, which can be larger than assumed in the text. A preprint by Nozieres and coworkers incidentally (not included) confirms our conclusion that spinons do carry current.

Work by Clarke and Strong (preprints (k), (l)) revises the considerations on c-axis conductivity given in the text from $\sigma \propto \omega$ at low temperatures to $\sigma \propto \omega^{2\alpha}$. Some data support this value. The sharp rise in ρ_c seen in some materials has been shown by Ong to be caused by the spin gap. This may be explained by the mechanism of Strong and PWA (reprint (s)). The ρ_c story is much clarified by the availability of c-axis infrared data. (See also chapters 4, 7.) (Reprint (n).)

I am happy enough to admit that the questions of localization and of spinon binding and hence of ab plane resistivity at low T remain open for further work;

the picture in the text is probably qualitatively correct. A recent preprint gives a first stab at this (preprint (n).) But the high-frequency and T regime has been much clarified by recent work in the infrared, referred to in preprint (m), and explained completely in Luttinger liquid terms; and by Ong's clear separation of chain and plane contributions to ρ_{ab} in YBCO.

Chapter 7 was the most tentative and incomplete of the chapters and I still feel that the question of blending superconductivity with Luttinger liquid concepts has much business yet to be taken care of.

The first major rethinking which has been done since mid-'93 is the restatement of the gap equation in terms essentially equivalent to those in chapter 7, but simplified by replacing the pair tunneling effect by an effective interaction incorporating $\chi_{pair} = \int G_1 G_1^* d\omega$. This interaction is short-range in time, like the conventional BCS interactions (short-range relative to Δ, at least) because of the convergence of the integral GG by the factor $\frac{1}{(x^2 - v_c^2 t^2)^\alpha}$ in G_1, as shown by reprint (j). This allows reformulation of the gap equation as used in Chakravarty et al. (reprint (h)) which is in a very much more perspicuous form.

Unfortunately, reprint (h) represents a wrong guess as to the nature of the "BCS"-type interactions $V_{kk'}$, at least for YBCO. It turns out that the experimental gap almost certainly changes sign in going from points near O, π to π, O in the unit cell. (Interference, x-ray, and measurements on vortex structure and spectra all now agree.) This however does not mean "d-wave" and the best measurements presently available do not agree with "d-wave" (i.e., $\cos 2\theta$) but have a more rapidly varying structure. A revised gap-structure paper has not yet been written.

Reexamination of the isotope effect leaves one dubious as to whether a positive isotope effect has ever been demonstrated on an optimally doped material. There is an interesting and very strong anomaly in the isotope effect of various impurity doped materials, or those showing a trend toward a commensurate superstructure, which has yet to be understood.

Thus all of the arguments given in chapter 7 against a structure with nodes have apparently evaporated. What remains puzzling is the total failure to observe depairing and scattering effects; in fact, Ong's thermomagnetic measurements suggest a quasiparticle mean free path in the range of .25 μ, which is quite outside the range of possible impurity scattering Fermi liquid quasiparticles. Again, this indicates that we have not yet understood the true nature of quasiparticles in the superconducting regime. Are they in some sense "naked spinons" with charge carried by backflow?

Another reprint to be noted is the analysis of c-axis electrodynamics, a strong support of our general point of view but a salutary reminder that the naive idea of restoring full coherence in the superconducting state is quantitatively impossible.

In general, it is clear that the details of the superconducting state remain complex and somewhat mysterious. The full explanation of the remarkably complex and detailed experimental picture is a task for the coming decades; this book can only outline the direction in which to start out. Every few months, it seems that new phenomena appear, or old ones require a new explanation. This book represents the only sound basis on which to attack these problems, and the reprint section will illustrate how much is done, as well as how much remains to be done.

NOTE ADDED IN PROOF

The reader may realize, if he has come this far, that I have said very little about efforts at numerical simulation. This work: exact solutions of small systems, quantum monte Carlo and variants, series expansions, and so on, has produced thousands of papers and consumed millions of research dollars. I believe, however, that the sizes of systems studied, the dynamical ranges of parameters such as energy or time, and possibly the sophistication of the algorithms used are all inadequate: at least one and probably two orders of magnitude from being useful. The history of the relevant "toy models" is instructive. The 1D and 2D Heisenberg Model, and the 1D Hubbard Model, have all been the subjects of fruitful numerical work, which, however, has in each case quantified and filled out a preexisting theoretical framework from the Bethe Ansatz or the non-linear sigma model, rather than indicating the nature of the solution.

The methods now in use could hardly trace out an accurate Fermi surface for a 2D model, much less give details of the low-energy excitation spectrum and of the $2k_F$ singularities of correlation functions. In the circumstances, most workers have either searched for a "magic bullet" to solve the whole high-T_c problem at one gulp, or reflected the local or personal preconceptions of the investigator. It was not worth my reporting on this unfortunate story.

Author Index

A note on the Indexes: No attempt at a full index of the supplementary material was made, since almost all of this has appeared in print elsewhere and much of it is referenced in the main text. General subject categories are indicated in the Subject Index.

Subject Index